DRUG RESISTANCE
IN BACTERIA

H. Umezawa

Werner Arber

Bruce Kline

Susumu Mitsuhashi

DRUG RESISTANCE IN BACTERIA

Genetics, Biochemistry, and Molecular Biology

Edited by
SUSUMU MITSUHASHI

JAPAN SCIENTIFIC SOCIETIES PRESS *Tokyo*
THIEME-STRATTON INC. *New York*

ISBN 4-7622-5334-X JAPAN SCIENTIFIC SOCIETIES PRESS
6-2-10 Hongo, Bunkyo-ku, Tokyo 113,
Japan
ISBN 0-86577-085-9 THIEME-STRATTON INC.
381 Park Avenue South, New York,
NY 10016, U.S.A.
ISBN 3-13-641401-2 GEORG THIEME VERLAG
Stuttgart 30, West Germany

Distributed in all areas outside Japan and other Asian countries by Thieme-Stratton Inc., New York, a subsidiary of Georg Thieme Verlag, Stuttgart.

Printed in Japan

Preface

Paul Ehrlich in 1907 described the trypanocidal activity of *p*-rosaniline, and in the same year his research group reported that *Trypanosoma brucei* became resistant by repeated exposure to the drug. Knowledge of drug resistance in microorganisms is therefore as old as the history of chemotherapy itself. Drug resistance of bacteria was reported by Morgenroth and Kaufmann soon after discovery of the antipneumococcal effect of ethyldihydrocupreine hydrochloride (optochin).

Many drug-resistant strains of bacteria reported in earlier papers were found primarily *in vitro* under experimental conditions. Although several of the studies produced interesting results, bacterial drug resistance still did not arouse clinical medical interest in those early years of research. However, with the increasing incidence of infections caused by resistant strains, durg resistance has now become a problem of prime importance and is at last receiving the attention it deserves from the clinical and pharmaceutical sciences.

Appearance in Japan of multiply resistant *Shigella* strains and their rapid increase in number have attracted the attention of microbiologists, initially from the standpoint of epidemiology and later from the standpoint of genetics. Conjugative resistance (R) plasmids were first identified by the following properties: (1) transmission of resistance by mixed cultivation, (2) interruption of the transmission with a fritted glass disk to separate the two parental cultures, and (3) spontaneous and induced loss of drug resistance from dug-resistant cells. A mixed incubation of a small number of bacterial cells carrying transmissible drug resistance results in the rapid acquisition of this resistance by a majority of the other incubated cells. R factors are transferable among all species of the family *Enterobacteriaceae*, the *Vibrio* group, *Pasteurella pestis*, *Bordetella bronchiseptica*, and among the *Pseudomonas* group. These properties of the R factor, *i.e.*, autonomous replication and its capability to transmit widely among various species of bacteria, are of great importance in public health and animal husbandry.

In 1963 we reported the presence of nonconjugative resistance(r) plasmids in *Staphylococcus aureus.* Thereafter mini-plasmids(r) were often found in both gram-positive and gram-negative bacteria. Multiple resistance conferred by the presence of both R and r plasmids has often been seen in gram-negative bacteria, and nonconjugative r plasmids are easily mobilized by the concomitant presence of R plasmid, resulting in the rapid spread of conjugative(R) and nonconjugative(r) plasmids among various bacterial species.

In the future, serious attention should be paid to the nonconjugative r plasmids as vectors in gene engineering by the genetic properties: (1) existance in a host cell as multiple copies, (2) wide host range, and (3) the presence of a resistance marker(s).

The studies of drug resistance in bacteria have enlarged our knowledge of drug resistance plasmids and the biochemical mechanisms of bacterial resistance. Drug resistance plasmids are revealing their mysterious roles in microbial evolution and their impact as biohazards. But genetic and molecular studies will open a new field of science capable of training plasmids as vehicles for useful genes.

We have met many individuals from all over the world who are trying to solve the problems caused by the new biohazards resulting from resistance plasmids and to utilize plasmids as helpful vectors in gene engineering.

The organizers of the Third Tokyo Symposium appreciate the financial support given by many pharmaceutical companies to defray the expenses of the foreign participants. We are also grateful for the good help and advice of the International Society of Chemotherapy, the Japan Society of Chemotherapy and the Japan Society of Microbiology.

October 1981

Susumu MITSUHASHI
Laboratory of Drug Resistance in Bacteria
Department of Microbiology
School of Medicine, Gunma University
Maebashi, Japan

Contents

Genetics and Molecular Biology (Replication and Others)

Genetics and Molecular Biology (Transposition)

Genetics and Molecular Biology (Vector and R Plasmid)

Genetics and Molecular Biology (*Pseudomonas*)

Biochemistry (β-Lactamase)

Biochemistry (Penicillin Binding Proteins and Outer Membrane)

Biochemistry (Resistance Mechanism)

Epidemiology

Biochemistry (New Drug)

GENETICS AND MOLECULAR BIOLOGY : REPLICATION AND OTHERS

THE INVOLVEMENT OF IS ELEMENTS OF E. COLI IN THE GENESIS OF TRANSPOSONS AND IN SPONTANEOUS MUTAGENESIS[1]

W. Arber and S. Iida

Department of Microbiology
Biozentrum of the University of Basel
Basel, Switzerland

INTRODUCTION

The worldwide, intensive study of drug resistance in bacteria has not only explained mechanisms of transmission of resistance genes, but it has also greatly helped to elucidate a longtime unexpected basic genetic principle: the mediation of DNA rearrangements by movable genetic elements. Resistance genes are often carried on such elements, which are then called transposons. In recent years it became obvious that many transposons are flanked on each side by homologous nucleotide sequences displaying themselves the properties of movable genetic elements. These then belong to the IS elements, defined as movable genetic elements not carrying known genes unrelated to the transposition process (1-3).

One may wonder whether new, IS-mediated transposons may arise any time as a consequence of IS transposition and comprise in principle any bacterial gene. We indeed succeeded recently to demonstrate the genesis of IS1-mediated transposons (4) carrying the genetic determinant for chloramphenicol acetyl transferase (*cat*) which provides resistance to chloramphenicol (Cm). IS elements are also known to promote the fusion of replicons and the formation of deletions, of inversions and of gene amplifications. Any of these different types of DNA rearrangements together with the possibility that IS elements can be excised from their site of residence is likely to contribute to microbial evolution.

[1] *Supported by Grant 3.479.79 from the Swiss National Science Foundation*

In this context it is of interest to know how many different IS elements a particular bacterial strain, e.g. *Escherichia coli* K12, carries, and with what frequencies DNA rearrangements occur in this strain. The study of IS transposition into the P1 prophage carried in the P1 lysogenic *E. coli* as a plasmid has proven to give at least partial answers to these questions (5, 6). These experiments revealed that transposition of several different IS elements from the host chromosome to the P1 DNA represents a very important source of spontaneous mutations on the P1 genome.

GENESIS OF IS1 MEDIATED TRANSPOSONS

We recently succeeded to demonstrate the natural *de novo* formation of IS1-flanked transposons carrying the *cat* gene. Since the details of these experiments have already been published (4), we limit ourselves here to explain the principle which is illustrated in Fig. 1. The *cat* gene is carried on pBR325 (7). This plasmid does not contain any nucleotide sequences from IS1 (8), but it can serve as a target for IS1 transposition. However, since it is not possible to directly monitor the transposition of a single IS1 into pBR325 we have chosen the following approach based on the fact that the genome of bacteriophage P1 contains one copy of IS1 (9). In transpositional events, this IS1 can mediate the formation of pBR325::P1 cointegrates (4, 10). These were isolated as specialized transducing P1 phages carrying the three determinants for resistance to Cm, ampicillin (Ap) and tetracycline (Tc). They arose with a frequency of about 10^{-8} per plaque forming P1 phage particle. The structural analysis revealed that these cointegrate genomes carried two copies of IS1 forming the junctions between P1 DNA and pBR325. This confirms that the residential IS1 of P1 was involved in the formation of the cointegrates.

Since the two IS1 elements on the cointegrates were carried in the same orientation, reciprocal recombination between them resulted in PBR325::IS1 segregants. We estimate the probability of their formation in *rec*$^+$ bacteria to be about 10^{-3} per generation. pBR325::IS1 segregates were purified by transformation (pBR325::IS1 transforms much more efficiently than the full-sized pBR325::P1 cointegrates). In this way, we obtained four different pBR325::IS1 derivatives carrying IS1 near the *cat* gene (4, 10, 11).

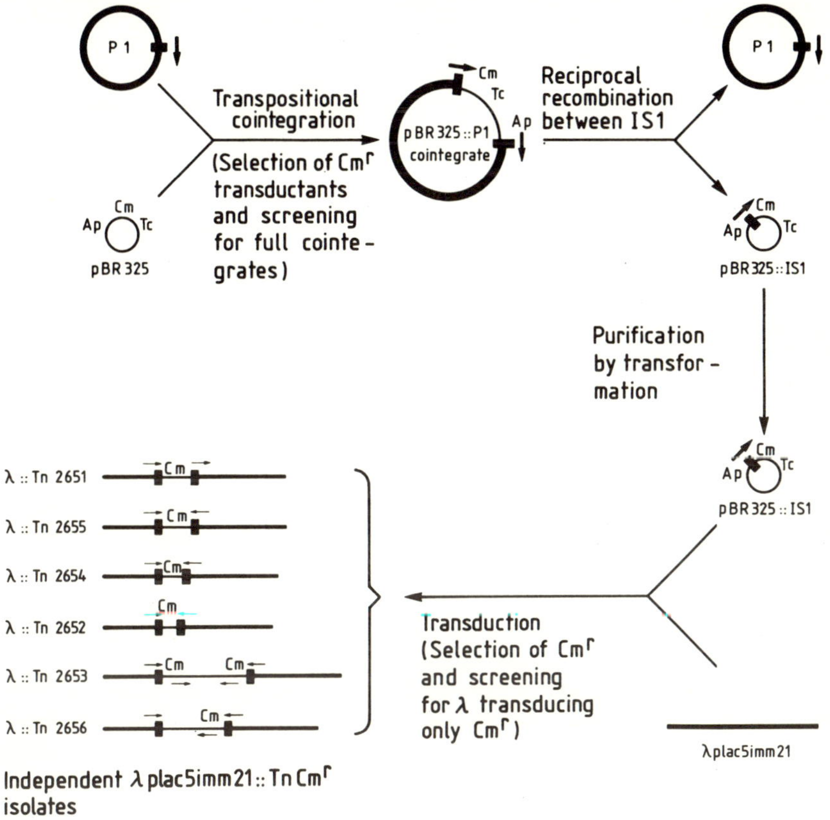

*Fig. 1. Schematic representation of the steps leading to the genesis of IS*1*-mediated Cm*^{r} *transposons. Cm identifies the gene* *cat* *providing resistance to chloramphenicol, Ap and Tc the determinants for resistance to ampicillin and tetracycline, respectively. Small arrows indicate the orientation of IS*1 *(wide box) and of the* *cat* *gene in Tn*2653 *and Tn*2656*. See text for comments on the steps leading to the formation of the 6 independent transposons and on their characterization.*

Two of these derivatives were used in further experimentation. This consisted in growing phage λp*lac*5*imm*21 (which does not contain IS1) in bacteria harbouring such a pBR325::IS1 derivative. From the resulting λ lysate plaque forming specialized transducing λ phage derivatives carrying only the *cat* gene were isolated. They arose with a frequency of about 10^{-9} per plaque forming particle. The investigation of the physical structures of six independent such Cm^r transducing λ genomes revealed that all carry the *cat* gene between two IS1 elements. In one case (Tn*2651*) the two IS1 elements are carried in the same orientation, in the five other cases (Tn*2652* to Tn*2656*) in inverted orientation (Fig. 1). The length of the DNA segment between the two IS1 elements differs for each isolate. Four of the isolates are likely to have been formed by acquisition of a second IS1 element to the side of the *cat* gene opposite of the IS1 resident in pBR325::IS1. The origin of the second IS1 has not yet been investigated. This could stem from transpositional duplication either of the residential IS1 of pBR325::IS1 or of a chromosomal IS1. On the other hand, the 0.5 kb long inverted repeats on pBR325 are likely to have been involved in the formation of the more complex structures of Tn*2653* and Tn*2656* (4, 8).

To demonstrate that the newly isolated elements comprising the *cat* gene between two IS1 actually are transposons, they were translocated to the genome of phage P1-15, a P1 derivative not carrying an IS1 element. In all cases studied, Cm^r transducing P1-15 derivatives revealed to carry the *cat* gene between two flanking IS1 elements, and the physical structures of these segments corresponded to those already present on the parental Cm^r transducing λ phage strains. This is taken as a proof that individual IS1 flanked transposons had evolved in the course of these experiments. The overall probability for their occurrence is obviously very low. Although it may formally not be correct to calculate this probability by simple multiplication of the probabilities determined in each individual step of the experiment, the value of 10^{-20} resulting from such a calculation may still be considered to reflect the approximate order of magnitude of the chance for an IS1-flanked transposon to be formed. The overall frequency for a new transposon to be formed, however, might be several orders of magnitude higher, if the formation of a stable cointegrate and its subsequent, independent resolution are not obligatory steps in the formation of pBR325::IS1.

TRANSPOSITION OF IS ELEMENTS IS A MAJOR CAUSE FOR SPONTANEOUS MUTATION OF THE PHAGE P1 GENOME

The IS1 element is carried in 6 to 10 copies in the chromosome of *E. coli* K12 strains (12). The chromosome carries also other IS elements, only some of which have already been identified and characterized. This is for example seen in investigations with the P1 prophage serving as a target for transposing IS elements. This experimental approach consists in the isolation and physical characterization of spontaneous mutations in the P1 genome (5, 6). For this purpose, lysogens for P1 or for its derivative P1-15 were kept at either 20° or 30° for prolonged times in the stationary phase which was, however, periodically interrupted by diluting the cultures once a week by a factor of 1000, thus allowing for short periods of exponential growth. Screening for defective prophages after about 10 weeks revealed that a few per mill of the tested subclones had acquired a mutation in their prophage, so that they were not any longer able to produce plaque forming phage particles. Prophage loss explained this phenomenon in only very rare cases. Rather, most studied subclones with a defect in phage production still carried the prophage. Their mutant phage genomes were then submitted to a physical characterization which yielded the results represented in Fig. 2.

Looking first at the results obtained with P1-15, all of 21 independently isolated mutations revealed to be caused by insertion of DNA sequences, varying in their length between about 700 and 1500 basepairs (6). Some of these sequences were identified as IS*2*, some as IS*30* (13) and one as IS1. Others have not yet been identified but differ from those mentioned as well as from IS*4* and IS*5*. We suspect that all of these IS elements had transposed from the bacterial chromosome to the P1-15 prophage, thereby abolishing an essential function for the vegetative growth of bacteriophage particles. An analysis of the sites of insertion shows that the IS elements are not carried at random along the P1-15 genome. Some areas which are known to contain essential genes for growth of phage P1-15 have not yet been seen to carry an IS element. Other areas represent hot regions for IS insertion. The most densely populated region extending over about 1700 basepairs carries mainly IS*2* and IS*30* elements. At least the former occupy a number of different insertion sites within this hot region (C. Sengstag, personal communication).

On the P1 prophage carrying a copy of IS1 as natural constituent (9), spontaneous mutation occurs roughly three times more frequently than on the P1-15 prophage. About 70% of the P1 prophage mutations analyzed were deletions and all of these started at the residential IS1 element (5). A majority of these deletions ended in a relatively narrow region on one side of the IS1 element (Fig. 2). Nearly 30% of spontaneously mutated P1 prophages carried an insertion, and the sites of these insertions in the P1 genome reflected the same distribution as seen with P1-15 (5).

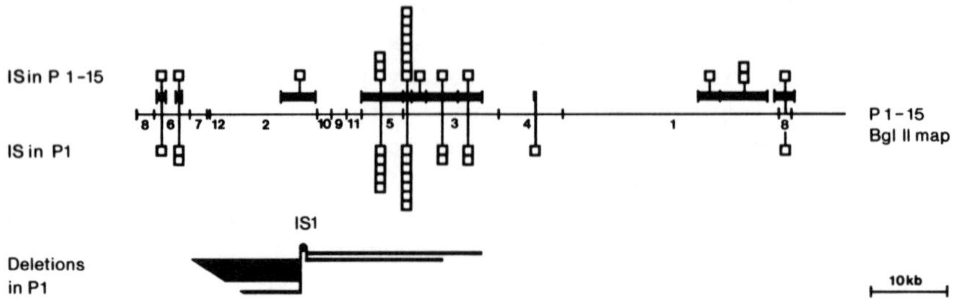

Fig. 2. Map positions of spontaneous insertion and deletion mutations in the phage P1-15 and P1 genome. Each small square represents an independently isolated insertion mutation, which is carried within the DNA segment indicated as determined by restriction cleavage analysis. For only a few isolates the precise location was determined by electron microscopical heteroduplex analysis. Deletions were only found with the P1 genome and they all start at the residential IS1 element of P1. Most of these deletions extend to the left of IS1 and stop at various sites in a "hot region of deletion end points". Some deletions, however, end at other sites either to the left or to the right of IS1. The P1-15 genome has no IS1, and no deletion was found among 30 studied spontaneous mutants. The BglII restriction cleavage map is that of the P1-15 genome. All DNA segments carrying insertions in either P1-15 or P1 DNA are homologous for the two closely related phages. Note that the phage genome is in fact circular and that the linearly drawn map overlaps in the region of BglII fragments 8 and 6.

Only a few % of all studied mutants did not show any structural alteration detectable by restriction cleavage analysis. This class has not yet been further analyzed, but it may contain point mutants, e.g. due to base substitution and frame shift mutations.

Besides furnishing interesting material for further studies of structure and functions of several IS elements resident in *E. coli*, these experiments allow the following conclusions with regard to the mechanisms of the formation of spontaneous mutations. In the P1-15 genome which does not carry any known residential IS element, transposition of a chromosomal IS element represents the most important source of spontaneous mutants. In the genome of phage P1 insertions occur also and with about the same frequency as in the P1-15 genome. However, because of its carrying a residential IS1, IS1-mediated deletion formation is here the prevalent cause of spontaneous mutagenesis, followed by transposition of chromosomal IS elements. Point mutations form only a minor class of all prophage mutations affecting vegetative phage growth. Because of the non random use of target sites in transposition, IS-mediated processes may or may not also form the major source for spontaneous mutations within a particular gene.

Among all insertion derivatives studied, only one insertion was longer than 2 kb and this was identified as the 5.7 kb γδ element. In contrast, we did not find any IS-flanked transposon. We take this as an indication that single IS elements transpose much more readily than IS-flanked transposons, provided that such elements are present on the *E. coli* chromosome. At least for IS1 one can assume that DNA segments flanked by two neighbouring IS1 transpose sometimes as units, similarly to what we have already discussed in the last section. It should be mentioned in this context that IS1-flanked transposons containing selectable genes from the *E. coli* chromosome have been obtained upon specific selection (14).

The question remains whether IS-mediated DNA rearrangement processes are also the most important cause for spontaneous mutagenesis in the *E. coli* chromosome. This might be considered as a likely implication of the results obtained with the P1 and P1-15 prophages. However, one might expect many *E. coli* rearrangement derivatives not to survive, and such genomes would thus not be available for experimental investigations.

ESTIMATION OF FREQUENCIES OF IS-MEDIATED DNA REARRANGEMENTS

Frequencies of transposition of IS elements and of other IS-mediated DNA rearrangements for which there is no direct selection are quite difficult to determine. The following crude estimations are based on the observed frequency of spontaneous mutagenesis of the P1-15 genome and on results of studies with other plasmids. Under the conditions where we found a few per mill of prophage mutants in a culture of *E. coli* K12(P1-15), the bacteria had grown for about 100 generations. However, their growth had not been continuous but periodically interrupted by a prolongated time in the stationary phase. By comparison, K12(P1-15) bacteria grown without interruption for the same number of generations in the exponential phase contain about 5-10 times less spontaneous prophage mutants. Many of these are also insertion mutants. It is thus likely that IS transposition occurs also in the stationary phase, or with increased probability in the transition between the two phases of growth. To which units should we then relate the number of transposition events: to the number of generations, to units of time or to the number of cells at the time of the analysis? Being aware of these difficulties, we will nevertheless give in the following frequencies of transposition and of other DNA rearrangements per generation. The P1 and P1-15 genomes have then about a chance of 10^{-5} to undergo spontaneous mutation per generation of the lysogenic bacteria. Since the chromosome of *E. coli* is about 40 times longer than the P1 prophage, the spontaneous mutation rate of the *E. coli* chromosome might be in the order of about 10^{-3} per generation. In a culture having gone through prolonged periods in the stationary phase, considerably more spontaneous mutants may be contained.

Since we concluded above that IS transposition is a major cause for spontaneous mutagenesis, at least for the P1-15 and P1 genomes, one may expect that in the average about one IS transposition occurs in roughly one cell of 1000 at each generation. Transposition of a particular IS element would then be perhaps 10 to 100 times less frequent. Since we have not detected any IS-flanked transposon in the studied prophage mutants, transposition of naturally occurring IS-flanked transposons in *E. coli* can be assumed to be less frequent than the transposition of single IS elements.

If more than one copy of an IS element is present in a cell, reciprocal recombination can occur between them. This can be promoted either by general recombination based on the *rec* pathways of *E. coli* or by a site-specific recombination which may be mediated by some of the residential movable genetic elements such as γδ (15). At least the system of general recombination is in principle available in any cell and one may suppose that recombination between two homologous IS elements mainly depends on the chance of their finding each other. As mentioned above, we estimated the probability of segregation of a cointegrate plasmid with two IS1 elements carried in the same orientation at about 10^{-3} per generation in *rec*$^+$ cells. This segregation, based on reciprocal recombination, can be considered as a kind of site-specific deletion formation. It occurs considerably more frequently than the deletion formation mediated by transpositional activities of the residential IS1 of P1.

Even if these extrapolations may not precisely reflect the reality, at least they make clear that IS-mediated DNA rearrangement processes occur only in minor parts of a bacterial population, rarely enough not to seriously affect either its overall viability or a certain degree of genetic stability under nonselective conditions, but frequently enough to allow evolutional alterations of a small proportion of the cells in a propagating clone.

IMPLICATIONS OF IS-MEDIATED DNA REARRANGEMENTS

These considerations touch the question of the function of IS elements in bacteria. This question is certainly not yet definitively solved. However, it is clear that movable genetic elements are important factors in evolution. We have shown above that a particular gene can become a component of an IS-mediated transposon. Since there is no strict limitation to the size of a transposon and since at least some IS elements may insert at many different sites on DNA molecules, one may expect many different transposons to be formed in the course of time. The entire population of such transposons may probably contain any segment of the *E. coli* chromosome.

Either by a transpositional event or by reciprocal recombination between homologous IS elements chromosomal genes forming part of a transposon may become integrated into a viral genome or a conjugative plasmid. By transduction or by conjugation (and perhaps also by transformation), these DNA segments may later be transferred to another bacterial strain. The presence of the flanking IS elements with their ability to transpose offers the chance to insert the transferred DNA into many different, nonhomologous regions of the host chromosome or into a plasmid harbored in the recipient bacteria, where at least some of the transferred genes may be expected to function. In addition, the observed possibility of excision of IS elements may lead to gene fusion which by chance may result in a new kind of polypeptide with novel structural and enzymatic properties. Such events are certainly much less frequent in nature than transposition of single IS elements, but they may be expected to occur frequently enough to be of extreme long-range importance. These mechanisms are also a likely explanation for the appearance of multiple drug resistance under selective conditions in bacteria previously not enabled to provide antibiotic resistance.

Certain types of point mutations, e.g. nucleotide substitution, certainly also play their important role in evolution as drivers of a stepwise modulation of gene activities. In contrast to these processes, transposition, particularly in conjunction with the possibility for horizontal gene exchange, should be considered not to be a random event, rather its occurrence must obey some rules with regard to space (e.g. the choice of transposition target is not random, see Fig. 2) and with regard to time. Indeed, since transposition is an enzyme-mediated process one may wonder whether particular environmental and physiological conditions, external or internal to the cell, influence the occurrence of transposition.

ACKNOWLEDGMENTS

The authors are indebted to their collaborators Helga Jütte, Solveig Schrickel, Patrick Caspers, Markus Hümbelin, Urs Karli, Jürg Meyer and Christian Sengstag for their appreciated help in the investigations outlined in this article.

REFERENCES

1. Campbell, A., Berg, D.E., Botstein, D., Lederberg, E.M., Novick, R.P., Starlinger, P. & Szybalski, W. (1979) *Gene 5*, 197-206
2. Calos, M.P. & Miller, J.H. (1980) *Cell 20*, 579-595
3. Iida, S., Meyer, J. & Arber, W. (1982) in *Mobile Genetic Elements* (Shapiro, J.A., ed.) Academic Press, New York in press
4. Iida, S., Meyer, J. & Arber, W. (1981) *Cold Spring Harbor Sympos. Quant. Biol. 45*, 27-37
5. Arber, W., Iida, S., Jütte, H., Caspers, P., Meyer, J. & Hänni, C. (1979) *Cold Spring Harbor Sympos. Quant. Biol. 43*, 1197-1208
6. Arber, W., Hümbelin, M., Caspers, P., Reif, H.J., Iida, S. & Meyer, J. (1981) *Cold Spring Harbor Sympos. Quant. Biol. 45*, 38-40
7. Bolivar, F. (1978) *Gene 4*, 121-136
8. Prentki, P., Karch, F., Iida, S. & Meyer, J. (1981) *Gene 14*, 289-299
9. Iida, S., Meyer, J. & Arber, W. (1978) *Plasmid 1*, 357-365
10. Iida, S., Meyer, J. & Arber, W. (1981) *Mol. Gen. Genet. 184*, 1-10

11. Iida, S., Marcoli, R. & Bickle, T.A. (1981) *Nature 294*, 374-376
12. Nyman, K., Nakamura, K., Ohtsubo, H. & Ohtsubo, E. (1981) *Nature 289*, 609-612
13. Iida, S., Hänni, C., Echarti, C. & Arber, W. (1981) *J. Gen. Microbiol., 126*, 413-425
14. York, M.K. & Stodolsky, M. (1981) *Mol. Gen. Genet. 181*, 230-240
15. Reed, R.R. & Grindley, N.D.F. (1981) *Cell 25*, 721-728

EXPRESSION OF *STREPTOCOCCUS MUTANS* PLASMID AND CHROMOSOMAL GENES IN *ESCHERICHIA COLI* K-12[1]

R. Curtiss III, E.K. Jagusztyn-Krynicka[2], J.B. Hansen, M. Smorawinska[2], Y. Abiko[3] and G. Cardineau

Institute of Dental Research and
Department of Microbiology
University of Alabama in Birmingham
Birmingham, Alabama, U.S.A.

I. INTRODUCTION

Streptococcus mutans is a principal etiologic agent of dental caries and is likely to be the most ubiquitous bacterial infectious disease agent world-wide (1). A thorough review of *S. mutans* has recently appeared (2). *S. mutans* owes its pathogenicity to its ability to attach to the pellicle-coated tooth surface in a sucrose-independent manner followed by a sucrose-dependent adherence and aggregation that is facilitated by the synthesis of water-insoluble glucans under the control of cell-surface associated glucosyltransferase enzymes. In addition to glucans, *S. mutans* synthesizes extracellular fructans by the polymerization of the fructosyl moiety of sucrose and is also able to synthesize intracellular glycogen-like polysaccharides. Either by catabolism of intra or extracellular polysaccharides or of free mono, di and trisaccharides, *S. mutans* produces copious amounts of lactic acid which cause demineralization of enamel and commence the process of tooth decay.

Much of what is known about *S. mutans* pathogenicity comes from isolation and characterization of some of the enzymes involved in the synthesis and metabolism of carbohydrates and from the isolation of mutants that are altered with regard to adherence, aggregation and acid production. Nevertheless, it

[1]*Research supported by grant DE-02670 from the National Institutes of Health, Bethesda, Maryland, U.S.A. J.B. Hansen was supported by NIH Training Grants DE-07026 and AI-07041.*
[2]*Present address: Institute of Microbiology, Univ. of Warsaw, Poland.*
[3]*Present address: Nihon University, Tokyo, Japan*

has been difficult to purify certain enzyme activities to homogeneity and the literature is fraught with contradictory information. Similarly, it is difficult to know how many genes might be involved in any given process since means to analyze *S. mutans* by conventional genetic techniques have been all but absent except for some recent as yet under-developed methodologic approaches which may be valuable in the future. Because of these considerations, we chose to analyze the biochemical and genetic basis of *S. mutans* caries pathogenicity by using recombinant DNA techniques. Since *S. mutans* and *Escherichia coli* have no DNA homology and exhibit no immunological cross reactions, we chose *E. coli* K-12 as the recipient host for *S. mutans* genes. Our rationale was that it would be easier to identify and characterize *S. mutans* gene products in *E. coli* and then, with specific antisera against *S. mutans* gene products made in *E. coli*, obtain specific known gene defects in *S. mutans* to evaluate the influence such mutations might have on pathogenicity.

II. EXPRESSION OF *S. MUTANS* GENETIC INFORMATION IN *E. COLI*

In our initial attempts, chromosomal DNA was isolated from the serotype c *S. mutans* strains PS14 and GS-5 and from the serotype g *S. mutans* strain 6715. DNA was recovered following mutanolysin (3) treatment of *S. mutans* cells and lysis by addition of sodium dodecyl sulfate. High molecular weight, chromosomal DNA purified by phenol extraction and ethanol precipitation was cleaved with the restriction enzymes *Eco*RI, *Hin*dIII, *Bam*HI and *Pst*I and these DNAs were ligated to the plasmid cloning vectors pBR322 (4) and pACYC184 (5), similarly cut with the same restriction enzymes. In addition, *S. mutans* cryptic plasmid DNA was recovered by standard methods and also used to form recombinant molecules with the pBR322 cloning vector (6). Following annealing and reestablishment of phosphodiester bonds with the addition of DNA ligase, recombinant molecules were introduced into suitable strains of *E. coli* K-12 by calcium chloride facilitated transformation.

Our initial desire was to determine whether *S. mutans* DNA would be expressed in *E. coli* and if so, whether it was able to substitute for *E. coli* gene defects. As indicated by the data in Table 1, the introduction of *S. mutans* DNA into χ1849 led to the recovery of clones that were able to grow in the absence of exogenously added purines to substitute for the Pur^- defect, in the absence of threonine and methionine to substitute for the Asd^- defect, and in the presence of galactose as sole energy source to substitute for the Gal^- defect. By using other *E. coli* K-12 strains, we found that some 30 to

Table 1. *Complementation of E. coli Gene Defects by Cloned S. mutans DNA*

E. coli mutant		*Complementation of defect*	*Restriction enzyme*
Genotype	*Phenotype*		
purE42	*Pur*$^-$	*Yes*	*HindIII, PstI*
Δ(gal-uvrB)	*Gal*$^-$	*Yes*	*BamHI, HindIII*
his-53	*His*$^-$	*No*	-
Δ(bioH-asd)	*Thr*$^-$ *Met*$^-$ *Dap*$^-$	*Yes*	*HindIII*
ilv-277	*Ilv*$^-$	*No*	-

40% of *E. coli* gene defects for purine, prymidine or amino acid biosynthesis or for inability to utilize carbohydrates could be restored by the addition of *S. mutans* genetic information in the form of recombinant plasmids.

A. *Cloning of pVA318 Plasmid DNA*

S. mutans serotype *c* and *e* strains occasionally contain a cryptic plasmid having a molecular mass of 3.6 Mdal (7,8). This one cryptic plasmid originally described by Dunny, et al. (7), and isolated again by Macrina and Scott from five different *S. mutans* strains isolated from all over the world, is wide-spread (9). We have found that some 10% of random *S. mutans* isolates obtained from children at the University of Alabama in Birmingham dental clinics also possess the same cryptic plasmid. We thus cloned the cryptic plasmid pVA318 (9) into the pBR322 cloning vector and transformed *E. coli* (6). By electronmicroscopy pVA318 had a contour length equivalent to 5.6 kb and based on buoyant density equilibrium centrifugation had a 32 to 34% guanine + cytosine content. The formation of the recombinant clones pYA656 and pYA658 (Fig. 1) by insertion of pVA318 into the *Hin*dIII site of pBR322 revealed that tetracycline resistance was expressed in one orientation but not in the other. This implied that a streptococcal gene sequence could restore a promoter function inactivated by separation of the "-35" RNA polymerase recognition site from the RNA polymerase binding site of the tetracycline resistance promoter in pBR322. Whether the pVA318 sequence in pYA658 provides the "-35" RNA polymerase recognition sequence or an entire promoter region is unknown. Although either pVA318 sequence orientation cloned into the *Hin*dIII site of pBR322 was completely stable, the cloning of

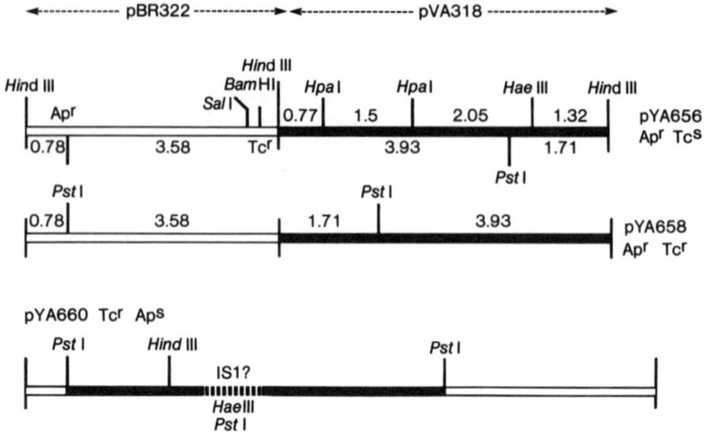

Fig. 1. Cryptic S. mutans plasmid pVA318 cloned into pBR322. *Linear maps, with HindIII ends, of three recombinant plasmids show the various restriction endonuclease sites and fragment sizes (kilobases). The black lines represent pVA318 DNA, the white lines represent pBR322 DNA. The antibiotic resistance phenotypes for ampicillin (Ap) and tetracycline (Tc) are indicated. For simplicity, only the sites for PstI and HindIII have been marked on the pYA658 and pYA660 maps. Also the multiple HaeIII sites on pBR322 have not been shown. This figure is reprinted from Hansen et al. (6).*

this plasmid using the *Pst*I enzyme led to gross instability with rapid loss of all or parts of the chimeric plasmid. In one stable isolate, pYA660, it appeared that an IS1-like element had inserted into the pVA318 DNA sequence. This was based on an 0.8 kb increase in size and new, adjacent *Pst*I and *Hae*III sites within the pVA318 sequence. pVA318, unlike pBR322, only has one *Hae*III site. Another stable *Pst*I cloned pVA318 derivative was also isolated after numerous tries. This plasmid, pYA661, has the same size as pYA656 and pYA658 but still could have some small alteration which permits stable maintenance in *E. coli* (6).

When the chimeric plasmid was introduced into a *polA*(Ts) strain and the temperature raised to 42°, the recombinant plasmids were rapidly lost irrespective of whether cloning was into the *Pst*I or *Hin*dIII site of pBR322. Thus, it would appear that the pVA318 replication machinery is inoperative in *E. coli* unless, of course, this plasmid also depends on a DNA polymerase I type of activity (6).

Examination of the proteins synthesized by the recombinant plasmids in minicells produced by strains harboring the recombinant plasmid indicated that the pVA318 sequence coded for a 20,000 molecular weight protein. This protein was evaluated for bacteriocin activity using a diversity of streptococcal and enteric strains. No such activity was found. Cold osmotic shock was also employed to determine whether this plasmid-specified protein was translocated to the *E. coli* periplasm. To the contrary, all protein remained in the *E. coli* cytoplasm (6). The function this protein may have in *S. mutans* biology is unknown.

B. *Expression of Genes for Carbohydrate Utilization*

As already stated, *S. mutans* genetic information functions in *E. coli* in many instances. One question of some interest was whether *S. mutans* genetic information that was involved in the transport of sugars in *S. mutans* would allow the transport of those sugars in *E. coli*. Thus, in one experiment, an *E. coli* that lacked the *malK* and *lamB* genes due to insertional inactivation of *lamB* by Mu*cts* followed by deletional loss of the prophage and adjacent genetic material was used to select Mal^+ clones able to grow on maltose. Mal^+ recombinant clones were obtained but remained lambda resistant as would have been expected. The ability to grow on maltose, however, implies that the product of the *malK* gene is a cytoplasmic membrane bound protein involved in the transport of maltose that can be replaced in a functional way by a comparable activity from *S. mutans*. Biochemical characterization of this clone, pYA507, has not been effected (J.C. Hsu and R. Curtiss III, unpublished).

The recovery of Gal^+ *E. coli* strains harboring *S. mutans* galactose utiliziation genetic information (Table 1) was of some interest since the *E. coli* K-12 mutant χ1849 contains a deletion of the entire *gal* operon (which encodes the enzymes of the Leloir pathway). Examination of these recombinant clones which grow relatively slowly on galactose as a sole carbon source indicated that UDP-galactose and UDP-glucose were still missing (M. Smorawinska, J.C. Hsu, Y. Abiko and R. Curtiss III, unpublished). This was indicated by the absence of ability to synthesize the exopolysaccharide colanic acid and by failure to now synthesize an LPS core with ability to confer sensitivity to bacteriophage Mu and resistance to bacteriophage C21. Furthermore, enzyme assays for the three enzymes of the Leloir pathway failed to detect any activities irrespective of the presence of cAMP or any potential inducers of the *E. coli gal* operon. On the contrary, our evidence indicates that the *S. mutans* genetic information encodes a series of proteins constituting the tagatose 6-phosphate pathway (10). In this pathway, external galactose is somehow converted by the *E. coli* recombinants to intracellular galactose 6-phosphate, quite possibly by an inefficiently operated phosphotransferase (PTS) system that must act in cooperation with the *E. coli* PTS enzyme I and HPr components. The galactose 6-phosphate is then converted to tagatose 6-phosphate by an isomerase, to tagatose 1,6-diphosphate by a kinase and to dihydroxyacetone phosphate and glyceraldehyde 3-phosphate by an aldolase. Experiments involving protein synthesis in minicells have revealed the production of several proteins by this clone and the proteins corresponding to the isomerase and kinase have been identified by using transposon mutagenesis and subcloning of the recombinant plasmid inserts. Although we have some suggestion that these genes are partially constitutive but mildly inducible in a coordinate way in *E. coli* the transposon mutants failed to show a strain polarity effect on transcription. It was somewhat surprising to find that the

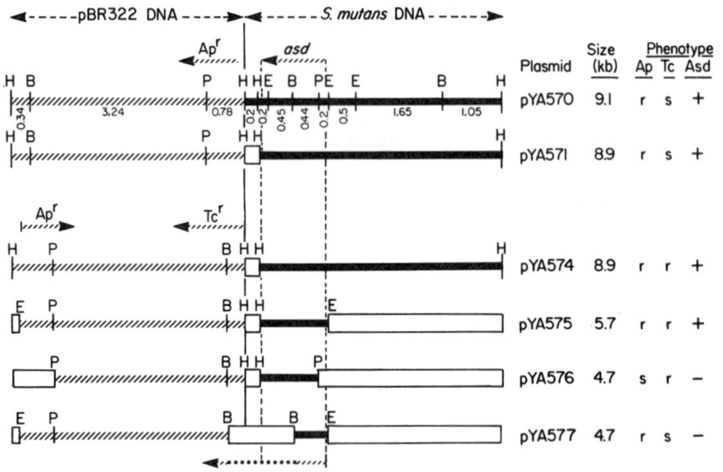

Fig. 2. S. mutans asd gene cloned into pBR322.

Linear maps of six recombinant plasmids show the various restriction endonuclease sites and fragment sizes (kilobases). The black lines represent S. mutans DNA, the broken lines pBR322 DNA and the open boxes are deletions made in subcloning experiments from the parental pYA570 plasmid. The location of genes and direction of transcription for ampicillin (Ap) and tetracycline resistances and for asd are indicated. pYA577 makes an inactive, fused asd gene product, and pYA576 lacks the asd promoter. Abbreviations for restriction endonucleases are B, BamHI; E, EcoRI; H, HindIII and P, PstI. This figure is reprinted from Jagusztyn-Krynicka et al. (12).

gene cluster does not constitute an operon with a single unit of transcription (E.K. Jagusztyn-Krynicka, J.B. Hansen and R. Curtiss III, unpublished).

C. *Expression of S. mutans Gene for Aspartic Acid Semialdehyde Dehydrogenase in E. coli*

E. coli strains with the deletion of the *asd* gene require threonine, methionine and diaminopimelic acid for growth (11). The introduction of the *S. mutans asd* gene permitted these *E. coli* Δ*asd* mutants to grow (Table 1) as rapidly as they would if supplied with optimal amounts of threonine, methionine and diaminopimelic acid (12). The initial recombinants formed by cloning with *Hin*dIII contained a DNA insert almost the same size as the pBR322 cloning vector (Fig. 2). Subcloning with a variety of restriction enzymes after determining the restriction enzyme map allowed a better definition of the *S. mutans asd* gene. The pYA571 and pYA574 plasmids have the large *Hin*dIII *S. mutans* insert in reverse orientations. Since pYA571 is Tc^s and pYA574 Tc^r, our results again indicate that the tetracycline resistance gene of pBR322 when inactivated by an insertion at the *Hin*dIII site can be expressed by information provided by *S. mutans* DNA. It is evident that the plasmid subclone, pYA575, exhibits ability to grow in the absence of the supplements mandated by the *asd* mutation in *E. coli* and confers resistance to both ampicillin and tetracycline. In pYA576, the removal of a small 210 basepair segment of *S. mutans* DNA between the *Eco*RI and *Pst*I sites of the insert DNA of pYA575 (along with other portions of the pBR322 vector including the β-lactamase promoter) led to an Asd^- phenotype while retaining tetracycline resistance (Fig. 2). This implied that this short segment of DNA missing from pYA576 might contain the promoter for the *asd* gene (12).

Minicells harboring the various *asd* recombinant plasmids were purified and used to evaluate protein synthesis. A surprising result was that the synthesis of the *asd* protein almost totally shut off the synthesis of β-lactamase encoded by the pBR322 cloning vector (12). Since minicells have a limited amount of RNA polymerase segregating into the minicells at the time of their production, these results implied that the *asd* promoter has a much greater affinity for the *E. coli* RNA polymerase than does the β-lactamase promoter. This is very surprising since the β-lactamase promoter has a very high affinity for *E. coli* RNA polymerase.

The level of expression of the *asd* gene can be readily seen when one uses whole *E. coli* cells producing the *S. mutans asd* gene product which has a molecular weight of 45,000. As seen in Fig. 3, this 45,000 protein is very prominent and can be seen by Coomassie Blue staining of proteins run by SDS-polyacrylamide gel electrophoresis.

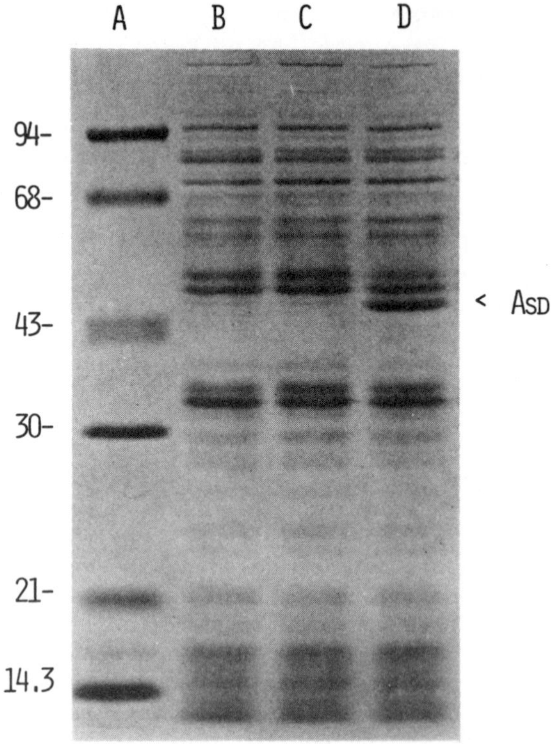

Fig. 3. SDS-polyacrylamide gel electrophoresis of proteins from E. coli containing the S. mutans asd gene.

Proteins solubilized from whole cells (265 μl of cells, O.D. at 600 nm = 0.6) by heating 5 min, 100°C in SDS and β-mercaptoethanol were electrophoresed according to Laemmli and Favre (13) and stained with Coomassie Blue. Size standard proteins, lane A, are from BioRad, and their molecular masses (in kilodaltons) are indicated on the left. Each standard protein band contains ~2 mg protein. Controls are E. coli strain χ1849 without any plasmid, lane B, and χ1849(pBR322), lane C. Lane D contains χ1849(pYA575): the asd protein is ~45,000 kilodaltons.

The strength of the *asd* promoter with regard to binding of *E. coli* RNA polymerase has been confirmed by RNA polymerase binding studies (G. Cardineau and R. Curtiss III, unpublished). The *S. mutans* DNA in pYA575 has two RNA polymerase binding sites: the 230 basepair segment adjacent to the tetracycline resistance gene and the 210 basepair *Eco*RI to *Pst*I site which contains the *asd* promoter. The former result proves that the *S. mutans* DNA in pYA574, pYA575 and pYA576 is providing an entire functional promoter region for the expression of the pBR322 tetracycline resistance gene. Nucleotide DNA sequencing of the 210 basepair *asd* promoter segment reveals a very unusual promoter (G. Cardineau and R. Curtiss III, unpublished). This control region contains a usual RNA polymerase recognition site, referred to as the "-35" sequence because of its location approximately 35 bases upstream from the start site for mRNA synthesis. This sequence contains the TTG sequence which is preceded further upstream by a T rich region (14,15). Approximately 15 basepairs downstream begins the coding sequence of five RNA polymerase binding sites. Three of these Pribnow boxes are overlapping and the second one has the perfect consensus sequence TATAATG (15-17). Following these three overlapping Pribnow boxes are two additional boxes in tandem. All five Pribnow boxes have TA at the first two positions and a T at the sixth position. Appropriate GAA sequences are seven, eight and nine bases downstream from the sixth position T in both the fourth and fifth Pribnow boxes and it is within these sequences that we believe that mRNA synthesis commences. Immediately following these potential transcription initiation sites is a nine base Shine-Dalgarno sequence (18) which is complementary to the 3'-terminus of *E. coli* 16S ribosomal RNA except for the replacement of an A for a G. Six bases further downstream is an ATG initiation codon followed by an open reading frame for the remainder of the DNA sequenced. It would thus appear that the *asd* promoter region is admirably well suited for expression in *E. coli*. The synthesis of the *S. mutans asd* gene product is constitutive in *E. coli*. It is not known how this coding sequence might be regulated in *S. mutans*. This is obviously an area requiring further investigation.

D. Cloning of S. mutans Genes for Cell Surface Proteins

Not only are *S. mutans* genes specifying enzymes involved in biosynthesis and metabolism expressed in *E. coli*, but also genes for enzymatic synthesis of glucans (J.P. Robeson, R. Barletta and R. Curtiss III, unpublished) and which specify proteins that are normally on the cell surface of *S. mutans* (R.G. Holt, Y. Abiko, S. Saito, M. Smorawinska and R. Curtiss III, unpublished). These proteins are, by and large, translocated across the cytoplasmic membrane of *E. coli* into the periplasm. In the case of one glucosyltransferase, it is transported across the cytoplasmic membrane without apparent modification. Thus, mechanisms for protein translocation across the cytoplasmic membrane may differ for Gram-positive and Gram-negative bacteria. Again, further work will be needed to elucidate these different processes.

ACKNOWLEDGMENTS

We thank our colleagues JeMin Charles Hsu, James P. Robeson, Raúl Barletta, Hettie Murchison, Sylvia Larrimore, Robert G. Holt and Shigeno Saito for helpful discussions and occasional assistance. We thank Pat Pierce for assistance with the preparation of this manuscript.

REFERENCES

1. Newbrun, E. (1978) Cariology. The Williams and Wilkins Co., Baltimore.
2. Hamada, S. & Slade, H.D. (1980) *Microbiol. Rev. 44*,331-384.
3. Yokogawa, K., Kawata, S., Nishimura, S., Ikeda, Y. & Yoshimura, Y. (1974) *Antimicrob. Ag. Chemother. 6*,156-165.
4. Bolivar, F., Rodriguez, R.L., Betlach, M.C. & Boyer, H.W. (1977) Gene *2*,75-93.
5. Chang, A.C.Y. & Cohen, S.N. (1978) *J. Bacteriol. 134*, 1141-1156.
6. Hansen, J.B., Abiko, Y. & Curtiss R. III (1981) *Infect. Immun. 31*,1034-1043.
7. Dunny, G.M., Birch, N., Hascall, G. & Clewell, D.B. (1973) *J. Bacteriol. 114*,1362-1364.
8. Macrina, F.L., Reider, J.L., Virgili, S.S. & Kopecko, D.J. (1977) *Infect. Immun. 17*,215-226.
9. Macrina, F.L. & Scott, C.L. (1978) *Infect. Immun. 20*, 296-302.
10. Hamilton, I.R. & Lebtag, H. (1979) *J. Bacteriol. 140*, 1102-1104.
11. Umbarger, H.E. (1978) *Ann. Rev. Biochem. 47*,533-606.
12. Jagusztyn-Krynicka, E.K., Smorawinska, M. & Curtiss, R. III (1982) *J. Gen. Microbiol.* In Press.
13. Laemmli, U.K. & Favre, M. (1973) *J. Mol. Biol. 80*,575-599.
14. Takanami, M., Sugimoto, K., Sugisaki, H. & Okamoto, T. (1976) Nature *260*,297-302.
15. Rosenberg, M. & Court, D. (1979) *Ann. Rev. Genetics 13*, 319-353.
16. Pribnow, D. (1975) *J. Mol. Biol. 99*,419-443.
17. Pribnow, D. (1975) *Proc. Nat. Acad. Sci. U.S.A. 72*,

784-788.
18. Shine, J. & Dalgarno, L. (1974) *Proc. Nat. Acad. Sci. U.S.A. 71*,1342-1346.

GENETIC STUDIES ON THE MAINTENANCE OF MINI F PLASMIDS

B.C. Kline, R.W. Seelke and J.D. Trawick

Department of Cell Biology
Section of Microbiology
Mayo Medical School
Mayo Clinic
Rochester, Minnesota, U.S.A.

S.B. Levy and J. Hogan

Department of Molecular Biology and Microbiology
Tufts University School of Medicine
Boston, Massachusetts, U.S.A.

INTRODUCTION

Plasmid F, the classic sex factor of *Escherichia coli*, has been used for several years as a model system in the study of DNA replication and partitioning. It is a large plasmid of 94.5 kilobases (kb). A nine kb *Eco*RI restriction fragment (f5) of F, when recombined with an appropriate marker, can form mini-F plasmids with normal F replication characteristics: low copy number, stable maintenance, incompatibility with other F plasmids, sensitivity to acridine orange, ability to replicate in a *polA* host, and ability to integratively suppress *dnaA* mutations (1,2). Another region of F, the f7 *Eco*RI fragment can also form mini-F plasmids but their characteristics are not typical for an F plasmid: the f7-derived plasmids are unstable at fast growth rates, require high levels of polymerase I, and are resistant to acridine orange (3). The f7 region will not be discussed further.

Our research has focused mainly on three areas: defining the essential region in f5 for replication (*rep*), mapping and characterizing the copy number control genes (*cop*) and determining the relationship of incompatibility (*inc*) to copy number control. In so doing, we hope to provide a factual basis for analyzing the mechanisms of plasmid maintenance.

Our findings may be summarized as follows:

(1) The essential replication region is between 44.1 and 46.19kb, a region which contains an origin of replication, at least two *inc* genes, a replication control region that probably overlaps these *inc* genes, and a locus for acridine orange sensitivity (*aos*).
(2) In the replication control region there is at least one *cop* gene which acts negatively to control copy number; mutations in this gene do not map within the known *inc* genes.

In the following sections we will give the data which supports the above conclusions and place them in the context of what other workers have found about F maintenance (Fig. 1).

THE ESSENTIAL REGION

Several mini-F plasmids partially deleted for f5 sequences have been constructed by us and others. Representative plasmids and their published properties are shown in Figure 2. The sequences common to all plasmids are 44.76 to 45.88kb suggesting that absolutely essential genes map within this region (4). However, if we include reasonably stable maintenance and low copy number as essential properties, then one must conclude that the 44.1 to 46.19kb region defines the sequences essential to normal replication (5). An example of a plasmid maintained by the essential region is pBK280. The sequences to the left of the 44.1kb site in pBK280 apparently are not essential to mini-F replication or control since pBAL-16a lacks these sequences yet it has a normal copy number (6).

Mini-F such as pJE2001 and PNZ430 are missing significant portions of the essential region; yet, they form replicons. Other workers have not been able to make mini-F with a deletion within 44.1 to 44.8kb (6,7). Nor have others been able to make mini-F out of the 44.1 to 45.88 *Pst*I fragment (6,7,8). By contrast, pJE2001 is such a plasmid but its existence requires that the mini-F *Pst*I fragment 45.88 to 47.3kb be present in the same cell (8). Ebbers and Eichenlaub speculated that the 45.88 to 47.3kb region produced protein A (Fig. 1) and that this protein prevented runaway replication of pJE2001; a replication that should kill host cells. The low copy number of several plasmids in Figure 2 lacking protein A contradicts their speculation. Hence, there is no

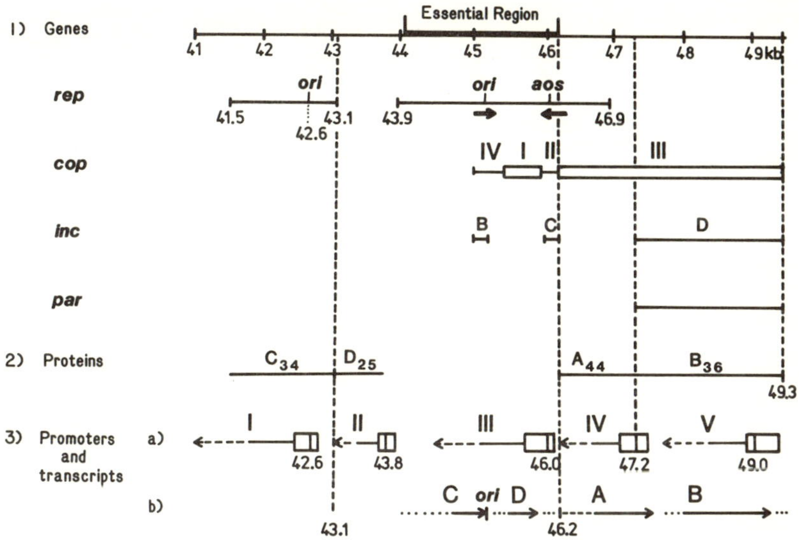

Fig. 1. Genes, gene transcripts and gene products identified in the Eco*RI f5 fragment. Gene symbols are defined in the text. Conditional mutations affecting replication were mapped by others (14,28). The arrows in the* rep *region are direct repeats which are specified in detail in Fig. 3. Copy number mutations and properties are expanded upon in Table 1. The subscript on the proteins represent mass in kilodaltons (7). In 3 (a) the boxes represent promoter map sites and the statistically most probable site is given by the line within each box. The vectors from the boxes represent direction and lengths of transcription (26). In 3 (b) solid vectors represent sites across which a transcript passes. The dotted portion of the vector signifies uncertainty for the promoter site except for transcript D. The most likely site for transcript D promoter based on sequence data analysis (7) and reinforced by our cloning results is about coordinate 45.2kb. The conclusions in 3 (b) are based on unpublished results of Trawick and Kline.*

satisfactory explanations for the existence of pJE2001. Likewise, the existence of pNZ430 is enigmatic unless the non-F, Tn*3* sequences inserted at 44.8kb provide an essential function.

The question naturally arises about the essentiality of the 44.1 to 46.19kb sequences to the maintenance of full size F plasmids. Several lines of evidence indicate these sequences are critical for maintenance of normal F. An explicit detailing of this evidence has been presented and can also be deduced by examining the gene locations mapped in Figure

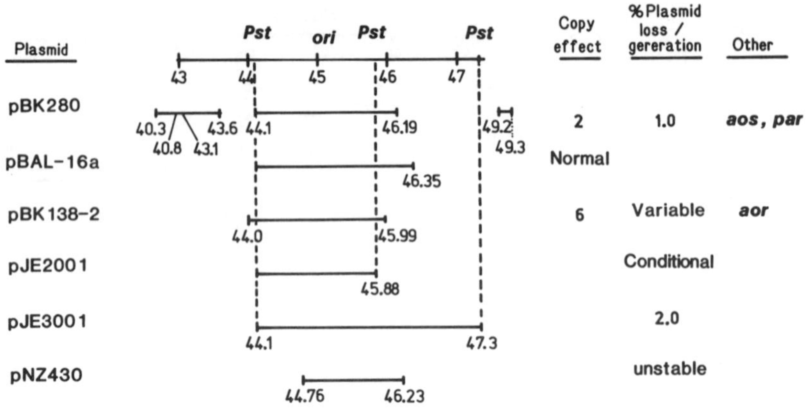

Fig. 2. Select mini-F replicon and their properties. The references for these plasmids are: pBK138-2 and 280 (5); pBAL-16a (17); pJE2001 and 3001 (9); and pNZ430 (4). Each plasmid uses non-F DNA for its phenotypic marker. Blank spaces on properties indicate no report available. Copy effects are same as defined in Table 1. For pBAL-16a, "normal" is the description given in reference (7). The verbal descriptions for plasmid loss per generation are explained as: (i) variable-*stable in some hosts, unstable in others; (ii)* conditional-*requires presence of complementary F sequences (45.88 to 47.3kb) for partial stability. The remaining instability is presumably due to incompatibility--see Fig. 1; and (iii)* unstable *extensive loss, no numerical value reported (4).* aos *and* aor *indicate plasmid genetic property of sensitivity and resistance to elimination by acridine orange (25).*

1 (5).

Although the essential region of F is critical to replication, other sequences to the left of this region also are important when they are present. Eichenlaub *et al.* (10) have shown that a frequently used origin of replication (*ori*) exists at coordinate 42.6kb. Manis and Kline (1) first showed that this origin is dispensible and then Figurski *et al.* (12) found a second origin at 44.4kb in plasmids deleted for the 42.6kb *ori*. The 44.4kb *ori* recently has been remapped to about 45.1kb by Eichenlaub *et al.* (13). The rules by which F selects one or the other *ori* for expression are unknown.

Curiously, Eichenlaub and Wehlmann (14) have found conditional replication mutants that are defective in the synthesis of protein C (Fig. 1). Deletion of the 40.8 to 43.1kb sequences from one mutant, sequences which contain the mutant gene, also caused a loss of the conditional replication phenotype. A satisfactory explanation for this behavior has not yet been made, but it does indicate that *rep* sequences outside of the essential region are not a trivial occurrence.

GENETIC COMPOSITION OF THE ESSENTIAL REGION

Figure 1 shows that the essential region contains *ori*, *cop*, *inc* and *aos* as identified by cloning or mutational analysis. Additionally, Morutsu *et al.* and Tolun and Helinski have sequenced the essential region and have found nine, 19 to 22 base pair direct repeats organized into two clusters as shown in Fig. 3 (6,15). Comparison of the gene maps in Figure 1 to the direct repeat maps suggests that the latter are integral components of *ori* and *inc* and perhaps *cop* and *aos* as well. More will be said about the direct repeats and these gene functions in subsequent sections. All of the known gene functions are found in the right half of the essential region. No gene functions in the left half have been found save for the general designation of *rep*, defined by the inability of deletion mutants of these sequences to form plasmids as detailed above.

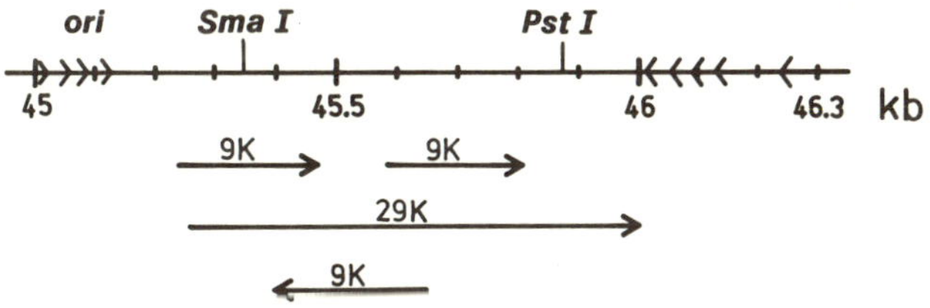

Fig. 3. Direct repeats and potential polypeptides in the replication/replication control sequences of mini-F. The direct repeats and their orientations are depicted as arrowheads (→→→→) on the map. The spacing of the repeats are approximately as shown. Sequences encoding theoretical polypeptides of various masses in kilodaltons (k) are indicated by the plain lines. The arrowheads represent the protein COOH terminus and, therefore, the 3'OH terminus of the transcripts. The information in this figure was taken from references 6 and 15.

MUTATIONS AFFECTING CONTROL OF F COPY NUMBER

Mini-F *cop* mutants have been produced by chemical mutagenesis, Tn*3* insertion and by *in vitro* generated deletions

(4,5,16). All mutations have been mapped and characterized more or less (Fig. 1 and Table 1). As seen in Table 1 the phenotypes of the mutant classes, I, II and III are not identical, the fourth class has not been sufficiently characterized. Considering phenotypic differences and mutation locations, possibly each of the four classes represents a *cop* gene; however, complementation analysis between mutants will be necessary to confirm or change this suggestion.

The phenomenon of plasmid incompatibility did represent a significant barrier to performing such an analysis. Recently, however, we solved this problem and we are now determining the number of *cop* complementation groups. First though, we used complementation analysis to study the mechanism for control of F replication as described in the next section.

GENETIC EVIDENCE THAT CONTROL OF F REPLICATION IS NEGATIVE

Manis and Kline first observed that *stable* cointegrates between *cop*$^+$ F plasmids can be easily made if at least one of the plasmids is deleted for the 40.8 to 43.1kb region. (17). We interpreted this to be a circumvention of the incompatibility barrier because genetic analysis of cointegrate structure indicated the presence of two functional F replicons.

What properties are expected of *cop*$^+$:*cop*$^-$ cointegrates? In a negative model, operator mutants should have a *cis* dominant Cop$^-$ phenotype and repressor mutants should have a *trans* recessive Cop$^+$ phenotype. In a positive control model, overproduction of a positive effector would be *trans* dominant. In Figure 4, we show the plasmid band intensities in cell lysates made from equal amounts of cells containing pML31 (*cop*$^+$), pBK84 (*cop50*) or the cointegrate pBK391A (pML31:pBK84). Clearly, the cointegrate has a band intensity similar to pML31. Thus, the cointegrate has a Cop$^+$ phenotype and the *cop50* mutation is *trans* recessive. Quantitative measurement of pBK391A copy number agrees with this conclusion (data not shown). Using a different *cop* mutant to form the *cop*$^+$ cointegrate, we found both *cop*$^+$ and *cop* replicons could be generated from the *cop*$^+$ cointegrate (data not shown). Hence, the data support the interpretation that region I mutations inactivate a negative copy number control system. Our results support those of Tsutsui and Matsubara (18) who made this same conclusion prior to our studies.

A protein repressor of replication has been identified in the λdv system (19) and small (100bp) RNA repressors of replication have been identified in *Col*E1 and *Col*E1-like

Table 1. Properties of Mini F Copy Number Mutants

Mutational sites	*Fig. 1 region*	*Mutagen*[a]	*Copy number*[b] *increase*	*Phenotype of cloned* **inc** *genes*[c]			*Acridine orange cures?*	*References*
				incB	**incC**	**incB+incC**		
None			*1.0*	+	+	+	*Yes*	*5,16,25*
45.3 -45.9	*I*	*NTG/EMS*	*8.0-28*	+	+	-	*Yes*	*5,25*
45.99 and 46.02	*II*	*Tn3*	*4.0-7.0*	+	-	*NT*	*No*	*5,25*
46.19-49.3	*III*	*Deletion*	*1.5-2.0*	+	+	+	*Yes*	*5*
45.15	*IV*	*Tn3*	*2.0*	*NT*	*NT*	*NT*	*Yes*	*4*

[a] *NTG, nitrosoguanidine; EMS, ethylmethane sulfonate.*

[b] *Copy numbers are expressed as multiples of the wild type value which is one or two copies per chromosomal equivalent.*

[c] *Sequences cloned to test* **inc** *determinants were: 44.1 to 44.88 (***incB***); 45.88 to 46.19 (***incC***); and 43.1 to 46.19kb (***incB+incC***). F'***lac*** was the test plasmid used in the incompatibility assays. NT: not tested.*

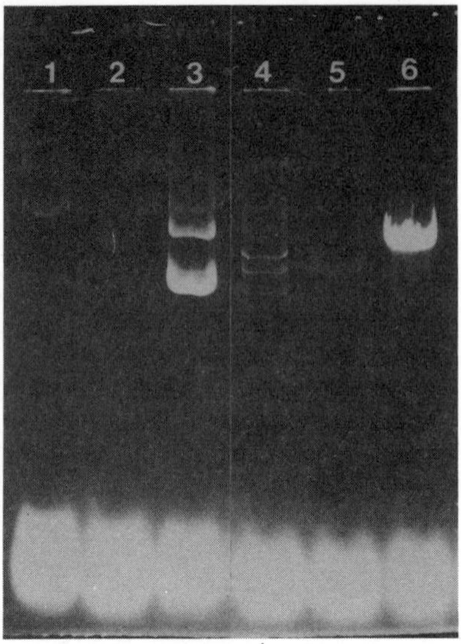

Fig. 4. Copy numbers of cop^+, cop^- *and* cop^+:cop^- *cointegrates of mini-F. Bacteria containing a* cop^+ *(pML31),* cop^- *(pBK84), or* cop^+:cop^- *(pBK391A) cointegrate were adjusted to equivalent cell densities and plasmid DNA was extracted and separated from chromosomal DNA by the method of Birnboim and Doly (29). In order to approximate a quantitative comparison of plasmid concentrations, equal volumes of plasmid-containing extracts were applied to the gel and electrophoresed. Details of the electrophoresis procedure, gel composition, buffers, etc. have been published (11). Lanes 1, 2 and 3 are untreated extracts with pBK391A, pML31 and pBK84, respectively. Lanes 4, 5 and 6 are the same extracts in the same order but treated with* Eco*RI. The number of* Eco*RI bands and their mobilities observed in the cointegrate are those expected based on the properties of pML31 and pBK84. The masses of pML31, pBK84 and pBK391A are 10.5, 7.6 and 18.1 Mdal, respectively.*

plasmids (20,21). Our transcriptional results (Fig. 1) strongly implicate the existence of the 29K transcript. This transcript is about 800bp in length; that is, quite unlike the RNA repressors. Hence, we propose that it is translated into the 29K protein which is a repressor of F replication. In fact, a 28K protein has been seen in extracts of minicells containing plasmids with the 44.1 to 45.88kb *Pst*I fragment (Levy *et al.*, unpublished observation).

INCOMPATIBILITY DETERMINANTS IN MINI-F PLASMIDS

To date, three *inc* determinants have been identified in

the 40.3 to 49.3kb region. These determinants are labelled *incB*, *incC* and *incD* (Fig. 1). By common agreement, *incA* has been retired from usage for reasons that have been discussed (22). Both *incB* and *incC* have been mapped precisely; they are, respectively, the direct repeat sequences found at the general locations of 45.1 and 46.1kb (5,6,15). Since the repeats are also implicated in replication and its control (Figs. 1 and 3), it is reasonable to speculate that these repeats determine incompatibility by titrating essential replication components or sites.

The available evidence suggests that incompatibility determined by *incD* reflects interference in plasmid partitioning. This conclusion rests on the observation that the same sequences that contain *incD* probably also contain the F partitioning (*par*) gene(s). *IncD* was originally mapped within the 46.19 to 49.3kb region (16,23) and more precisely mapped to the 47.5 to 49.4kb region (5,22). Note this region makes a 36K protein, protein B (Fig. 1). The evidence that *incD* is related to *par* is as follows: (i) Ogura *et al.* (64) have converted a Par$^-$ plasmid to Par$^+$ by inserting *in vitro* the 47.3 to 49.3kb *Pst*I/*Eco*RI fragment of F into the Par$^-$ plasmid; (ii) Nordstrom and coworkers (personal communication) have converted Par$^-$ R1 plasmids to Par$^+$ by recombining *in vitro* the 46.19 to 49.3kb region of F with the Par$^-$ mutant; and (iii) plasmids with the 46.19 to 49.3kb region deleted are unstable and exhibit a segregation kinetic (about 1% loss per generation) expected for *par* plasmids (5). This same deletion (region III, Fig. 1) also cause a Cop$^-$ phenotype. Whether or not this has anything to do with partitioning is not clear.

COP AND *INC* GENE INTERACTIONS AND ACRIDINE ORANGE SENSITIVITY

While *par* and *incD* genes seem to be synonymous, the situation with *cop* and *incB*/*incC* genes is much more complex and no simple relationship is completely apparent. Clearly, the Tn*3* insertions at 45.15, 45.99 and 46.04kb imply that *incB* and *incC* are involved in copy number control. For the 45.99 and 46.04kb insertions, Seelke *et al.* have actually shown that the *incC*$^+$ locus becomes *incC*$^-$ (5). What complicates this apparently simple correlation is that the putative 29K protein repressor may also interact with the *incB* and *incC* determinants.

The above conclusion is based on the following set of observations. In 1978, Manis and Kline found that cloning the 43.1 to 46.19kb region from *cop*$^+$ plasmids generated pSC101:F recombinants that were incompatible when tested against F'*lac* plasmids (16). Further, they found that when chemically-induced *cop* mutants (region I, Fig. 1) were used as the source of the 43.1 to 46.19kb sequences in making pSC101:F recombinants, the recombinants were compatible with F'*lac*. We know that these mutations map within the gene for the putative 29K protein, that is, between *incB* and *incC*.

The silencing of the *incB*$^+$ and *incC*$^+$ determinants, which were also contained in the pSC101:F recombinants, is not due to secondary mutations in these sequences. This conclusion is based on the observation that when these *inc* genes were cloned individually from the same *cop* mutants, that is cloned as the 44.1 to 45.88kb and 44.88 to 46.19kb regions, they formed *inc*$^+$ recombinants (5). Thus, we must conclude that in some way there is an interaction between the putative repressor of replication and the *incB*/*incC* direct repeats.

Analysis of *ori*, *cop* and *inc* genes revolves around the direct repeats defined in Fig. 3; so also does the sensitivity of F replication to acridine orange. This conclusion is based on the observations made by Wechsler and Kline (25) that the same Tn*3* insertion at 45.99 and 46.04kb used to induce *cop* mutations were subsequently found not only to be *incC*$^-$ but also to be resistant to acridine orange curing. Tn*3* insertion at 46.19kb did not cause acridine resistance, so the locus for this sensitivity cannot extend rightward to this coordinate. The leftward extent of the drug sensitivity locus is unknown. The exact step that acridine inhibits in F replication is unknown; however, the step of chain elongation is probably not involved (25). Given the above implications for the involvement of direct repeats in control, that is, initiation of replication, it is reasonable to suggest that acridine orange may inhibit initiation of F replication.

We have described a diversity of genotypes and phenotypes for replication and copy number control processes that map in the 45.0 to 46.19kb region. But as yet these descriptions have not been fitted to a detailed model. In order to achieve greater understanding of the way these genes are integrated, we and others (26) have begun to map the promoters and transcripts in the f5 region.

PROMOTERS AND TRANSCRIPTS IN MINI-F

Wehlmann and Eichenlaub (26) have used the R-loop mapping technique with transcripts generated *in vitro* from the f5 *Eco*RI restriction fragment. They found five units of expression (I to V, Fig. 1). We have performed a similar analysis but instead used cloning techniques to place F promoters next to plasmid-borne promoterless antibiotic resistance genes

(Trawick and Kline, unpublished). Subsequently, we selected for antibiotic resistant cells harboring the recombinant plasmids. Thus far, we have found four promoters (A through D, Fig. 1). Presumably the A and B promoters determine expression of proteins A and B while promoter D determines expression of the putative 29K repressor. The function of promoter C is unknown. An obvious possibility is that C determines synthesis of an RNA primer of replication since this transcript potentially impinges on the 45.1kb *ori* sequences. It is not clear why the *in vitro* and *in vivo* results of the two groups do not agree; however, we feel that the *in vivo* results are less subject to artifactual effects.

ACKNOWLEDGMENTS

This research was supported in part by grants from the National Institutes of Health to B.C.K. (GM25604) and from the National Science Foundation to S.B.L. (PCM75-03540 and 80-13480) and in part by the Mayo Foundation. We are especially indebted to Dr. K. Matsubara for sharing information with us that has lead to the hypothesis that F copy number is controlled by a protein repressor of 29,000 molecular weight. We are also indebted to Dr. S. Mitsuhashi and his many colleagues who have provided the opportunity and means for us to present this brief summary on F maintenance.

REFERENCES

1. Lovett, M.A., and D.R. Helinski. 1976. J. Bacteriol. *127:* 982-987.
2. Timmis, K., F. Cabello, and S.N. Cohen. 1975. Proc. Natl. Acad. Sci. U.S.A. *72:*2242-2246.
3. Lane, D., and R.C. Gardner. 1979. J. Bacteriol. *139*:141-151.
4. Berquist,PP.L., R.A. Downard, P.A. Caughey, R.C. Gardner, and H.E.D. Lane. 1981. J. Bacteriol. *147*:888-899.
5. Seelke, R.W., B.C. Kline, J.D. Trawick, and G.D. Ritts. 1982. Plasmid *7*:000-000, in press.
6. Murotsu, T., K. Matsubara, H. Sugisaki, and M. Takanami. 1981. Gene, in press.
7. Wehlmann, H., and R. Eichenlaub. 1980. Mol. Gen. Genet. *180*:205-211.
8. Ebbers, J., and R. Eichenlaub. 1981. J. Bacteriol. *147*: 736-743.
9. Kline, B., and S. Palchaudhuri. 1980. Plasmid *4*:281-291

10. Eichenlaub, R., D. Figurski, and D.R. Helinski. 1977. Proc. Natl. Acad. Sci. U.S.A. *74*:1138-1141.
11. Manis, J.J., and B.C. Kline. 1977. Mol. Gen. Genet. *152*: 175-182.
12. Figurski, D., R. Kolter, R. Meyer, M. Kahn, R. Eichenlaub, and D.R. Helinski. 1978, in Microbiology-1978 (Schlessinger, D., ed.) American Society for Microbiology, Washington, D.C.
13. Eichenlaub, R., H. Wehlmann, and J. Ebbers. 1981, in Molecular Biology, Pathogenicity, and Ecology of Bacterial Plasmids (Levy, S.B., Clowes, R.C., and Koenig, E.L., eds.) pp. 327-336, Plenum Press, New York.
14. Eichenlaub, R., and H. Wehlmann. 1980. Mol. Gen. Genet. *180*:201-204.
15. Tolun, A., and D. Helinski. 1981. Cell *24*:687-694.
16. Manis, J.J., and B.C. Kline. 1978. Plasmid *1*:492-507.
17. Manis, J.J., and B.C. Kline. 1978. Plasmid *1*:480-491.
18. Tsutsui, H., and K. Matsubara. 1981. J. Bacteriol. *147*: 509-516.
19. Morutsu, T., and K. Matsubara. 1980. Mol. Gen. Genet. *179*:509-519.
20. Itoh, T., and J. Tomizawa. 1980. Proc. Natl. Acad. Sci. U.S.A. *77*:2450-2454.
21. Veltkamp, E., and A.R. Stuitje. 1981. Plasmid :76-99.
22. Kline, B., and D. Lane. 1980. Plasmid *4*:231-232.
23. Palchaudhuri, S., and K. Maas. 1977. Proc. Natl. Acad. Sci. U.S.A. *74*:1190-1194.
24. Ogura, T., T. Miki, and S. Hiraga. 1980. Proc. Natl. Acad. Sci. U.S.A. *77*:3993-3997.
25. Wechsler, J., and B.C. Kline. 1980. Plasmid *4*:276-280.
26. Wehlmann, H., and R. Eichenlaub. 1981. Plasmid *5*:259-258.
27. Thompson, R., and P. Broda. 1973. Mol. Gen. Genet. *127*: 255-258.
28. Gardner, R.C., P.A. Caughey, D. Lane, and P.L. Berquist. 1980. Plasmid *3*:179-192.
29. Birnboim, H.D., and J. Doly. 1979. Nucleic Acids Res. *7*:1513-1523.

γδ-MEDIATED MOBILIZATION OF PLASMIDS

N. Goto, S. Horiuchi, A. Shoji, and R. Nakaya

Department of Microbiology
Tokyo Medical and Dental University School of Medicine
Bunkyo-ku, Tokyo, Japan

The γδ sequence of the F plasmid has been ascribed a role in many types of *rec*A-independent recombination events. It is the DNA sequence located between coordinates 2.8 and 8.5 kb on the map of F. It is one of three sequences on F which serves as attachment sites for recombination between F and the *E. coli* chromosome leading to the formation of Hfr strains.

In 1977, Bolivar and others reported that the F plasmid is able to mobilize pBR322 during conjugation at a low frequency (1), and subsequently it was found that pBR322 has invariably an insertion of γδ after mobilization (2).

In the present paper, we describe findings that some small recombinant plasmids mobilized by F were sometimes associated with a γδ insertion located in a restricted area of the plasmid.

I. CONSTRUCTION OF PLASMIDS

pNR5302 was obtained by *Eco*RI-digestion and self ligation of R*ts*1, which is a well-known temperature-sensitive Km-resistant R plasmid (3). pNR5302 consists of 18.5 and 3.2 kb *Eco*RI fragments of R*ts*1 (4). Meyer and others found that the Km-resistant transposon Tn*2680* resides in the 18.5 kb fragment (personal communication).

We constructed another plasmid containing the replication gene of R*ts*1, using ColE1-*kan* as a source of Km-resistance gene which is a hybrid plasmid between ColE1 and the Km-resistance *Eco*RI fragment of R6-5 carrying Tn*903* (Miki, T., personal communication). pNR5311 was thus formed by recombining the 3.2 and 7.2 kb *Eco*RI fragments of R*ts*1 and R6-5, respectively. pNR5311 was then subcloned using the restriction endonuclease *Mlu*I (5), and pNR5411 was obtained.

II. MOBILIZATION OF THE PLASMIDS

The three plasmids were not self-transmissible. We found, however, that when harbored in Hfr and F'*lac* strains, pNR5311 and pNR5411 were able to transfer the Km-resistance by conjugation at relatively low frequencies but not pNR5302 (Table 1). pNR5321 is a plasmid which was isolated from a transconjugant of pNR5311 transfered by F. It carries a γδ insertion in the original plasmid pNR5311 (see below). Apparently pNR5321 could be mobilized at frequency of 10 to 30 times higher than that of its progenitor pNR5311. F'*lac* mobilized these plasmids more efficiently than Hfr.

III. PHYSICAL MAPS OF THE PLASMIDS

Physical maps of pNR5311, pNR5411 and pNR5321 are shown in Figure 1. The size of the inverted repeats of Tn*903* was adopted from the base sequence data reported by Oka and others (6). Compiling the map of R6-5 (7), pNR5302 (4), pNR5311 and pNR5411, the replicator gene of R*ts*1 was located between coordinates 1.65 and 3.2 kb on the map of pNR5311. pNR5321 had an insertion of 5.7 kb segment at coordinate 3.7 kb. The size, which was also confirmed by an electron microscopic study, and the restriction patterns of the segment strongly suggest that the inserted segment is the γδ sequence (2).

Table 1. *Transfer Frequency of the Plasmids Carried by Hfr and F'lac Strains*

Plasmid	*Transfer frequency*		*Ratio*
	Hfr[a]	*F'lac*[a]	*(F'lac/Hfr)*
pNR5311	6.1×10^{-7}	6.1×10^{-6}	*10*
pNR5411	8.1×10^{-7}	5.3×10^{-6}	*6.6*
pNR5320	$< 1 \times 10^{-8}$	$< 1 \times 10^{-8}$	-
pNR5321	6.5×10^{-6}	2.0×10^{-4}	*31*

[a]*Donor strain used. Hfr: W1895 (HfrC) and F'lac: JC3272 (F42.1).*

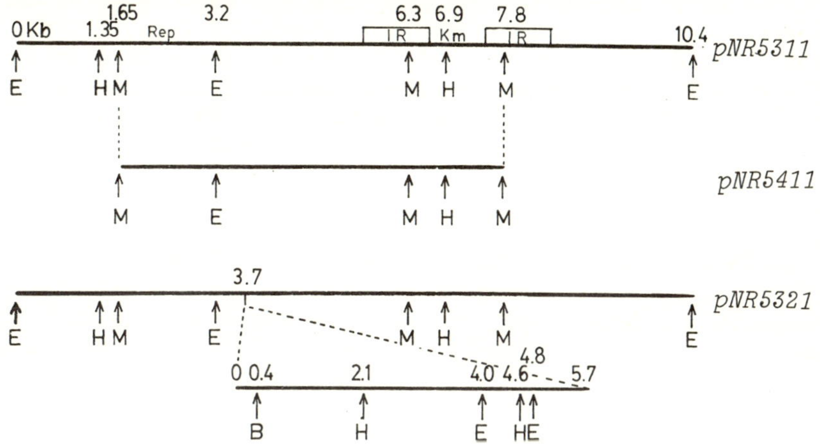

Fig. 1 Restriction map of the plasmids. The head of each arrow indicates the cleavage site of the restriction enzymes; B: BamHI, E: EcoRI, H: HindIII, and M: MluI. Rep and Km are the replication and the kanamycin-resistance genes, respectively. Coordinates are from one of the two EcoRI sites of pNR5311. In pNR5321, the line at 3.7 kb designates the insertion of the 5.7 kb fragment shown at the bottom. In pNR5311, IR in the rectangular represents inverted repeats of Tn903 (7).

IV. INSERTION OF THE γδ SEQUENCE INTO pNR5311 AND DISTRIBUTION OF THE INSERTION SITES

Conjugation was further carried out using W1895 [HfrC (pNR5311)] as a donor and 48 independent transconjugants were isolated. Plasmid DNAs were purified and examined for the insertion of γδ. As seen in Table 2, pNR5311::γδ plasmids were found in 95% of the transconjugants at 37 C. Only one of 20 plasmids had no insertion and was subsequently found to have a Tn*903* in inverted orientation.

These results seem to imply that the γδ sequence (and in some cases Tn*903* as well) is involved in mobilization of pNR5311. The frequency of the γδ-insertion was remarkably low at either 42 C or 30 C, suggesting that one or more other mechanisms which do not require the insertion of γδ may be involved in the mobilization of pNR5311 by F.

A total of 21 pNR5311::γδ were examined and the sites of insertion were located on the map of pNR5311 (Fig. 2). Most of the sites were mapped in a rather restricted area. The zone II, whose size is 43% of the total nonessential region of pNR5311 DNA, contained more than 90% of the insertion sites (Table 3). Statistical analysis of the observed distribution of γδ indicated the distribution within the 3.4 kb sequence is not random with a 99% probability.

We then purified the DNA fragments II, III and III + I by enzyme digestion followed by preparative agarose gel electrophoresis and determined for their buoyant densities in a CsC1 gradient. The density of the zone I fragment DNA was

Table 2. *Insertion Frequency of γδ into Mobilized Plasmids*

Conjugation temperature	*No. of plasmid DNAs Examined/γδ-inserted*		*Insertion frequency*
37 C	*20*	*19*	*0.95*[a]
42 C	*18*	*2*	*0.11*[b]
30 C	*10*	*2*	*0.20*

[a] *The plasmid without γδ had an inverted Tn903.*

[b] *Two out of 16 plasmids without γδ had an inverted Tn 903.*

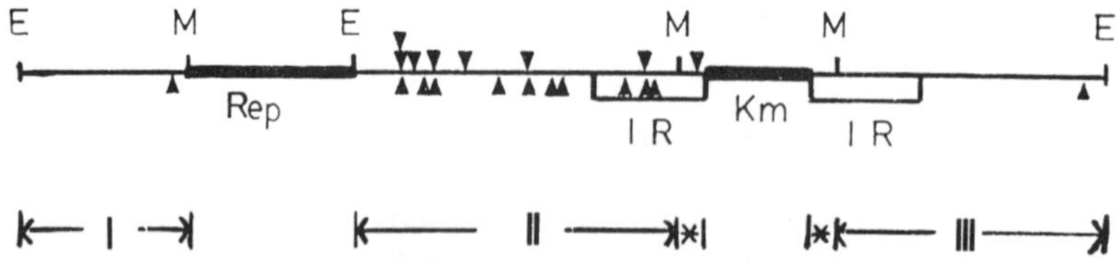

Fig. 2. Location of the insertion sites of γ to δ (▼) and δ to γ (▲) orientations, respectively, on pNR5311:: γδ plasmids. The thick lines indicate essential regions. Other symbols are described in the legend of Fig. 1. I, II and III are three nonessential regions each flanked by an EcoRI and an MluI site. The asterisks designate the small segments flanked by a MluI site and the inner end of the inverted repeats of Tn903.

Table 3. *Distrubution of γδ-insertion Sites in pNR5311*

Zone[a]	*Size (kb)*	*Number of γδ-sites*[b] *Expected/Observed*		*%GC*
I	*1.65*	*4*	*1*	*49*
*II**	*3.4*	*9*	*19*	*47*
*III**	*2.9*	*8*	*1*	*57*

[a]*See Fig. 1. II* and III* designate *-segment plus zone II and III, respectively.*

[b]*Expected values were calculated assuming that the 21 γδ-sites were distrubuted at random in each zone.*

calculated from the densities of III and III + I by taking into account their relative sizes (Table 3). It appears that the observed distribution of the insertion sites reflects the presence of AT-rich regions as have been reported in the case of IS*1* insertion into *gal*- (8) and into phage P1 (9).

Reed and others analyzed the base sequences of three different pBR322::γδ plasmids and found that the sequences were flanked by five base-pair direct repeats (10). Among a total of 15 base pairs of such sequences, 11 pairs were AT. Putting their findings and our present results together, it is possible to postulate that γδ is inserted preferably into AT-rich regions, resulting in the duplication of five base-pair sequence.

REFERENCES

1. Bolivar, F. et al. (1977) Gene, *2*, 95-112
2. Guyer, M.S. (1978) J. Mol. Biol. *126*, 347-365
3. Terawaki, Y., Rowns, R., and Nakaya, R. (1974) J. Bacteriol. *117*, 687-695
4. Goto, N. Manuscript in preparation
5. Sugisaki, H. and Kanazawa, S. (1981) Gene, in press
6. Oka, A., Sugisaki, H., and Takanami, M. (1981) J. Mol. Biol. *147*, 217-226
7. Andrés, I. et al. (1979) Mol. Gen. Genet. *168*, 1-25. 8
8. Kühn, S., Fritz, H. J., and Falkow, S. (1975), Proc. Natl. Acad, Sci. USA *70*, 3623-3627
9. Meyer, J. and Iida, S. (1979) Mol. Gen. Genet. *176*, 209-219
10. Reed, R.R. et al. (1979) Proc. Natl. Acad,.Sci. USA *76*, 4882-4886

INITIATION OF DNA REPLICATION OF ANTIBIOTIC RESISTANCE PLASMID R6K[1]

Manabu Inuzuka, Noriko Inuzuka, Akira Taketo and Donald R. Helinski*

Department of Biochemistry
Fukui Medical School
Fukui, Japan
and
** Department of Biology*
University of California at San Diego
La Jolla, California, U.S.A.

INTRODUCTION

Plasmid DNA has proved to be an interesting model system for the study of replication of closed circular duplex DNA. Replication and maintenance of this element depends both on host and plasmid functions. Especially replicon-specific genes and gene products play an important role on the stable inheritance of the plasmids in host cells. Detailed analysis of the DNA replication can be obtained by the application of recombinant DNA technique, rapid method for the determination of nucleotide sequence and establishment of an *in vitro* system for plasmid replication. These approaches are carried out in the analysis of the initiation of replication of R6K. Activity of all three replication origins by a plasmid-encoded initiation protein and interaction of this protein with one of the origin regions for the initiation of replication will be considered in this article.

PLASMID R6K

R6K is conjugative, 38 kilobase pairs(kb) in size, a member of incompatibility group X, and specifies resistance to ampicillin and streptomycin. This plasmid exists at 10-15 copies per chromosome in *E. coli* and does not require DNA polymerase I for replication(1). This plasmid employs a very

[1] *This work was supported by Ministry of Education of Japan, National Science Foundation and National Institute of Health of U.S.A.*

unique replication mode which has been revealed in the presence of three replication origins, designated *ori*α, β and γ, and an asymmetric terminus(Fig. 1)(2-5). In *E. coli*, at least two of origins, *ori*α and β, exhibit a sequential, bidirectional replication toward an asmmetric terminus(2,3). An *in vitro* system has been developed for the replication of R6K (6,7). This can give direct evidence that the plasmid-coded protein, designated π protein, is required for the initiation of R6K DNA replication(7). Recombinant DNA technique has dissected the basic replicon into the *ori*γ and the structure gene of π protein, *pir*(8). Nucleotide sequence of the *ori*γ region reveals a striking feature of the presence of 7 tandem direct repeats consisting of 22 base pairs(bp). An eighth 22 bp repeat is located near the putative promoter for the *pir* gene(9).

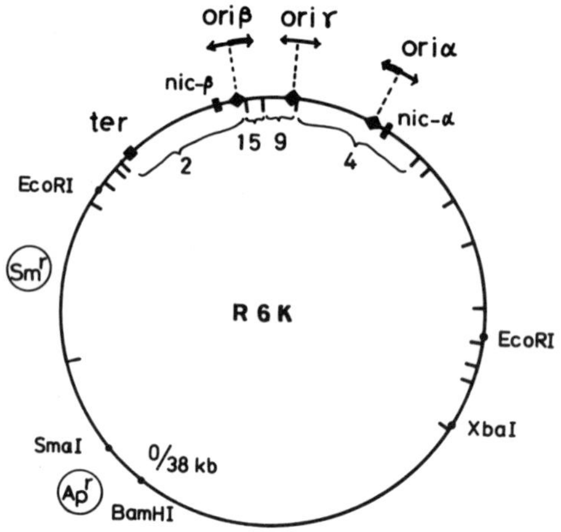

Fig. 1. Physical and genetic map of plasmid R6K. Ter refers to the terminus of replication. 4,9,15 and 2 refers to specific HindIII fragments in the replication regions, and (|) indicates other HindIII sites.

π PROTEIN ACTIVATES ALL THREE ORIGINS FOR REPLICATION

*Ori*γ plasmid was isolated as a π-dependent *ori* plasmid *in vivo*(8). However, an attempt to isolate *ori*α and *ori*β as a functional origin was unsuccessful even when π protein was supplied *in trans*. As mentioned earlier, three origins are active *in vivo* and *in vitro*. This discrepancy raised the question that the π protein requirement may be limited to the replication from *ori*γ or that the other protein(s) encoded by the *ori*α and β regions may be required *ori*α- and *ori*β-functions in addition to π protein. To this question, the following *in vitro* replication system can give a clear answer, since this system does not synthesize any template DNA-encoded protein.

A miniR6K derivative, pRK419, produced only π protein as an R6K-encoded protein that is 35,000 daltons of molecular weight(10). Replicative intermediates were synthesized in the *in vitro* system containing gene products of pRK419 according to the procedure previously described(5). Electron microscopic analysis of 87 molecules has shown that the replication starts from *ori*α, β and γ at the frequencies of 39 %, 30 % and 31 %, respectively(10). We can conclude therefore that π protein is required for the initiation of replication from any one of the three origins and is only one protein encoded by R6K genome as the initiation protein.

INTERACTION OF π PROTEIN WITH THE *ORI*γ REGION FOR INITIATION

Replication of R6K DNA *in vitro* is strongly inhibited by excess R6K DNA. This inhibition is brought back to the normal state with the addition of π protein fraction, but it is not affected by the concentration of the host proteins(7). These results indicate that π protein activity can be titrated by the excess target DNA which binds to π protein or π protein-complex. In other words, a binding site of π protein on R6K genome for the initiation can be detected by this procedure.

First, the origin regions(*ori*α, β and γ) were cloned onto a pACYC177 vector, and their replication properties were examined. As shown in Fig. 2, pNI1 and pMI234 containing the *ori*α and β regions, respectively, could not replicate in *E. coli*, even when π protein was supplied *in trans*. Furtheremore, insertion of *Hind*III fragment 15 of R6K at the *Hind*III site between fragment 4 and 9 in pNI22(i.e. pNI24) inactivated the origin activity. Thus, replication from *ori*α and β requires a *cis* interaction of the *ori*γ region.

To test the binding site of π protein, plasmid DNA carrying DNA segment indicated in Fig.2 was used as a competitor DNA in the *in vitro* R6K DNA synthesis system. pACYC177 and pNI1 did not inhibit the DNA synthesis even when 6 times more DNA than the template R6K DNA were added (Table 1). On the other hand, pNI8 and pNI21 DNA had a strong inhibitory activity. These results indicate that π protein does not interact with the *ori*α region and that the binding site of π protein or π protein-complex locates on *Hind*III-*Bgl*II fragment(277 bp) in the *ori*γ region. This binding activity of π protein was not necessary to requir the origin activity. Involvement of the direct repeats on π binding site is also suggested, because pNI8 carries the *pir* promoter region containing eighth repeat

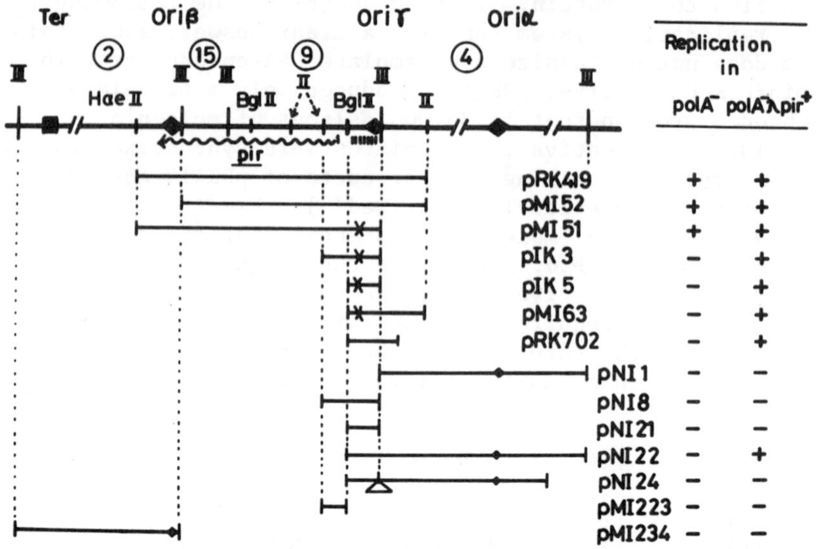

Fig. 2. MiniR6K derivatives and plasmids carrying the R6K origin regions. Plasmids shown below from pNI1 are pACYC177 derivatives. This vector cannot replicate in polA⁻ cells(11). Deletion mutation of 6th repeat among 7 direct repeats in the HindIII-BglII fragment on pMI51 to pMI61 is shown(X) and insertion site for HindIII fragment 15 is indicated(Δ). Replication properties in polA⁻ π pir⁺ cells are shown.

in addition to 7 direct repeats region that is also present in pNI21.

ROLE OF DIRECT REPEAT SEQUENCE ON THE π BINDING ACTIVITY

Region of π binding site contains 7 direct repeats consisting of 22 bp. A mutant of a miniR6K derivative, pMI51, that deleted the 6th repeat was isolated and found to have approximately one half of copy number when compared with that of the parental plasmid *in vivo*. From this deletion mutant, low molecular weight of *ori* plasmids that are dependent on π protein were derived as shown in Fig. 2. Using this plasmid DNA as competitor DNA, binding activity was examined as described before. Deletion mutant DNA had less π binding activity, one half to one third, than that of intact wild DNA (Table 1). These results clearly show that at least the 6th repeat in the *ori*γ region is involved in the binding of π protein or π protein-complex for replication. Furtheremore, the results suggest that less copy number of the deletion

Table 1. *Competitive Inhibition of R6K DNA Synthesis In Vitro by Plasmid DNA Containing π Binding Site*

Competitor DNA	Relative DNA synthesis[a]			
	Molar ratio of competitor DNA to template DNA			
	X 1.5	*X 2*	*X 3*	*X 6*
R6K	*29 %*		-	-
pACYC177	-		-	*100 %*
pNI1	-		*73 %*	*104 %*
pNI21	-		*69 %*	*29 %*
pNI8	-		*8 %*	*1 %*
pIK3		*62 %*		
pIK5		*69 %*		
pMI61		*86 %*		
pRK702		*28 %*		

[a] *Amount of the incorporation of TMP into R6K DNA in the standard reaction mixture without any competitor DNA was 53 pmoles. This refers to 100 %. pACYC177 is a vector DNA and can not replicate under this condition.*

mutant *in vivo* is due to less binding activity of π protein which is an initiation protein.

CONCLUDING REMARKS

An R6K-encoded initiation protein, π, is required for the activity of all three origins of R6K replication *in vitro*. No other protein encoded by the *ori*α and β regions is necessary for the initiation except for *E. coli* replication proteins. For the initiation from any one of the three origins, π protein interacts with only the *ori*γ region in the *in vitro* system. This conclusion is also supported by the *in vivo* data which states that the *ori*α or β segment lacking the *ori*γ region can not replicate even when π protein is supplied *in trans*. Involvement of the direct repeats in the binding site of π protein has been revealed. It is also suggested that π protein interacts with the putative promoter region of the *pir* gene. At present it is not clear whether π protein itself or π host protein complex binds to R6K DNA at the specific site. Considering the above results, it is very interesting that an initiation protein of phageλ, *o* protein, binds to the 4 direct repeats in the origin region (12).

To explain the regulation of initiation of R6K DNA replication, we proposed the following working model. That is, π protein interacts with the origin as a positive regulatory element for the initiation and also interacts with the π promoter region resulting in the autogenous expression of *pir* gene(8). In this article, we have proved the interaction of π protein and R6K DNA. However, it remains to be revealed by more investigation how R6K DNA replication initiates at multiple origin sites and how the copy number is constantly maintained.

REFERENCES

1. Kontomichalou, P., Mitani, M. & Clowes, R.C.(1970) J. Bacteriol. *104*, 33-44
2. Lovett, M.L., Sparks, R.B. & Helinski, D.R.(1975) Proc. Natl. Acad. Sci. USA *72*, 2905-2909
3. Crosa, J.H., Luttrop, L.K., Heffron, F. & Falkow, S.(1975) Mol. Gen. Genet. *140*, 39-50
4. Crosa, J.H.(1980) J. Biol. Chem. *255*, 11075-11077
5. Inuzuka, N., Inuzuka, M. & Helinski, D.R.(1980) J. Biol. Chem.*255*, 11071-11074
6. Inuzuka, M. & Helinski, D.R.(1978) Biochemistry *17*, 2567-2573
7. Inuzuka, M. & Helinski, D.R.(1978) Proc. Natl. Acad. Sci. USA *75*, 5381-5385
8. Kolter, R., Inuzuka, M. & Helinski, D.R.(1978) Cell *15*, 1199-1208
9. Stalker, D.M., Kolter, R. & Helinski, D.R.(1979) Proc. Natl. Acad. Sci. USA *76*, 1150-1154
10. Inuzuka, M. & Inuzuka, N. FEBS Letters, in press
11. Chang, A.C.Y. & Cohen, S.N.(1978) J. Bacteriol. *134*, 1141-1156
12. Tsurimoto, T. & Matsubara, K (1981) Nucleic Acid Res. *9*, 1789-1799

GENETICAL DETERMINANTS FOR THE REGULATION OF DNA REPLICATION OF IncFII PLASMID,R6-5

H.Danbara[1] and M.Yoshikawa

The Institute of Medical Science
The University of Tokyo
Minato-ku,Tokyo,Japan

J.K.Timmis and K.N.Timmis

Department of Medical Biochemistry
University Medical Center
Geneva,Switzerland

INTRODUCTION

A fundamental property of living cells is the ordered transmission of a complete set of genetic elements from parental to progeny cells at cell division.Plasmids belonging to incompatibility group FII (IncFII) such as R1(R1drd-19),R6-5 (R6) and R100 (NR1,R222,R100-1) have been used widely as a good replicon model for the studies on the regulation mechanisms of DNA replication ,particularly,because their replication is regulated stringently.The basic replicon of these IncFII plasmids,designated as a RepA replication region,has been generated in a convenient molecular size by genetic engineering to allow the investigation on molecular levels and now has become,like ColE1 and F plasmids,one of the most defined replicons.We have already reported the genetical determinants required for the replication of R100-1 plasmid (1-3) and recently found the genetical elements for copy number control and incompatibility of R1 and R6-5 (4-9). In this communication,isolation and genetical characterizations of high as well as low copy number mutants derivatives of mini-R6-5 plasmid are described and the role of the genetical elements,so far identified,on the copy number control and incompatibility mechanisms is discussed.

1. *H.D. gratefully acknowledges receipt of a postdoctoral fellowship from the Alexander von Humboldt Stiftung.*

I.NATURES OF *RepA* REPLICON OF *IncFII* PLASMIDS

R1,R6-5 and R100 plasmids have been reported to share extensive sequence homology in heteroduplex analysis(10).Genetical studies have demonstrated that these plasmids are a composite replicon consisting of the regions of RepA,RepB,Stb(Par),RepC and RepD.RepA replication region is known to be a basic replicon of IncFII plasmids.RepB,Stb(Par),RepC and RepD regions seem to be required for the stable inheritance of RepA replicon,although mechanisms are not known yet(1,3,7,8,11-14). RepA replicon consists of RepA determinant and other determinants required for copy number control and incompatibility. Polynucleotide sequences of RepA replication region of R1,R6-5 and R100 plasmids are reported to be essentially identical(9, 17). We have constructed miniplasmids of R6-5 which contain only RepA replicon and have a convenient size for the studies on molecular levels(6,9). These mini-R6-5 plasmids consist of two essential PstI fragments generated by the digestion of R6-5 plasmid with restriction endonuclease PstI and replicate stringently like parent R6-5 plasmid. P-6, one of the two essential PstI fragments of mini-R6-5 plasmid contains the determinants,Cop/Inc,required for copy number control and incompatibility.Another PstI fragment,P-4,contains the origin of vegetative replication,OriV,of R6-5 plasmid(18). RepA determinant which is a genetical element required for replication of R6-5 plasmid straddles the junction of P-6 and P-4 fragments(6). The Genetical structure of RepA replication region of R6-5 plasmid is shown in Fig.1.

II. COPY NUMBER MUTANT DERIVATIVES OF R6-5 PLASMID

A. DIRECT ISOLATION OF HIGH COPY NUMBER MUTANTS

To understand the mechanisms of the regulation of the DNA replication of IncFII plasmids,high copy number mutants have been already isolated from wild type plasmids,R1 and R100,which are a composite plasmid containing not only RepA replication region but also other regions such as RepB,Stb(Par),RepC and RepD and also IS elements(19-21).This happened to provoke difficulties to characterize the mutants because of their structural complexities and intramolecular instability.We have taken an advantage to use a miniplasmid of R6-5 which contains only RepA replication region and isolated high copy number mutants.Cells carrying RepA replication region of R6-5 with a determinant for ampicillin resistance derived from Tn3 were treated with hydroxylamine *in vivo* and spread on plate containing 800 μg of ampicillin per ml,on which cells carrying the wild type ampicillin resistant miniplasmid of R6-5 can not grow.Cells carrying high copy number mutants and expressing elevated ampicillin resistance were isolated and characterized.Some properties of high copy number mutant derivatives were shown in Table 1.In consistent with the report by Uhlin and Nordström(20), incompatibility of all of the copy number mutants thus isolated was altered. One class,exemplified by pKT410,showed a weaker incompatibility, whereas others , exemplified by pKT411 and pKT419,showed stronger incompatibility towards a tester IncFII plasmid than that of the parent,pKT401.Reconstruction experiments of miniplasmid consisting of one essential PstI fragment from a high copy number mutant and another from R6-5 revealed that only P-6 fragment from copy number mutants was able to generate a new miniplasmid that exhibited high copy number and altered incompatibility.

B.INDIRECT ISOLATION OF HIGH COPY NUMBER AND LOW COPY NUMBER MUTANTS

The finding obtained above by the characterization of high copy number mutants that incompatibility of all of the copy number mutants isolated was altered either to be strengthened or to be weakened suggested us another strategy to isolate copy number mutant derivatives of R6-5 plasmid.Thus a mutant with altered incompatibility was expected to be a copy number mutant. P-6 fragment of RepA replicon of R6-5 plasmid,on which the determinants for IncFII incompatibility is located,had already been cloned into PstI site of a vector plasmid pBR322(4).This hybrid plasmid between pBR322 and P-6 fragment of R6-5,conferring resistance to tetracycline(TC),was treated with hydroxylamine *in vitro* and tranformed a streptomycin(SM) resistant derivative of C600 strain by selection with TC and SM resistance. These transformants were then replica-plated onto a nutrient agar plate containing TC,SM and chloramphenicol(CM).Cells of CR34 strain carrying an R100 plasmid derivative which was TC sensitive and highly transmissible by conjugation had been spread on the same plate before replica-plating.Two types of TC and SM resistant transformants of C600 strain were observed with respect to their ability to coinherit R100 plasmid; one showed decreased coexsisting ability and another increased coexsisting ability when these abilities were compared with that of transformant carrying the parent hybrid plasmid. DNA of these hybrid plasmids was isolated and introduced into a recA strain carrying a CM resistant miniplasmid of R6-5.Segragation tests revealed that hybrid plasmids which coexsisted with R100 plasmid stably showed weaker incompatibility,whereas hybrid plasmids which excluded R100 plasmid more frequently than the parent

Table 1. *Properties of Copy Number Mutants of R6-5*

Mutant[a]	*Copy effect*	*Incompatibility*[b]			*Mutation*[e]
		Miniplasmid		*P-6/ vector hybrid*[d]	
		N-1[c]	*Pref.loss*		
pKT401 (A)	*1*	*8.8*	*Parent*	+++	*Wild*
pKT410 (A)	*16*	~*70*	——	-	*CopA/W.Inhibitor*
pMY1123 (B)	*4*	~*70*	——	-	*CopA/W.Inhibitor*
pKT411 (A)	*4*	*4.7*	*Test*	+++	*CopT/W.Target*
pKT419 (A)	*4*	#	*Mutant*	+++++	*CopT/W.Target*
pMY1124 (B)	*0.5*	*8.7*	*Test*	+++++	*CopA/S.Inhibitor (CopT/W.Target)*
pMY1125 (B)	*0.25*	*9.7*	*Test*	+++++	*CopA/S.Inhibitor (CopT/W.Target)*

a. *Copy number mutants were isolated by direct(A) or Indirect(B) method as described in the text.*

b. *Test plasmids are CM resistant mini-R6-5 derivative, pKT421, in case of pKT401, pKT410, pKT411 and pKT419 or KM resistant mini-R6-5 derivative, pKT071, in case of pMY1123, pMY1124 and pMY1125.*

c. *Number of cell doublings required to reduce the proportion of mixed plasmid state cells by one order of magunitudes(# Indicates incompatibility too severe to be able to determine).*

d. *P-6 fragment of the mutant and parent miniplasmids were cloned into pACYC177 in case of pKT401, pKT410, pKT411 and pKT419 or into pBR322 in case of pMY1123, pMY1124 and pMY1125 (+++ and +++++ indicate severe and severer incompatibility, respectively.- indicates compatible).*

e. *W. and S. indicate weaker and stronger, respectively.*

showed severer incompatibility towards the tester IncFII plasmid than that of the parent hybrid plasmid. Miniplasmids were constructed by ligating a P-6 fragment of hybrid plasmids which exhibited altered incompatibility, P-4 fragment of the parent R6-5 plasmid and P-CM fragment which carried CM resistance determinant of S-a plasmid generated by PstI endonuclease digestion. Tn3 element was transposed from pSC101::Tn3 to the CM resistant miniplasmids thus constructed. CM and AP resistant miniplasmid with Tn3 inserted outside P-6+P-4 were used and characterized further. Minimum inhibitory concentration to ampicillin exhibited by these miniplasmids was determined to estimate the copy number of these plasmids. Two types of AP and CM resistance miniplasmids were obtained with respect to their copy number. One class of miniplasmid containing P-6 fragment with strengthened incompatibility showed a copy number lower than that of the parental miniplasmid. Another class of miniplasmid containing P-6 fragment with a weakened incompatibility was shown to exhibit a copy number higher than that of the parent miniplasmid.

C. CHARACTERIZATION OF COPY NUMBER MUTANT DERIVATIVES

Pritchard (22) has proposed that regulation of DNA replication is controlled by attachment of a negatively trans-acting control element to a target site. Based upon this hypothesis, pKT410 and pMY1123 appear to have been mutated in a replication inhibitor and produced inhibitor with weaker activity than that of the parent. pKT411 appeared to have been mutated in the inhibitor target to exhibit a weaker activity. The result that pKT4 19 was preferentially lost and the parent tester plasmid remained would be explained as follows; pKT419 has a reduced target activity and the inhibitor produced by this mutant can not bind to its mutated target site efficiently, whereas the inhibitor produced by the parent can bind to this mutant target more efficiently than to the wild type target. These results suggested that a part of polynucleotide sequences required for both inhibitor and inhibitor target would be common and located in a same region on P-6 fragment. Polynucleotide sequence analysis of P-6 fragment of the parent, target-type copy number mutants and also inhibitor-type copy number mutants has confirmed this supposition and revealed that the target sequence lies within the possible structure gene for the inhibitor(9, G.Brady and K.N. Timmis submitted). pMY1124 and pMY1125 are low copy number mutants. The marked feature of these low copy number mutants is their ability to exclude the tester plasmid preferentially. This result suggests that the mutation of these two low copy number mutants resides in the inhibitor gene. This mutation strengthened the inhibitor activity and simultanously weakened the target activity.

III. MODEL FOR COPY NUMBER CONTROL AND INCOMPATIBILITY MECHANISMS OF R6-5 PLASMID

Genetical elements and their possible role for the copy number control and incompatibility mechanisms of R6-5 plasmid are shown in Fig.1.CopB determinant would produce a 9.5K small basic peptide(6,9).CopA determinant could produce either 7.2K small peptide or a short untranslated 91 nucleotides RNA (9). This RNA molecule has been identified in R100 plasmid(23,see Rownd et al. this volume),whereas a 7.2K peptide has not.CopT determinant is presumably a DNA sequence for inhibitor binding. Sequence analysis of two independently isolated target-type mutants showed that nucleotide deoxyguanosine(G) was changed to deoxyadenosine(A) in both mutants and these mutation sites were localized within CopA determinant(9).The results described above and other evidences such as RNA polymerase binding experiments(24,R.Lurz,H.Danbara,B.Rückert and K.N.Timmis, submitted) seem to have substantiated a model for the mechanism of regulation of DNA replication of R6-5 plasmid shown in Fig.1.

At least three genetical determinants on P-6 fragment are involved in copy number control and incompatibility.The products of CopB and CopA genes are negatively acting elements which interact with the third element, that is ,the CopT target sequence ,to regulate the frequency of initiation of plasmid replication.

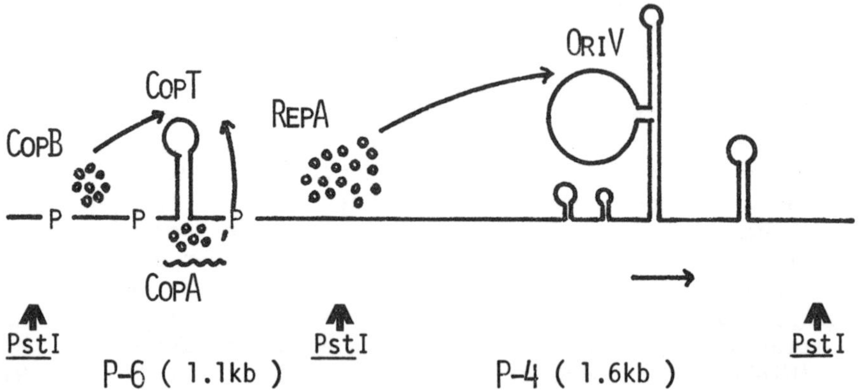

Fig.1. Structure of RepA replication region and possible mechanisms of copy number control and incompatibility of R6-5 plasmid. P; RNA polymerase binding site.

ACKNOWLEDGMENTS

Parts of this study were done at Max-Planck Institute for Molecular Genetics in W.Berlin.We are very grateful to R.Lurz, G.Brady,T.Hashimoto-Gotho,M.Bagdasarian and C.Franklin for continuous discussions and critics. We also thank D.Vogt for valued technical assistance.

REFERENCES

1. Yoshikawa M.(1974) J.Bacteriol.*118*,1123-1131
2. Yoshikawa M.(1975) J.Bacteriol.*124*,1097-1100
3. Yoshikawa M.,Danbara H. and Sasakawa C.(1975) in Microbial Drug Resistance (Mitsuhashi S. and Hashimoto H.eds.) pp.51-60,Univ.of Tokyo Press,Tokyo
4. Timmis K.N.,Andrés I. and Slocombe P.M.(1978) Nature (London) *273*,27-32
5. Timmis K.N.,Andrés I.,Slocombe P.M. and Synenki R.M.(1979) Cold Spring Harbor Symp.Quant.Biol. *43*,105-110
6. Andrés I.,Slocombe P.M.,Cabello F.,Timmis J.K.,Lurz R., Burkardt H.J. and Timmis K.N.(1979)Molec.gen.Genet. *168*,1-25
7. Danbara H.,Timmis J.K.,Lurz R. and Timmis K.N.(1980) J.Bacteriol. *144*,1126-1138
8. Danbara H.,Timmis J.K. and Timmis K.N.(1980) in Antibiotic Resistance -Transposition and Other Mechanisms(Mitsuhashi S.,Rosival L. and Krcméry V. eds.)pp.105-112, Springer-Verlag,Berlin,Heiderberg,New York
9. Danbara H.,Brady G.,Timmis J.K. and Timmis K.N.(1981) Proc.Natl.Acad.Sci.USA. *78*, 4699-4703
10. Sharp P.A.,Cohen S.N. and Davidson N.(1973) J.Mol Biol. *75*,235-255
11. Hashimoto H. and Mitsuhashi S.(1970) in Progress in Antimicrobial and Anticancer Chemotherapy (Umezawa ed.) pp. 545-551,Univ.of Tokyo Press, Tokyo
12. Timmis K.N.,Cabello F. and Cohen S.N.(1978) Molec.gen. Genet. *162*,121-137
13. Miki T.,Easton A.M. and Rownd R.H.(1980) J.Bacteriol. *141*,87-99
14. Nordsröm K.,Molin S. and Aagaard-Hansen H.(1980) Plasmid *4*,215-227
15. Ike Y.,Hashimoto H. and Mitsuhashi S.(1981) J.Bacteriol. *147*,578-588
16. Timmis K.N.,Danbara H.,Brady G. and Lurz R.(1981) Plasmid *5*,53-73

17. Rosen J.,Ryder T.,Inokuchi H.,Ohtsubo H. and Ohtsubo E. (1980) Molec.gen.Genet. *179*,527-537
18. Synenki R.M.,Nordheim A. and Timmis K.N.(1979) Molec.gen.Genet. *168*,27-36
19. Morris C.F.,Hashimoto H.,Mickel S. and Rownd R.(1974) J.Bacteriol. *118*,855-866
20. Uhlin B.E. and Nordström K.(1975) J.Bacteriol.*124*,641-649
21. Luibrand G.,Blohm D.,Mayer H. and Goebel W.(1977) Molec.gen.Genet. *152*,43-51
22. Pritchard R.H.(1978) in DNA Synthesis -Present and Future (Molineux I. and Kohiyama M.ed.) pp.1-26 Plenum,New York
23. Rosen J.,Ryder T.,Ohtsubo H. and Ohtsubo E.(1981) Nature (London),*290*,794-797
24. Slocombe P.M.,Ely S. and Timmis K.N.(1979) Nuc.Acid Res. 7,1469-1484

STUDIES ON VARIOUS PHENOTYPES OF DRUG RESISTANCE FACTOR, Rts1[1]

Akira Kaji, Hisashi Yoshimoto
Tatsuo Yamamoto and Sheldon Finver

Department of Microbiology, School of Medicine,
University of Pennsylvania
Philadelphia, Pennsylvania, U.S.A.

Rts1 is multiphenotypic drug resistance factor which confers host bacteria resistance to kanamycin. Its multiphenotypes include temperature sensitive host cell growth (Tsg) (*1, 2*), temperature sensitive formation of ccc plasmid DNA (Tsc) (*3, 4*), temperature dependent T4 phage restriction (*5-7*) and plasmid elimination. Some but not all of these phenotypes are dependent on the presence of cAMP (*8*). In this article we wish to review briefly the representative information for these phenotypes. We shall discuss their relationships and our recent preliminary attempts to clone some of these genes.

The temperature dependent restriction of T4 phage growth is largely due to an endonuclease coded by Rts1. This endonuclease is unique in its specificity which requires glucosylated T4 DNA. The requirement for glucosylation is dependent on an appropriate Mg^{2+} concentration. As shown in Table 1 this endonuclease converts T4 DNA to slower sedimenting DNA fragments. It does not degrade T4 DNA into acid soluble oligonucleotides. Other DNA such as T7 DNA is insensitive to this endonuclease (Table 1). This enzyme has been purified 500 fold and shown to be unstable at 42°C reflecting its *in vivo* property (*6*). The major reason for restriction of T4 phage growth by Rts1 is due to efficient inhibition of new T4 DNA formation caused by the fragmentation of incoming T4 DNA by this enzyme. The fragmented DNA appears to serve as template for T4 RNA synthesis because early T4 meassenger RNA was still made even though T4 progeny was not made (*7*).

The second phenotype of Rts1 to be reviewed here is its temperature sensitive growth effect (tsg). A peculiar nature of this effect is that at 42℃ the total number of *E. coli* harboring Rts1 increases linearly while viable counts stay constant (Fig. 1). The simplest explanation for this observation is that every cell division yields one non-viable cell and one viable cell. The viable cell will undergo a similar cell division giving another dead cell. Repetition of such cell division will result in accumulation of dead cells without changing the number of viable cells.

The third phenotype of interest is Tsc; temperature sensitive for-

[1] *Supported by U.S.P.H.S. GM 12053*

Table 1. *Substrate Specificity of Rts1 Endonuclease*

DNA and cell extracts used	*DNA cleaved* [a]
	%
Nonglucosylated T4[³H] DNA	*0*
Nonglucosylated DNA + crude JC7623, 32 ℃ extract	*0*
Nonglucosylated DNA + crude JC7623, 42 ℃ extract	*4.9*
Nonglucosylated DNA + crude JC7623/Rts 1, 32 ℃ extract	*7.0*
Nonglucosylated DNA + crude JC7623/Rts 1, 42 ℃ extract	*9.1*
Glucosylated T4 DNA + crude JC7623/Rts 1, 32 ℃ extract	*96.5*
T7[³H] DNA [b]	*0*
T7[³H] DNA + crude JC7623/Rts 1, 32 ℃ extract	*4.8*

The reaction was carried out and reaction products were analyzed by SDS-sucrose gradient centrifugation.

[a] *Cleaved DNA is defined as DNA sedimenting above Fraction 19 after 10 min of digestion at 32 ℃.*

[b] *T7DNA was centrifuged at 35,000 rpm for 3hr.*

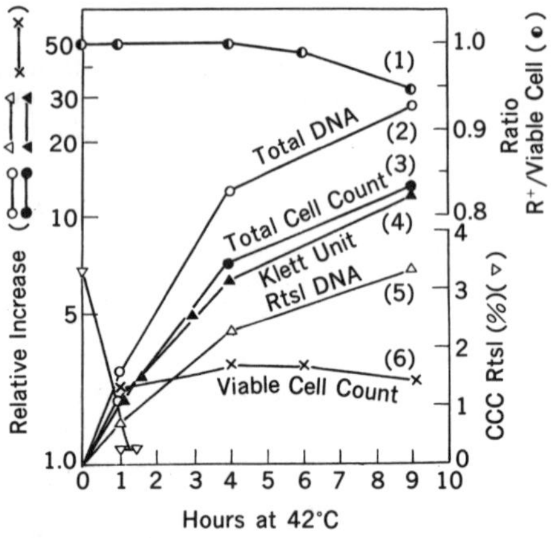

Fig. 1. Time course of the change of (1) percentage of Rts1-containing cells, (2) total DNA, (3) total cell counts, (4) Klett units of the culture, (5) the radioactive ccc Rts1 DNA, (6) viable counts, and (7) newly synthesized ccc DNA of E. coli 20S0 (Rts1) cells during growth at 42 ℃.

mation of covalently closed circular (ccc) DNA. As shown in Fig. 1 within 1 hr, after the temperature shift up to 42 ℃, no synthesis of ccc Rts1 DNA was observed. On the other hand, absolute amount of Rts1 DNA, measured by DNA-DNA hybridization, increased at least five fold during the incubation period at 42 ℃ (4). It should be pointed out that

Table 2. *Loss of R Factors from Host Cells*

Culture	*Incubation temperature (°C)*	*Incubation period (hr)*	*Viable count (cells/ml)*	*Percent colonies sensitive to kanamycin or ampicillin (%)*
R28K		*0*	0.42×10^3	*0*
	27	*24*	2.2×10^9	*0*
	42	*24*	5.8×10^8	*0*
Rts1		*0*	0.98×10^3	*0*
	27	*24*	1.3×10^9	*0*
	42	*24*	2.5×10^4	*77*

Cultures of E. coli 20S0/(Rts1) and (R28K) were diluted to 0.5 to 1 × 10^3 *cells per ml and incubated in trypticase soy broth at 27 ℃ or 42 ℃ for 24 hr without shaking.*

the pre-formed Rts1 DNA does not disappear at 42 ℃ nor coexisting other plasmid was influenced by Rts1. These observations point out the specific nature of inhibition of ccc DNA formation.

The fourth phenotype of Rts1 to be discussed here is its temperature dependent elimination. When inoculated at low concentration, Rts1 is lost from the host cells during the overnight growth at 42 ℃ but not at 32 ℃ (Table 2). This phenomenon is termed temperature dependent instability in this communication and should be distinguished from instability which was observed at 32 ℃ with plasmids derived from Rts1 as discussed later.

THE ROLE OF AN 8 Mdal Bam H1 RESTRICTION ENZYME FRAGMENT FOR Tsg EXPRESSION

A series of Rts1 miniplasmids was prepared by ligation of Bam H1 fragments of a smaller Rts1 derivative (pAK8) (*9*). These miniplasmids were screened for Tsg phenotype and Tsg^+ and Tsg^- miniplasmids were digested by Bam H1 endonuclease. The digests were subjected to electrophoretic separation as shown in Fig. 2. As summarized in Table 3, the 8 Mdal Bam H1 fragment is present in all miniplasmids which retained the Tsg^+ phenotype of Rts1. This correlation suggests that expression of the Tsg phenotype requires the presence of a locus on the 8 Mdal fragment. It is also noted in this table that all the Bam H1 miniplasmids tested contained the 18.6 and 14.1 Mdal fragments. Since miniplasmids were selected for kanamycin resistance, one of these fragments must represent the kanamycin resistance and the other replication region. In the experiment shown in Fig. 3 kanamycin resistance fragment was cloned into pBR322 and shown to be the 14.1 Mdal fragment. This suggests that the 18.6 Mdal fragment contains the replication region. It is still possible that all three DNA fragments described above may play a role in expressing Tsg. For example, the structural gene for Tsg may reside in either or both of 14.1 and 18.6 Mdal fragment, while the controlling gene for Tsg

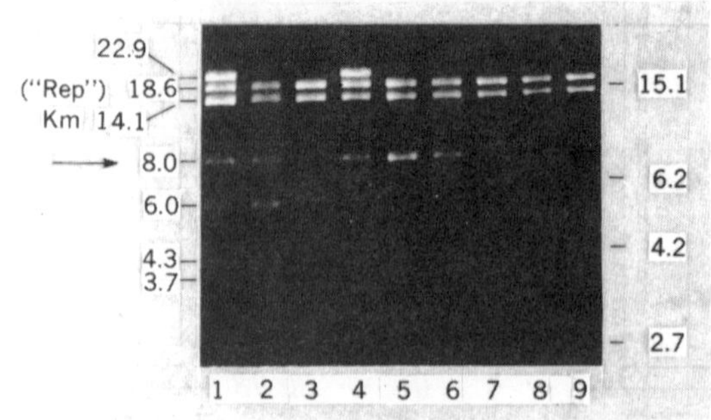

Fig. 2. Bam, H1 endonuclease digestion of Rts1 miniplasmids derived from Bam H1 digest of pAK8. Plasmid DNA was digested with endonuclease Bam H1 and subjected to agarose gel electrophoresis. The molecular weights of Marker DNA are shown on the right. Lane 1, pAK8, (tsg$^+$); lane 2, pFY505 (tsg$^+$); lane 3, pFY545 (tsg$^-$); lane 4, pFY551 (tsg$^+$); lane 5, pFY553 (tsg$^+$); lane 6, pFY556 (tsg$^+$); lane 7, pFY560 (tsg$^-$); lane 8, pFY601 (tsg$^-$); lane 9, pFY603 (tsg$^-$).

Table 3. *Fragment Composition of the Bam H1 Derived Rts1 Miniplasmids*

		Bam H1 Fragments, Mdal									
Plasmid	*Tsg*	*1.1*	*3.7*	*4.3*	*6.0*	*8.0*	*14.1*	*18.6*	*22.9*	*Total MW*	*Inc*
pAK8	+	+	+	+	+	+	+	+	+	*78.7*	*T*
pFY505	+	-	+	+	+	+	+	+	-	*54.7*	*T*
pFY551	+	-	-	-	-	+	+	+	+	*63.6*	*T*
pFY553	+	-	-	+	-	+	+	+	-	*45.0*	*T*
pFY556	+	-	-	-	-	+	+	+	-	*40.7*	*T*
pFY545	-	-	-	-	-	-	+	+	-	*32.7*	*T*
pFY560	-	+	-	-	-	-	+	+	-	*33.8*	*T*
pFY601	-	-	-	+	-	-	+	+	-	*37.0*	*T*
pFY603	-	-	-	-	-	-	+	+	-	*32.7*	*T*

The presence of restriction enzyme DNA fragments of Rts1 miniplasmids was determined by comigration with digested pAK8 DNA as shown in Fig. 2. pAK8 fragment molecular weights were determined previously (9).

may reside in the 8 Mdal fragment of *vice versa.*

The Tsg characteristic is apparently transferable to pBR322. Thus, in the experiment shown in Table 4 such recombinant plasmid pPF6 gave Tsg$^+$ effect. In this experiment pBR322 and the miniplasmid, pFY556 were digested with Bam H1 and the recombinant plasmids of pBR322 were formed with the T4-ligase. The bacteria harboring such plasmids were selected for ampicillin resistance and sensitivity to tetracycline and kana-

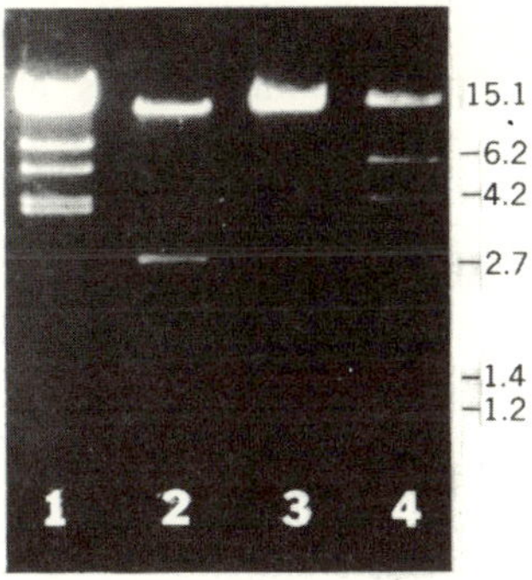

Fig. 3. Identification of the Bam H1 fragments carrying the kanamycin resistance and pAK8 replication regions. Plasmid DNA was digested with Bam H1 and subjected to agarose gel electrophoresis as described. Lane 1, 1.2 μg of pAK8; lane 2, 0.3 μg of pFK004; lane 3, 0.3 μg of pFY603; lane 4, 0.25 μg of λ Hind III digest. pFK004, kanamycin resistant pBR 322 recombinant; pFY603, see Table 3.

Table 4. *Temperature Sensitive Growth of E. coli Strains Carrying an Rts1-derivative Miniplasmid or its pBR322-clone*

Plasmid		*Grown at*	
		32 °C	*42 °C*
pFY505	*Tsg⁺ Mini*	*0.898*	*0.058*
pFY545	*Tsg⁻ Mini*	*0.941*	*0.940*
pPF6	*Tsg⁺ pBR332 clone*	*0.980*	*0.067*
pPF3	*Tsg⁻ pBR322 clone*	*1.050*	*0.671*

Data was shown by $O.D._{540}$ after 8 hr shaking.
Cultures of E. coli with plasmid (10^4 cells/ml were grown for 8hr in trypticase soy broth (TSB).

mycin. The miniplasmids pFY556 consists of only three Bam H1 DNA fragments (8 Mdal, 14.1 Mdal and 18.6 Mdal). The recombinant should contain either *rep* or *tsg* gene. As expected, not all the recombinant plasmids were Tsg^+. For example, pPF3 was Tsg^- recombinant plasmid.

COVALENTLY CLOSED CIRCULAR DNA SYNTHESIS OF Rts1 MINI-PLASMIDS

As discussed in the introduction, synthesis of Rts1 DNA can occur at

42 ℃, but formation of covalently closed circular (ccc) molecules is markedly inhibited (*3*). To determine if this characteristic (temperature sensitive ccc DNA formation; Tsc) is correlated with Tsg, cultures harboring Tsg^+ and Tsg^- Rts1 miniplasmids were grown at 32 ℃ or 42.5 ℃, pulse labeled with H^3-thymidine and subjected to alkaline sucrose density gradient analysis. As shown in Fig. 4 the Tsg^+ miniplasmids such as pFY551 gave lowered ratio of ccc DNA to the host DNA at 42.5 ℃ while pFY601 (Tsg^-) miniplasmid formed ccc molecules irrespective of the incubation temperature. Summary of experiments on other Tsg^+ and Tsg^- miniplasmids are given in Table 5. It appears that Tsg^+ and Tsc^+ correlate at this stage of analysis.

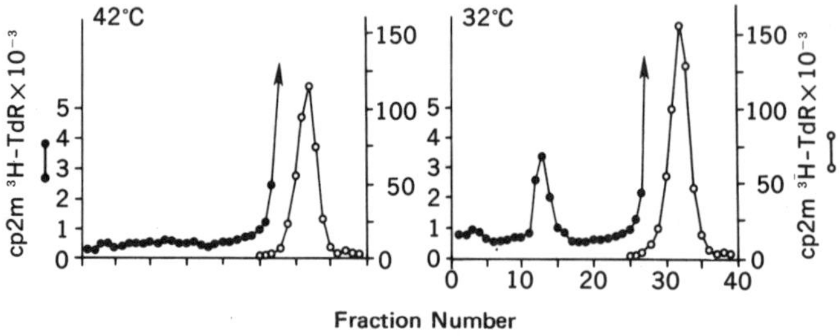

Fig. 4. Temperature sensitive formation of covalently closed circular DNA of Tsg^+ *miniplasmids. Cultures were pulse labeled with* ^3H-*thymidine, lysed, and subjected to centrifugation through 5-20% alkaline sucrose gradients. Sedimentation is from right to left. E. coli 20S0 harboring pFY551 (*Tsg^+*). ○, host DNA; ●, plasmid DNA.*

Table 5. *Thermosensitivity of CCC Formation in Strains Harboring* tsg^+ *and* tsg^- *Rts1 miniplasmids*

Plasmid	*tsg*	*CCC Plasmid Ratio (% of Total DNA)* 32 ℃	42.5 ℃	*42.5 ℃ / 32 ℃*
pFY545	-	*1.35*	*1.25*	*0.93*
pFY560	-	*1.13*	*1.56*	*1.38*
pFY601	-	*2.19*	*1.71*	*0.78*
pFY551	+	*1.29*	*0.22*	*0.17*
pFY553	+	*1.30*	*0.36*	*0.28*
pFY556	+	*1.13*	*0.36*	*0.32*

The percentage of ccc plasmid to total DNA was calculated by using the values determined by alkaline sucrose sedimentation analysis as performed in Fig. 4.

STRINGENT COPY NUMBER CONTROL IN Rts1 MINIPLASMIDS

As shown in Table 2, all Rts1 miniplasmids tested retained the stringent control of its copy number. Since copy number of Rts1 is approximately 1-2 per host genome, it appears that no increase of copy number is observed even though the molecular weight of the plasmid is reduced to about 1/3 to 1/4 of the original plasmid DNA. These results are in accordance with the concept that functional incompatibility locus is closely related to the region for proper plasmid replication (*10-12*).

LOSS OF INS$^+$ pBR322 RECOMBINANT PLASMID

During the studies on the recombinant plasmids between pBR322 and Rts1 DNA fragments, it became clear that some of the recombinant plasmids are very unstable. For example, pFK801, pFK813 and pFK820 were lost from the host bacteria within 60 generations at 32 ℃ (Fig. 5). Although the stability is increased at 42 ℃, the loss still occurred even at this temperature. The reason for faster loss at 32 ℃ is found to be due to slower growth of bacteria harboring Ins$^+$ plasmids. At 32 ℃, the R$^-$ cells grew slightly faster while the growth of R$^+$ and R$^-$ cells were almost identical at 42 ℃ (Fig. 6). Therefore, the rapid loss of Ins$^+$ plasmid at 32 ℃ is partially due to overgrowth of R$^-$ cells at 32 ℃.

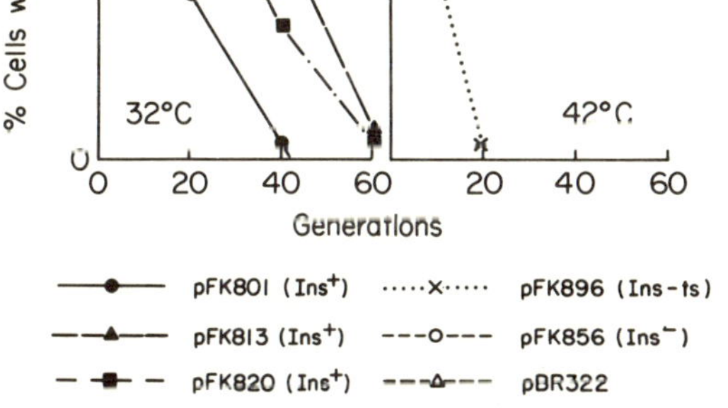

Fig. 5. Time course of loss of Ins recombinant plasmid. E. coli 20S0 harboring the recombinant plasmids were inoculated at the density of 2 × 10^3 /ml in trypticase soy broth. Samples were taken at various times and ampicillin resistant colonies and total number of colonies were counted.

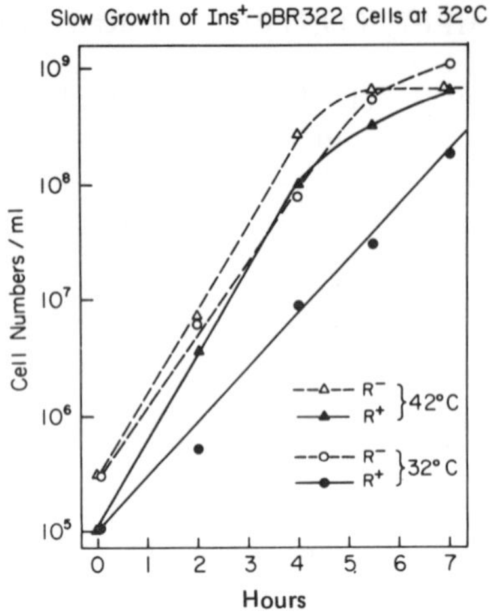

Fig. 6. Kinetics of growth of E. coli harboring Ins⁺ pBR322 recombinant plasmids. E. coli 20S0 with or without pFK813 were grown in trypticase soy broth without antibiotics.

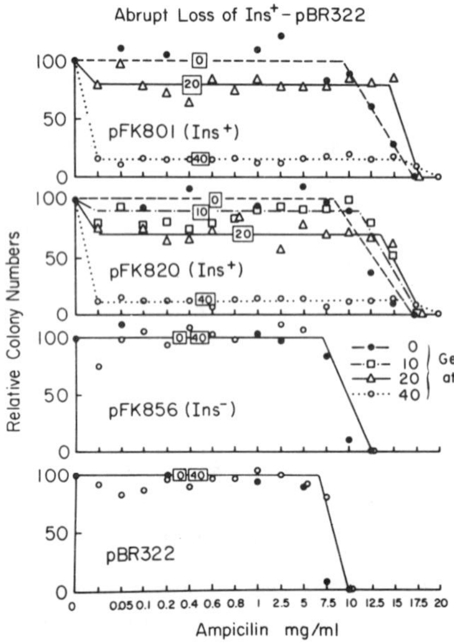

Fig. 7. The level of ampicillin resistance of E. coli with the Ins⁺ pBR322 during the growth in the absence of ampicillin. E. coli harboring recombinant plasmids were grown in trypticase soy broth as in Fig. 6. At various times samples were plated on plates containing various amounts of ampicillin as shown in the figure. The colonies grown on the ampicillin plates are expressed as relative numbers to those grown on the plates without ampicillin.

Vector pBR322 is a multicopy plasmid and the replication region of this plasmid was derived from Col E1, the copy number of which is approximately 24 per cell (*13*). If the loss of Ins^+ pBR322 recombinant DNA occurs only randomly during cell division, the probability of such event is approximately $\frac{1}{2^{24}}$ (*14*).

To obtain some notion about the mechanism of loss of Ins^+ pBR 322 recombinant plasmid, experiments illustrated in Fig. 7 were performed. In this experiment, relative plasmid copy number was determined on the basis of the level of ampicillin resistance. It has been shown that the level of bacterial resistance to ampicillin is directly correlated with plasmid copy number (*15*). After R^+ cells were grown without ampicillin for some generations, the culture was exposed to various concentrations of ampicillin to determine whether part of the ampicillin resistant population has decreased number of the recombinant plasmid. It is clear from this figure that the loss of plasmid is *not* gradual. The plasmid is present in full number or none is present.

ISOLATION OF Ins DNA FRAGMENT

Figure 8 shows the restriction enzyme analysis of Ins^+ recombinant plasmids as compared with that of Ins^- pBR322 recombinants. It is clear from this figure that unstable recombinant plasmids (Ins^+) always contained the 1.2 Mdal DNA fragment which is present in Bam H1 digests of Rts1 DNA. It should be pointed out that the Ins^+ recom-

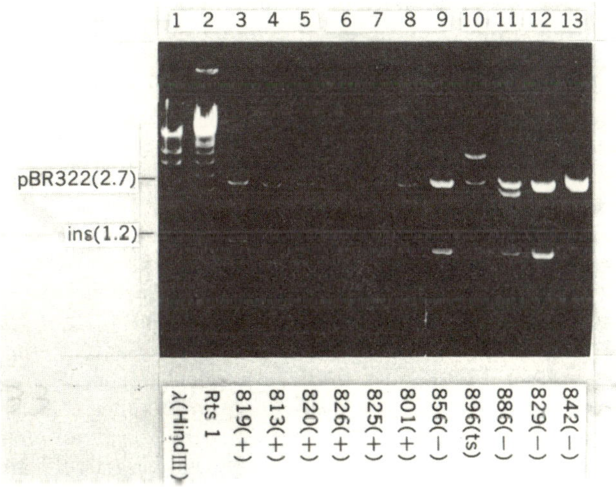

Fig. 8. Localization of Ins gene in 1.2 Md Bam H1 Fragment of Rts1. Agarose gel electrophoresis of Bam H1 digests of recombinant plasmid between pBR322 and fragments of Rts1 DNA is shown. (+) denotes Ins^+. Note the presence of 1.2 Mdal DNA with Ins (+) plasmids.

binant plasmid was lost even at 32 ℃ as discussed in the preceding section. In contrast, temperature sensitive instability was conferred by the DNA fragment which has approximately molecular weight of 6.0 Mdal. This 6.0 Mdal fragment was also observed among the Bam H1 digests of Rts1 DNA. From these observations, we make tentative conclusions that the Ins^+ DNA is the 1.2 Mdal Bam DNA fragment. We do not understand the exact reason why the Ins^+ DNA confers pBR322 instability even at 32 ℃ while Rts1 itself is stable at 32 ℃.

To explain the above observation we postulate that within Rts1, a DNA element exists which counteracts the Ins DNA activity. This hypothetical element is tentatively called *stb* (stability) element. Many of Rts1 miniplasmids prepared by Eco R1 are extremely unstable even at 32 ℃ (unpublished observation). Such instability can be explained on the basis of (presumed) destruction of the hypothetical *stb* gene by Eco R1 enzyme. On the other hand, the temperature dependent instability of a recombinant plasmid pFK896 reflects the characteristics of Rts1. We therefore suggest that the Rts1's temperature dependent instability is at least partially related to the action of an element residing in this fragment of DNA. (6 Mdal DNA, see Fig. 8)

DISCUSSION

In this article we attempted to review our work on the multiphenotypic drug resistance factor Rts1. The world's 1st naturally occurring temperature sensitive plasmid Rts1 was originally discovered in Japan and we feel that it is very appropriate to present a review of this fascinating plasmid in this conference held in Tokyo.

It appears that understanding of molecular basis of T4 phage restriction is at the beginning stage. The isolation and identification of the Rts1 coded restriction enzyme gives a basis for further studies. The molecular mechanism for specificity of glucosylated T4 DNA is not understood. Nor do we have the pure enzyme to study this reaction from an enzymological view point. Its uniqueness for preferring glucose residue certainly warrants further studies of this enzyme. In addition, work by other laboratory suggests that additional mechanisms, may operate for restriction of T4 phage. This mechanism, probably at the initial stages of T4- life cycle, would also have to be elucidated further (*5*).

Perhaps the most intriguing Rts1 phenotype is its temperature sensitive growth effect (Tsg). No molecular mechanism of this phenotype has been proposed. We do not even know whether it is exerted through heat labile protein as most of the temperature sensitive phenotypes are. In this connection, it is interesting to note that Tsg could so far not be separated from Tsc, temperature sensitive ccc DNA formation. It is possible that accumulation of non-ccc DNA may be toxic to the host cells. Although the data in this communication suggest that the expression of Tsg by Rts1 and its miniplasmid depends on the Bam H1 8 Mdal DNA fragment, our preliminary data (not shown) with Sal 1 restriction enzyme suggest that the *tsg* gene can be localized within a 2-3 Mdal fragments. It is important to note that the possibility exists that the *tsg* may reside in the 14.1 (Km) or 18.6 (rep) fragments and expression of it in Rts1 may depend on the presence of the 8 Mdal fragment.

From the view point of molecular biology of plasmid DNA replication, the Tsc character of Rts1 is most intriguing. It should be pointed out that Rts1 is the first and so far the only plasmid reported to be of capable of replication without involving ccc form. For this reason it is often thought that Rts1 is an exotic, unusual plasmid. On the contrary, it is quite possible that all plasmid could have two pathways of replication. Survey of the literature indicates that no evidence has ever been presented to rule out the possibility that a minor pathway of replication representing about 10-20% of total replication may go through the replication not involving ccc form of DNA. Replication without ccc form has been established with T7 and adenovirus DNA.

Lastly, we should point out possible significance of *ins* gene. It is clear that some DNA fragment exerts *cis* effect of making other plasmids unstable. Since the elimination is dependent on cell growth, it appears that this piece of DNA influences proper partitioning of plasmid DNA during cell division. If we understand the molecular basis of this elimination, it would be one step closer to the eventual control of undesirable drug resistance. On the other hand, many interesting questions can be asked from the view point of plasmid biology. Does the effect of *ins* counteracted by *par* or *stb* gene which is believed to function for proper segregation of plasmid (*12, 16*)? What is the mechanism of proper segregation of plasmid? What is the nature of *ins* DNA and the DNA segment which counteract the effect of *ins*? How is the Ins described in this paper related to the temperature dependent *Ins* which represents the instability of original Rts1? Is the *ins* similar to the occasionally observed general phenomena that recombinant pBR322 with inserts become labile? The answers to these questions would certainly give fundamental information not only for the maintenance of Rts1 but also for the important problems of plasmid instability in general.

ACKNOWLEDGMENT

This work was supported by U.S.A. PHS GM-12053.

REFERENCES

1. Terawaki, Y., Kakizawa, Y., Takayasu, H., and Yoshikawa, M. (1968) Nature *219*, 284-285
2. DiJoseph, C.G., Bayer, M.E., and Kaji, A. (1973) J. Bacteriol. *115*, 399-410
3. DiJoseph, C.G. and Kaji, A. (1974) Proc. Nat. Acad. Sci. U.S.A. *71*, 2515-2519

4. Yamamoto, T. and Kaji, A. (1977) J. Bacteriol. *132*, 90-99
5. Sasakawa, C., Tsuchiya, T., and Yoshikawa, M. (1980) Mol. Gen. Genet. *177*, 243-250
6. Ishaq, M. and Kaji, A. (1980) J. Biol. Chem. *255*, 4040-4047
7. Raghavan, N., Isaq, M., and Kaji, A. (1980) J. Virol. *35*, 551-554
8. Yamamoto, T., Yokota, T., and Kaji, A. (1977) J. Bacteriol. *132*, 80-89
9. Yamamoto, T., Finver, S. Yokota, T., Bricker, J., and Kaji, A. (1981) J. Bacteriol. *146*, 85-92
10. Danbara, H., Timmis, J.K., Lurz, R., and Timmis, T.N. (1980) J. Bacteriol. *144*, 1126-1138
11. Meacock, P.A. and Cohen, S.N. (1980) Mol. Gen. Genet. *174*, 135-147
12. Miki, T., Easton, A.M., and Rownd, R.H. (1980) J. Bacteriol. *141*, 87-99
13. Clewell, D.B. (1972) J. Bacteriol. *110*, 667-676
14. Nordstrom, K., Molin, S., and Aagaard-Hansen, H. (1980) Plasmid *4*, 215-227
15. Uhlen, B.E. and Nordstrom,K. (1977) Plasmid *1*, 1-7
16. Nordstrom, K., Molin, S., and Aagaard-Hansen, H. (1980) Plasmid *4*, 332-349

POSITIVE REGULATION OF THE *xyl* OPERON ON TOL PLASMID[1]

T. Nakazawa, S. Inouye, and A. Nakazawa

Department of Biochemistry
Yamaguchi University School of Medicine
Ube, Yamaguchi, Japan

I. DEGRADATIVE PLASMIDS IN PSEUDOMONADS

Members of the bacterial geneus *Pseudomonas* have capacities to utilize a wide range of compounds as carbon and energy sources (1). Thus the question once arose whether the genome of *Pseudomonas* species is fit in size to code for so many catabolic functions. It is now known that some of these functions are specified by genes borne on plasmids.

The first such plasmid described was a plasmid for salicylate utilization (SAL). A wide variety of such degradative plasmids, some of which are self-transmissible but some are not, has been identified (2). Thus plasmids confer on host pseudomonads the ability to catabolize naphthalene (NAH), camphor (CAM), *n*-alkane (OCT), and toluene, xylenes and toluates (TOL).

Degradative plasmids carry a whole array of functional genes responsible for successive catabolism of substrates. The two plasmids most extensively studied, OCT and TOL, are known to carry the structural genes of catabolic enzymes as well as the regulatory genes. These plasmids would provide a useful experimental system for the detailed analysis of the genetic organization of catabolic pathways in non-enteric bacteria.

We have mapped some *xyl* genes specifying toluene and xylene metabolism on the TOL plasmid (3,4), and cloned a regulatory gene separately from the *xyl* operon (5). It was demonstrated that the regulatory gene produces a positive factor which is activated specifically by the inducer.

[1] *This work has been supported by Grants-in-Aid for Scientific Research from the Ministry of Education, Science and Culture, Japan.*

II. STRUCTURE AND FUNCTION OF THE TOL PLASMID

A. *The Degradative Pathway Encoded by TOL*

The TOL plasmid of *Pseudomonas putida(arvilla)* mt-2 codes for a set of enzymes responsible for the degradation of toluene, benzyl alcohol, benzaldehyde, and benzoate via the *meta*-cleavage pathway (Fig. 1) (6). This set of enzymes also catalyzes the metabolism of methylated derivatives of these aro-

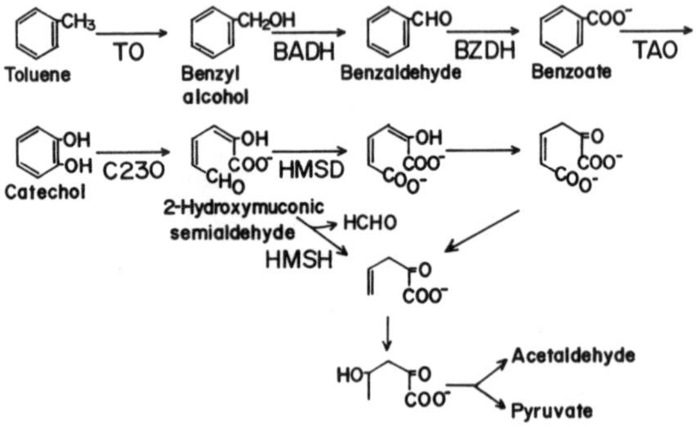

Fig. 1. The degradative pathway encoded by the TOL plasmid. Enzyme abbreviations: TO, toluene oxygenase; BADH, benzyl alcohol dehydrogenase; BZDH, benzaldehyde dehydrogenase; TAO, toluate oxygenase; C230, catechol 2,3-dioxygenase; HMSH, 2-hydroxymuconic semialdehyde hydrolase; HMSD, 2-hydroxymuconic semialdehyde dehydrogenase.

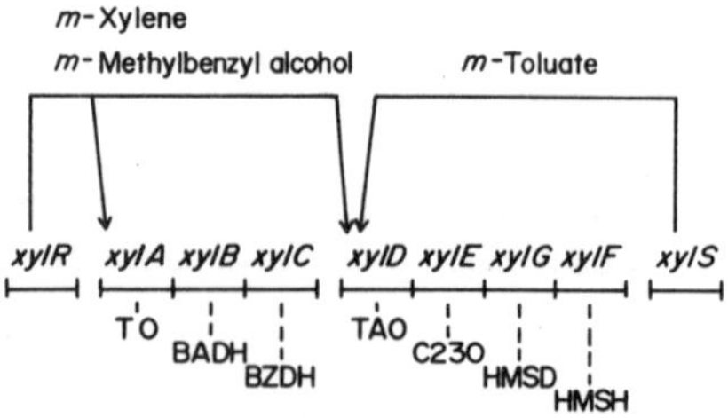

Fig. 2. A model for the regulation of early enzymes of the TOL-encoded pathway. The product of xylR combines with m-xylene or m-methylbenzyl alcohol, and induces both xylABC and xylDEGF. The product of xylS combines with m-toluate and induces only xylDEGF. Enzyme abbreviations are in Fig. 1.

matic compounds. These enzymes are induced by a specific inducer, and a model for the regulation of their synthesis has been proposed (7) (Fig. 2). The enzymes are organized in at least two regulatory blocks (operons). One block specifies the oxidation of the methyl group (*xylABC*), and the other controls the further oxidation and degradation of the carboxylic acids (*xylDEGF*). Two regulatory genes, *xylR* and *xylS*, are involved: *xylR* plays a role in the induction of both regulatory blocks by the hydrocarbons and the alcohol metabolites (and probably also the aldehyde metabolites), whereas *xylS* is responsible for the induction of only the second block by the carboxylic acids. The positive regulation for the synthesis of enzymes by the product of *xylR* has been demonstrated by constructing a partial diploid of the TOL genes (3,8).

B. *Physical and Functional Mapping of TOL*

1. Mapping of RP4 and TOL Recombinants. To understand the molecular basis of the regulation of *xyl* genes, recombinant plasmids of TOL and RP4 were isolated in vivo in *Pseudomonas aeruginosa*, which is transmissible to, and replication-proficient in, *Escherichia coli* (4). The induction of the TOL pathway enzymes in cells of *P. putida* carrying one of the recombinants, pTN2, is similar to that of the wild-type TOL plasmid. Cells of *E. coli* carrying pTN2 also showed induction of benzyl alcohol dehydrogenase and catechol 2,3-dioxygenase, which are encoded by the *xylB* and *xylE* genes, respectively, although the enzyme levels were much lower than in *P. putida*.

The pTN2 plasmid was purified from a strain of *E. coli*, and the cleavage sites for the restriction endonucleases *Xho*I, *Bam*HI, *Hin*dIII, and *Eco*RI were mapped (3). The plasmid contains the entire length of the RP4 plasmid (about 54 kilobase pairs [kb]) and the TOL segment (about 56 kb). The TOL segment is inserted about 12 and 5 kb away from the *Eco*RI and *Bam*HI cleavage sites of RP4, respectively. Similar recombinants of RP4 and TOL were reported in many laboratories, but the length and the inserted position of the TOL segment appear to be different (P. Broda, personal communication).

Downing and Broda previously reported a cleavage map for *Xho*I and *Hin*dIII of the TOL plasmid (117 kb) (10). The physical map of TOL was constructed based on these and our own results of RP4 and TOL recombinants (Fig. 3). Some deletion and insertion mutants of pTN2 were isolated, and their physical and functional maps were constructed (3). The results allowed us to estimate regions containing some *xyl* genes on the TOL plasmid.

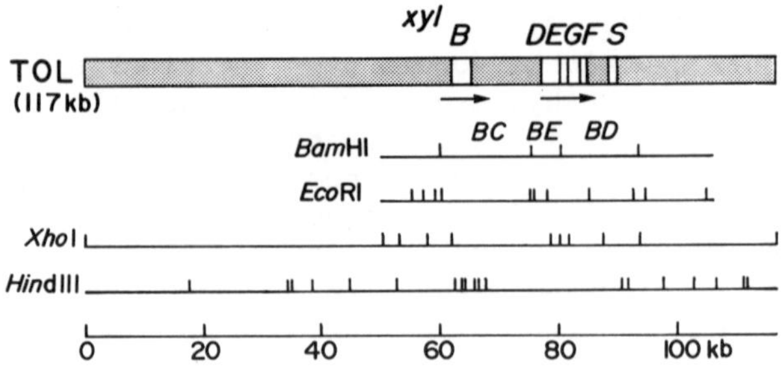

Fig. 3. A physical and genetic map of the TOL plasmid. For each of the restriction endonucleases indicated on the left, the relative locations of cleavage sites are marked by vertical lines. The open boxes on the TOL map indicate the regions of TOL genes. The arrows under the map indicate the direction of transcription.

2. *Molecular Cloning of TOL Genes in E. coli.* Two structural genes, *xylB* and *xylE* were cloned onto pBR322 in *E. coli*. The *xylB* gene (benzyl alcohol dehydrogenase) was mapped on a 2.9-kb region within the *Bam*HI BC fragment, whereas the *xylE* gene (catechol 2,3-dioxygenase) was mapped in a 1.8-kb region within the *Bam*HI BD fragment (Fig. 3) (4). The directions of transcription of these genes were deduced from the expression of the cloned genes which had been ligated in orientations opposite pBR322 at its *Bam*HI site within the tetracycline gene. The *xylB* and *xylE* genes in pTN2 are inducibly expressed in *E. coli* by the specific inducer, whereas they are not inducible in the pBR322-TOL hybrids. The map positions of *xylD* (toluate oxygenase system), *xylG* (2-hydroxymuconic semialdehyde dehydrogenase) and *xylF* (2-hydroxymuconic semialdehyde hydrolase) were also determined (Fig. 3) (5).

Catechol 2,3-dioxygenase of *P. putida* mt-2 has a molecular weight of 140,000 in the native form, and consists of a 35,000 dalton polypeptide (12). The *xylE* product formed in *E. coli* is indistinguishable from that formed in *P. putida* with respect to catalytic activity, molecular weight and antigenicity. The DNA sequence corresponding to the coding region of the enzyme was determined, and the complete amino acid sequence was deduced. The sequence was in perfect agreement with the sequences of the amino- and carboxy-termini which had been determined for the enzyme protein (Nakai et al. manuscript in preparation).

C. Positive Regulation of xylDEGF Operon by xylS

1. Molecular Cloning of xylDEGF Operon. The *xylDEGF* operon including the complete operator-promoter region was cloned in *E. coli* (Fig. 4). The *Hin*dIII fragment HA was inserted into pACYC184 to produce pTS66. The plasmid contains *xylDEGF* operon and *xylS* as evidenced by the inducible synthesis of the enzymes by *m*-toluate (5). Deletion of a part of fragment BC did not affect the expression of the operon (pTS69).

The operon was separately cloned from *xylS* (pTS71) by purifying BE fragment containing *xylD* followed by ligation to the *Bam*HI-digested pTS11, which has a part of fragment BD on pBR322 and contains *xylE*, *xylG*, and *xylF*. The enzymes encoded by pTS71 were not induced by *m*-toluate (5).

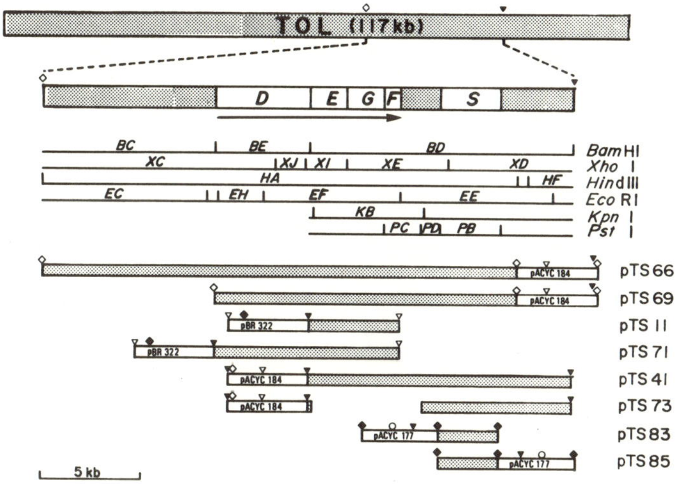

Fig. 4. A physical and genetic map of the TOL segment around the xylDEGF operon. The open boxes on the map indicate the region of TOL genes. The arrow indicates the direction of transcription. For each restriction endonuclease indicated on the right, the cleavage sites are marked by vertical lines. The positions of the inserted TOL segments in derivatives of pACYC184, pBR322 and pACYC177 are shown in the lower part of the figure. Stippled boxes indicate the TOL segment, and open boxes indicate the region of vectors. Symbols for restriction sites are as follows: ▼, BamHI; ◇, HindIII; ▽, EcoRI; ◆, PstI; ○, XhoI.

2. Molecular Cloning of xylS. Since the above results suggested the location of *xylS* in the region present in pTS66 but absent from pTS71, the region was further analyzed. The BD fragment was inserted into pACYC184 to produce pTS41. Then pTS73 was made after the KB fragment was removed by digesting pTS41 DNA with *Kpn*I. Because the double transformants of pTS71 and pTS73 showed induction of the enzymes by *m*-toluate, pTS73 contained *xylS*. The non-induced levels of the enzymes encoded by pTS71 were not repressed by the presence of pTS73. Therefore, the *xylS* product is a positive regulator which has no repressor function. The *xylS* gene was further mapped on the PB fragment (3.0 kb) by constructing pTS83 and pTS85 (Fig. 4) (5). Both the plasmids conferred the inducibility on cells containing pTS71.

3. Positive Regulation of Operons in Degradative Plasmids. Catabolic functions of degradative plasmids are controlled by regulatory genes carried by the plasmid. Thus the *alk* operon in OCT is regulated by *alkR* (11), while the *xyl* operons in TOL is controlled by *xylR* and *xylS*. The positive control seems to be a regulatory system common in degradative plasmids, in which the regulatory factor functions only when an inducer (substrate) is present.

REFERENCES

1. Stanier, R.Y., Palleroni, N.J. & Doudoroff, M.(1966) *J. Gen. Microbiol. 43*,159-271
2. Chakrabarty, A.M.(1976) *Ann. Rev. Genet. 10*,7-30
3. Nakazawa, T., Inouye, S. & Nakazawa, A.(1980) *J. Bacteriol. 144*,222-231
4. Inouye, S., Nakazawa, A. & Nakazawa, T.(1981) *J. Bacteriol. 145*,1137-1143
5. Inouye, S., Nakazawa, A. & Nakazawa, T.(1981) *J. Bacteriol. 148* (in press)
6. Worsey, M.J. & Williams, P.A.(1975) *J. Bacteriol. 124*,7-13
7. Worsey, M.J., Franklin, F.C.H. & Williams, P.A.(1978) *J. Bacteriol. 134*,757-764
8. Franklin, F.C.H. & Williams, P.A.(1980) *Mol. Gen. Genet. 177*,321-328
9. Nakazawa, T., Hayashi, E., Yokota, T., Ebina, Y. & Nakazawa, A.(1978) *J. Bacteriol. 134*,270-277
10. Downing, R.G. & Broda, P.(1979) *Mol. Gen. Genet. 177*,189-191
11. Fennewald, M., Benson, S., Oppici, M. & Shapiro, J.(1979) *J. Bacteriol. 139*,940-952
12. Nozaki, M.(1979) *Topics Curr. Chem. 78*,145-186

CLONING OF REPLICATOR REGIONS OF COPY NUMBER MUTANT pNR300

S. Horiuchi, R. Nakaya, N. Goto,
N. Okamura, and A. Shoji

Department of Microbiology, Tokyo Medical and Dental University School of Medicine, Tokyo, Japan.

The control of replication of plasmids has been studied as a means of understanding the basis for the autonomy of these interesting genetic elements and as model systems for replication of DNA molecules. The control of replication of IncFII group plasmids has been extensively studied (1, 2, 3). It is now known that the control of replication of plasmids involves at least two plasmid determinants, a trans-acting replication inhibitor molecule, and a cis-acting target of this inhibitor molecule (4).

The plasmid pNR113 is a self-transmissible plasmid with a molecular size of 113.3 kilobase pairs, confers resistance ampicillin(Ap), kanamycin(Km), chloramphenicol(Cp), streptomycin(Sm), sulfonamide(Sa), and tetracycline(Tc), and belongs to the FII incompatibility group (Table 1). We obtained a copy number mutant pNR300 from pNR113. The number of copies of pNR300 in Proteus mirabilis Pm171 was estimated to be 18-23 per host chromosome. We constructed a composite plasmid with miniplasmids from pNR113 and pNR300 put together, and measured the number of copies of this composite plasmid to know how the replication gene from pNR113 affect the gene that determines the copy number of pNR300.

1. Physical map of pNR113

There are 15 EcoRI cleavage sites on pNR113 (Fig. 1). Of the 15 EcoRI fragments, 10 fragments are comparable in molecular size to those of EcoRI fragments of NR1, suggesting a close relatedness in molecular composition between pNR113 and NR1. NR1 has a molecular size of 90 kb and confers resistance to Hg, Cp, Sm, Su, and Tc. On the other hand, pNR113 has a molecular size of 113 kb and confers resistance to Cp, Sm, Sa, Ap, Km, and Tc. The resistance genes to Ap and Km were

Table 1. *Plasmids Used*

Plasmids	*M.W. (kb)*	*Copy No.*	*Derivation*
pNR113	*113.3*	*1*	*Natural isolates*
pNR1139	*20.4*	*ND*[a]	*EcoRI fragments of the size 9.7kb, and 13.2kb*
pNR1140	*4.8*	*6-7*	*PstI fragments of the size 1.1kb, 1.7kb and 2.0kb*
pNR300	*113.3*	*18-23*	*Copy number mutant of pNR113*
pNR302	*48.9*	*4-5*	*Deletion mutant of pNR300*
pNR308	*15.1*	*42-68*	*PstI fragments of the size 1.1kb(double), 1.4kb, 1.7kb, 2.8kb, and 7.0kb*
pNR333	*4.8*	*172-441*	*PstI fragments of the size 1.1kb, 1.7kb and 2.0kb*
pNR3040	*19.9*	*10-13*	*Composite plasmid*[b]

[a]*Not done*

[b]*(pNR308:pNR1140)*

Table 2. *Characteristics of Cloned and Composite Plasmids*

Plasmid	*Resistance marker*	*M.W. (kb)*	*Length (μm)*	*Copy number*
pNR1140	*Km*	*4.8*	*1.3 ± 0.1*	*6-7*
pNR308	*Ap*	*15.1*	*4.9 ± 0.1*	*42-68*
pNR3040[a]	*Ap Km*	*19.9*	*6.4 ± 0.1*	*10-13*

[a]*pNR308:pNR1140*

located on the EcoRI fragments No. 5, 7 of pNR113. Since we did not examine the DNA homology between NR1 and pNR113 by heteroduplex method, it is not clear how close these two plasmids are related with each other. However, it is conceivable from the present results that pNR113 was originated from R plasmid like NR1 by substitution of EcoRI fragments H or I of NR1 with a transposon-like DNA segment of 13 kb confering resistance to Ap and Km.

2. *Identification of specific sites essential for replication.*

In restriction endonuclease cleavege experiments on the various self-cloned plasmids EcoRI fragment No. 2 of pNR113 was shown to be required for autonomous replication. PstI endonuclease was used to subclone EcoRI fragment No. 2. Two Pst1 fragments of sizes 1.7 kb and 1.1 kb must be ligated in native orientation to be able to mediate autonomous replication which is in accordance with the findings of Miki et al (2) (Fig. 2).

3. *Incompatibility of plasmids cloned from the wild type plasmid and the copy number mutant plasmid.*

Incompatibility was tested by constructing the cells harboring both plasmids. After initial selection on doubly selective plates, cells were grown for 40 generations in Penassay broth at 37°C and the number of cells carrying either or both plasmids was then determined by replica plating. As expected, both pNR1139 and pNR1140 were incompatible with each other. pNR1140 was also incompatible with pNR308. However, both pNR302 and pNR308 were compatible with each other. These results indicate that the miniplasmids cloned from the copy number mutant plasmids exhibit reduced incompatibility. When pNR308 was coexisted with pNR333, pNR308 was preferentially excluded in 25% of the cells, regardless of its being an incoming or resident plasmid.

The copy number mutants could be grouped into three distinct classes based on their copy number and degree of incompatibility. Class 1 mutants have very high copy number, and show weak incompatibility. Class 2 mutants have moderate copy number, and show weak incompatibility. Class 3 mutants have low copy number (4-5) with greatly reduced incompatibility.

4. *Determination of the copy number of both plasmids coexisting in the same host cells.*

The experiments described above showed that the plasmids cloned from a copy number mutant coexist well in the cells. We then measured the copy number of both plasmids coexisting in the cells. Coexisting plasmids had essentially the same number of copies as when existed separately in the host cells.

5. *The copy number of the composite plasmid pNR3040 (pNR308:pNR1140)*

Because of pNR1140 and pNR1139 are incompatible, the copy numbers of these plasmids when coexisting in the same host cells could not examined.

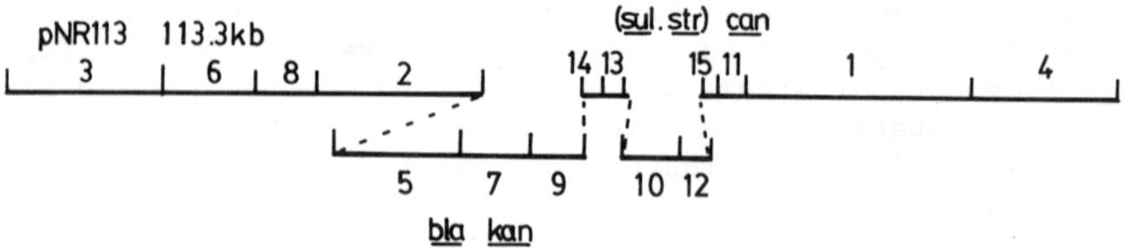

Fig. 1. Physical map of the pNR113 cleaved by EcoRI endonuclease.

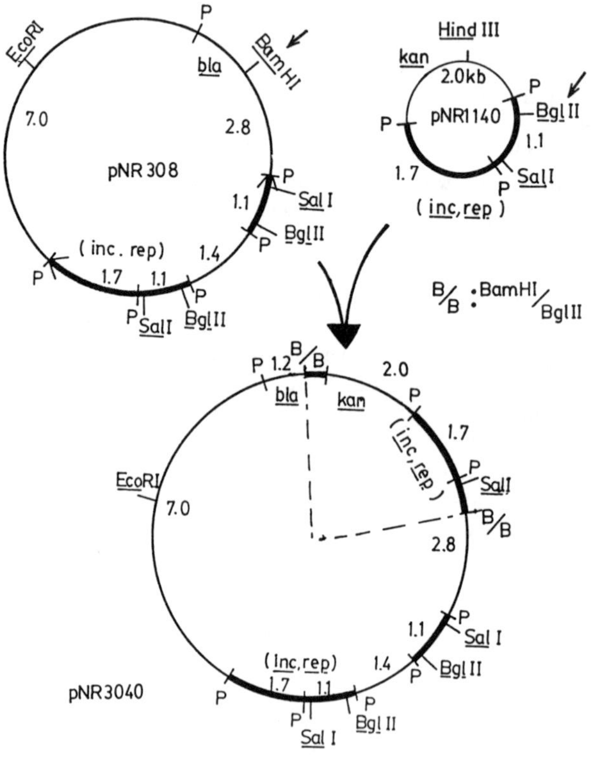

Fig. 2. Construction of the composite plasmid pNR3040, pNR1140, and pNR308 were digeted by BglII and BamHI, respectively, and ligated. pNR3040 was selected with kanamycin and ampicillin.

We then constructed a composite plasmid by recombining the DNAs of pNR308 and pNR1140 (Fig 2). This composite plasmid pNR3040 showed an intermediate copy number between that of pNR308 and pNR1140 (Table 2). It was, however, 4-5 times less than that of pNR308. These results suggest that the replication gene from pNR1140 is dominant for the control of replication of this plasmid, although the number of copies of pNR3040 was 2 times higher than that of pNR1140. Since pNR1140 gene products inhibited the pNR308 rep gene function, it seems obvious that cis-acting target of the inhibitor on pNR308 genome is still sensitive to the wild-type inhibitor. This strongly suggests that the increase of the copy number of pNR300 is due to a mutation of the trans-acting replication inhibitor molecule.

REFERENCES

1. Thomas,C.M. (1981) Plasmid *5*,277-291
2. Miki,T., Easton,A.M., & Rownd,R.H. (1980) J. Bacteriol *141*, 87-99.
3. Danbara,H., Timmis,J.K., Lung,R., & Timmis,K.N. (1980) J. Bacteriol. *144*,1126-1138.
4. Rownd,R.H., Easton,A.M., Barton,C.R., Womble,D.D., Mckell, J., Sampathkumar, P., & Luckow,A.V. (1980) in ICN-UCLA Symposia on Molecular and Biology (Bruce, A., ed.) *19*, 311-334, Academic Press, New York.
5. Timmis,K.N., Danbara,H., Brady,G., & Lung,R. (1981) Plasmid *5*, 53-75.

GENETICS AND MOLECULAR BIOLOGY : TRANSPOSITION

GENETIC AND PHYSICAL CHARACTERIZATION OF CITRATE UTILIZATION TRANSPOSON Tn*3411* FROM A NATURALLY OCCURRING CITRATE UTILIZATION PLASMID

N. Ishiguro and G. Sato

Department of Veterinary Public Health
Obihiro University of Agriculture and Veterinary Medicine
Obihiro, Hokkaido, Japan

C. Sasakawa[1], H. Danbara and M. Yoshikawa

Department of Bacteriology
Institute of Medical Science
University of Tokyo
Tokyo, Japan

Lack of citrate utilizing ability is a key character of identification of *Escherichia coli* in the family *Enterobacteriaceae*. However, plasmid-mediated citrate utilizing ability (abbreviated as Cit^+) has been demonstrated in Cit^+ *E. coli* (1,2) and also on thermosensitive R plasmids isolated from *Salmonella* strains (3,4). Most of the Cit^+ plasmids derived from Cit^+ *E. coli* strains were conjugative, and were classified into at least 3 groups (IncW, IncI1, and Inc unidentified groups) under incompatibility property. The findings of the genetic determinant for citrate utilization distributed on the different plasmids suggested the possibility that Cit^+ determinant may be located on an element transposable to new sites on the other replicons. Therefore, the potential transposability of Cit^+ determinant from naturally occurring Cit^+ plasmids derived from Cit^+ *E. coli* has been investigated.

I. ISOLATION OF CITRATE TRANSPOSON, Tn*3411*

Among Cit^+ plasmids tested for transposability of Cit^+ determinant, the determinant of pOH3001 has transposed to λ *b*519

[1]*Present address: Washington University, St Louis, Missouri.*

Table 1. *Cit$^+$ plasmids constructed in this study*

Plasmids	*Phenotype*[a]
pOH2	*pBR322 carrying citrate transposon (Tn3411), Cit$^+$ Ap Tc*
pOH3	*pBR322 carrying citrate transposon (Tn3411), Cit$^+$ Ap Tc*
pOH4	*R100.1 carrying citrate transposon (Tn3411), Cit$^+$ Cm Sm Su*
pOH5	*pBR322 carrying SalI digested A fragment of pOH2, Cit$^+$ Ap*
pOH6	*pBR322 carrying SalI digested A fragment of pOH2, Cit$^+$ Ap*
pOH7	*Self-cloning pOH2 with BamHI, Cit$^+$ Ap*

[a]*Phenotype symbols: Cit$^+$, citrate utilization; Ap, ampicillin resistance; Cm, chloramphenicol resistance; Sm, streptomycin resistance; Su, sulfonamide resistance; Tc, tetracycline resistance.*

*b*515*c*I857*S*7 (abbreviated as λ*bb*) (5). *E. coli* YC3031, lysogenic for λ*bb* and carrying a Cit$^+$ plasmid, was heat-induced and the λ*bb* lysates obtained were used to lysogenize *E. coli* YC11 23 (*recA* and rif^r mutant of C600) using the Cit$^+$ selection medium (2). Cit$^+$ transductants on Cit$^+$ selection plates were obtained at 2 x 10^{-9} per plaque forming unit of λ*bb* lysate. These Cit$^+$ transductants showed resistance to λ*c*I857 (abbreviated as λ) but not to λ*vir* phage and producing a Cit$^+$ transducing particle upon heat-induction. The transposability of the Cit$^+$ determinant from λ*bb* phage with citrate transposon (λ*bb*::Tn*3411*) to pBR322 dimer or monomer was examined by the preparation of plasmid DNA, and the resulting plasmid DNA was used for transformation to *recA*-deficient strain.

As shown in Table 2, the Cit$^+$ determinant was transposed from λ*bb*::Tn*3411* to pBR322 dimer at 1.6 x 10^{-4}. However, the transposition frequency of Cit$^+$ determinant from λ*bb*::Tn*3411* to target plasmid monomer was 10^4 times lower than to pBR322 dimer. Two Cit$^+$ derivatives of pBR322 monomer with Tn*3411* were designated as pOH2 and pOH3 (Table 1). In addition, both Cit$^+$ and drug resistance markers of Cit$^+$ transconjugant of R100.1, pOH4, obtained were cotransferred, and the evidence of transposition of Tn*3411* to the target plasmid R100.1 was demonstrated physically. This transposable element coding for citrate utilizing ability was assigned the number Tn*3411* from the above genetic evidence.

Table 2. *Transposition frequency of citrate transposon Tn3411*

Donor replicon of Tn3411	*Target replicon*	*Selection*[a]	*Transposition frequency of Tn3411*
λbb::Tn3411 lysogenized	*pBR322 dimer*	*Cit*	1.6×10^{-4} *(/Tet)*[b]
λbb::Tn3411 lysogenized	*pBR322 monomer*	*Cit*	1.6×10^{-8} *(/Tet)*[b]
pBR322 dimer::Tn3411	*R100.1*	*Cit*	2.9×10^{-7} *(/Cml)*[c]

[a] *Cit, described in Table 1, legend a.*
[b] *Transposition frequency of Tn3411 is the ratio of the number of* Cit^+ *transformants appearing on Simmons citrate agar plates to the number of* Tet^r *transformants of pBR322.*
[c] *Expressed as the number of* Cit^+ *transconjugants per* Cml^r *transconjugants of R100.1.*

II. ENDONUCLEASE DIGESTION AND CLEAVAGE MAP OF Tn*3411*

Phage and plasmid DNAs were digested with *Eco*RI endonuclease and *Hin*dIII endonuclease enzymes, respectively, and electrophoresed in an agarose gel (Fig. 1). The insertion of Tn*3411* has occurred in the *Eco*RI-C fragment of λ*bb* because of the increase in its size (Fig. 1, compare lane B with C). From the digestion patterns of phage DNAs and two Cit^+ plasmids, pOH2 and pOH3 with *Hin*dIII endonuclease, it was found that the insertion site of Tn*3411* was *Hin*dIII-digest C fragment of λ*bb* (Fig. 1, compare H with I), and the third largest band in *Hin*dIII-digest of λ*bb*::Tn*3411* was identical in size to the largest band (5.7 kb) in *Hin*dIII-digest of pOH2 and pOH3 (See lanes I, J and K). From these results, Tn*3411* possesses no site susceptible to *Eco*RI endonuclease but contains 2 restriction sites for *Hin*dIII. Three fragments were generated from pOH2 by digestion with either *Sal*I or *Bam*HI, demonstrating that Tn*3411* contained 2 recognition sites for *Sal*I and *Bam*HI as well as *Hin*dIII. However, no recognition site for *Pst*I was found in Tn*3411*. From the results of these restriction analyses, an endonuclease cleavage map of pOH2 was constructed as shown in Fig. 3.

III. ELECTRON MICROSCOPE OBSERVATION OF Tn*3411*

Heteroduplex molecules formed between λ*bb* and λ*bb*::Tn*3411* revealed a characteristic double stem-loop structure (Fig. 2A). The insertion of Tn*3411* was found to be located at map posi-

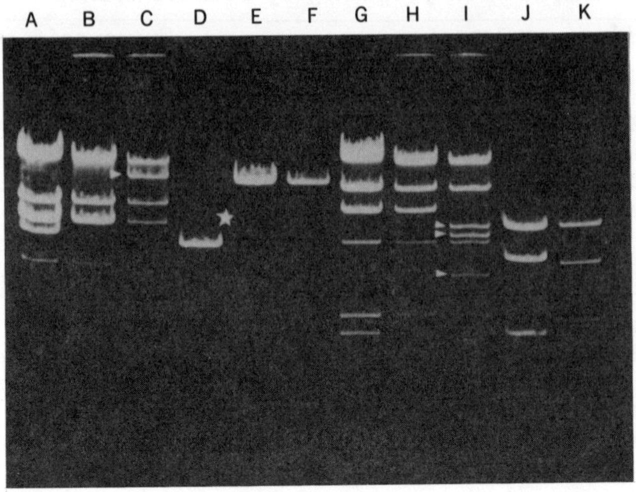

Fig. 1. Agarose gel electrophoresis of various phage DNAs and plasmid DNAs cleaved by EcoRI and HindIII. Electrophoresis was carried out on 0.8 % gel at 80 V for 4 h. Digestion with EcoRI: (A) λ , (B) λbb, (C) λbb::Tn3411, (D) pBR322, (E) pOH2, (F) pOH3. Digestion with HindIII: (G) λ (H) λbb, (I) λbb::Tn3411, (J) pOH2, (K) pOH3. Molecular weight standards were HindIII-cleaved phage λDNA (lane G). Arrows indicate new constructed fragments on λbb phage transposed with Tn3411. Faint band marked by asterisk is derivative from pOH2.

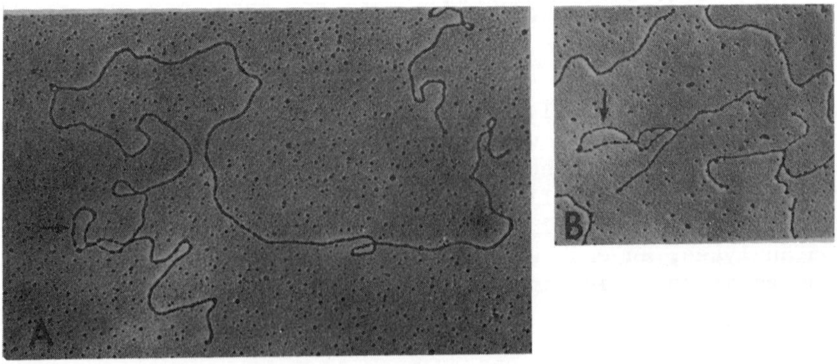

Fig. 2. DNA/DNA heteroduplexes between λbb and λbb::Tn3411 (A). DNA/DNA heteroduplexes between EcoRI-cleaved pOH2 and pBR322 (B). Arrows indicate Tn3411.

tion 83 % on the *λbb* gemome. A similar double stem-loop structure was observed in the heteroduplex molecules between pOH2 and pBR322 (Fig. 2B). The dimensions of 2 single-stranded loops constructed in heteroduplex molecules between pOH2 and pBR322 are presented in Fig. 3. Tn*3411* is a 7.4 kb long transposable element possessing 0.19 ± 0.07 kb external inverted repeats (Fig. 3, IR-1 and IR-4) and additional internal inverted repeats (Fig. 3, IR-2 and IR-3) of 0.2 ± 0.06 kb.

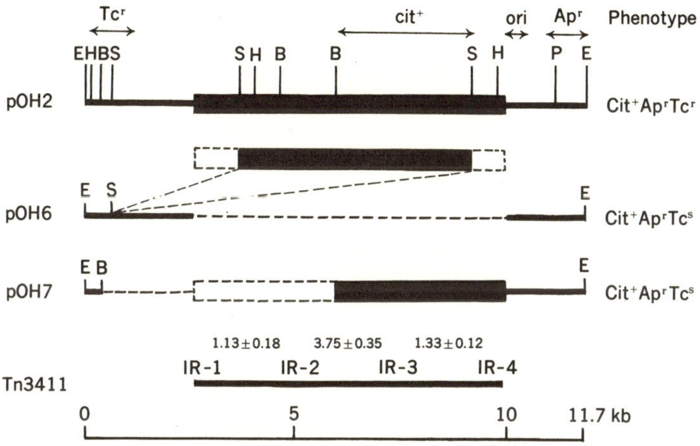

Fig. 3. Restriction endonuclease cleavage of pOH2 and schematic representation of pOH6 and pOH7.
Solid lines indicate fragments contained in the derivative plasmids. Closed boxes indicate the inserted regions of Tn3411 into vector plasmid pBR322. Dotted lines and boxes indicate the region deleted in the derivatives. Distances between cleaved sites are in kb. Vertical lines indicate the cleavage sites with endonuclease: B, BamHI; H, HindIII; E, EcoRI; P, PstI; S, SalI. Phenotype symbols are described in Table 1, legend a. Distances between 4 inverted repeats (IR) on Tn3411 were the mean measurements from separate heteroduplex molecules between pOH2 and pBR322. Inverted repeats on Tn3411 are designated serially as IR-1, IR-2, IR-3 and IR-4.

IV. CLONING OF CITRATE UTILIZATION DETERMINANT

The identification of a genetic DNA fragment for citrate utilizing ability on Tn*3411* was carried out by inserting the endonuclease-cleaved DNA fragments of pOH2 into vector plasmid pBR322. Two types of Cit$^+$, Ap^r and Tc^s transformant, pOH5 and pOH6 consisting of pBR322 and a *Sal*I-generated fragment of Tn*3411* were obtained (Table 1). The schematic representation of pOH6 is shown in Fig. 3. Since the orientation of the inserted *Sal*I-cleaved fragment (5.5 kb length) of Tn*3411* in pOH6 is different from that of pOH5, the phenotype of citrate utilization encoding on this fragment is expressed regardless of the orientation of the cloned *Sal*I fragment on pBR322. The Cit^+ self-cloning plasmid, pOH7, derived from pOH2 with *Bam*HI was also obtained (Table 1). The schematic structure of pOH7 is shown in Fig. 3. These results suggested that the structural genes corresponding to citrate utilizing ability were located on the 3.3 kb DNA fragment between the *Bam*HI and *Sal*I cleavage sites on Tn*3411* (Fig. 3).

DISCUSSION

A new transposon, Tn*3411*, encoding citrate utilizing ability was initially transposed from a naturally occurring citrate plasmid pOH3001 to λ*bb* phage genome. Transposition frequency of Tn*3411* from λ*bb*::Tn*3411* to pBR322 dimer is very high compared with that of the monomer. This distinction of transposition frequency of Tn*3411* to target replicon of dimer and monomer was not clearly explained by this study. There are two very diverse interpretations: one is that the dimer structure of pBR322 may generate a unique insertional hot spot for the transposition of Tn*3411*; the other is that this transposon Tn*3411* may preferentially insert into the replication origin of pBR322. Tu and Cohen (6) reported that Tn*3* can insert preferentially into AT rich regions of target replicon. On the other hand, in two Cit^+ derivatives of pBR322 monomer with Tn*3411*, pOH2 and pOH3, the insertion site of Tn*3411* on pBR322 was actually located in the nearest region of replication origin of pBR322. It is not yet clear whether Tn*3411* can preferentially insert into AT rich DNA regions existing on pBR322 dimer but not on the monomer, or replication origin of pBR322. Tn*3411*, identified as a transposable element, is demonstrated to be a characteristic structure possessing two pairs of inverted repeats with single-stranded loops under microscopic observation. The Cit^+ determinant in Cit^+ plasmid pOH3001 transposed to λ*bb* and further to pBR322 as a form of Tn*3411*, which is flanked by external inverted repeats (IR-1 and IR-4). However, the transposability of the internal Tn*3411* region (3.75 + 0.35 kb) which is flanked by the internal inverted repeats (IR-2 and IR-3) remains to be solved. Neither the Cit^+ transducing phage nor the Cit^+ transformant of pBR322 carrying internal DNA regions and flanked by a pair of internal inverted repeats was found on the Cit^+ selection plate. To examine whether the internal fragment has intrinsic transposability, further experiments are needed.

Test for citrate utilizing ability is both important and valuable for identification in the family *Enterobacteriaceae*. Tn*951*, encoding for *lac* operon which is homologous to the *E. coli lac* operon, was isolated in *Yersinia enterocolitica* (7). The potential transposition of biochemical characteristics such as lactose transposon Tn*951* and citrate transposon Tn*3411* may give rise to confusion in identification of the family *Enterobacteriaceae*.

REFERENCES

1. Ishiguro, N.& Sato, G.(1979) *J. Hyg. 83*,331-344
2. Ishiguro, N., Hirose, K., Asagi, M. & Sato, G.(1981) *J. Gen. Microbiol. 123*,193-196
3. Ishiguro, N., Hirose, K. & Sato, G.(1980) *Appl. Environ. Microbiol. 40*,446-451
4. Smith, H.W., Parsell, Z. & Green, P.(1978) *J. Gen. Microbiol. 109*,305-311
5. Young, R. & Grillo, D.S.(1980) *Mol. Gen. Genet. 178*,681-689
6. Tu, C.P.D. & Cohen, S.N.(1980) *Cell, 19*,151-160
7. Cornelis, G., Ghosal, D. & Saedler, H.(1978) *Mol. Gen. Genet. 160*,215-224

CHLORAMPHENICOL TRANSPOSONS FROM *SALMONELLA NAESTVED* AND *ESCHERICHIA COLI* OF DOMESTIC ANIMAL ORIGIN

N. Terakado, T. Sekizaki and K. Hashimoto

Biological Products Division, National Institute of Animal Health

S. Yamagata and T. Yamamoto

Faculty of Pharmaceutical Sciences, Chiba University

An interesting feature of drug resistance among enteric bacteria isolated from domestica animals is that many *Salmonella* strains have recently become resistant to chloramphenicol (Cm), and most of them carry conjugative R plasmids with Cm resistance (14,16,17). From laboratory evidences with transposons (2), it seems likely that the transposable natures of the Cm-resistant gene may play an important role in such widespread occurrence of Cm-resistant *Salmonella*. This paper deals with the demonstration of Cm transposons from *Salmonella* and *Escherichia coli* of domestic animal origin.

TRANSPOSABILITY OF CHLORAMPHENICOL-RESISTANT GENES FOUND ON R PLASMIDS FROM *S. NAESTVED* AND *E. COLI*

To examine the transposability of the Cm-resistant gene, two strains were examined. *S. naestved* strain AHI-21, of calf origin, harbors a conjugative R plasmid of group H, pTE21, which encodes resistance to Cm, tetracycline, streptomycin and sulfadimethoxine. *E. coli* strain AHI-1, of pig origin, also harbors a conjugative R plasmid of group I, pTE1, which encodes resistance to Cm and trimethoprim. First, both R plasmids, pTE21 and pTE1, were transferred by conjugation to *recA* strain AB2463, carrying pMK1, which is a composite plasmid of ColE1 and a kanamycin transposon (Tn*5*). Transformation was then carried out with DNA preparations extracted from the transconjugants AB2463 (pMK1,pTE21) and AB2463 (pMK1,pTE1). When selection was carried out with both kanamycin and Cm, transformants of C600 carrying a single plasmid DNA of approximately 9.4 megadaltons (Mdal) were obtained.

It was also found that C600 transformed with these 9.4-Mdal DNAs acquired the parental resistance character (Km Cm) and produced active colicin. When a *polA* strain, P3478, was used as the recipient, however, no resistant transformants could be obtained. Since the ColEl plasmid requires an active *polA* gene for its replication system (8), these results indicated that the Cm-resistant gene was transposed from both pTE21 and pTEl to the pMKl plasmid DNA. Thus, the 9.4-Mdal plasmids obtained in this experiment were designated pMKl::Tn*3351* and pMKl::Tn*3352*, respectively.

ELECTRON MICROSCOPIC ANALYSIS OF PMKl::TN*3351* AND PMKl::TN*3352*

As described above, both pMKl::Tn*3351* and pMKl::Tn*3352* had a molecular mass of about 9.4 Mdal. On the other hand, their parental pMKl was estimated to be about 7.7 Mdal; this estimate was slightly lower than that reported by Yamamoto and Yokota (18). It was therefore suggested that the transposable Cm resistant gene might be presented in the 1.7-Mdal element.

To confirm this hypothesis, pMKl::Tn*3351*, pMKl::Tn*3352* and pMKl DNAs were denatured and then self-annealed and examined by electron microscopy. All of the DNA samples showed the presence of inverted repeat structures which are characterictic of Tn*5*. The sizes of Tn*5* structures obtained in all preparations were identical and were calculated to be about 3.5 Mdal. However, the sizes of single-stranded ColEl loops observed in pMKl::Tn*3351* and pMKl::Tn*3352* were found to be about 1.7 Mdal larger than that of pMKl; 5.9 Mdal for pMKl::Tn*3351* and pMKl::Tn*3352* and 4.2 Mdal for pMKl. Figure 1 shows a pMKl::Tn*3351*/ColEl heteroduplex molecule. As is clear from this figure, a single-stranded insertion loop of about 1.7 Mdal was observed within the double-stranded ColEl loop region. However, no inverted repeat structure was seen except for that of Tn*5*. A similar result was obtained in a pMKl::Tn*3352*/ColEl heteroduplex. From these results, we concluded that the insertion loop observed under electron microscopy consisted of the Cm transposon.

RESTRICTION ENZYME CLEAVAGE OF PMKl::TN*3351* AND PMKl::TN*3352*

To examine the similarity between Tn*3351* and Tn*3352*, both plasmid DNAs were digested with three restriction endonucleases, and the resulting DNA fragments were subjected to agarose gel electrophoresis. Both DNAs were cleaved by *Eco*Rl into two, by *Pvu*ll into seven, and by *Pst*l into eight sections (Fig. 2). Moreover, the cleavage patterns were found to be identical, suggesting that they have homologous DNA sequences. On the other hand, the numbers of cutting sites on the pMKl DNA supplied as controls were found to be one for *Eco*Rl, five for *Pvu*ll, and six for *Pst*l. These results indicated that both Tn*335I* and Tn*3352* have a single cleavage site for *Eco*Rl and two each for *Pvu*ll and *Pst*l.

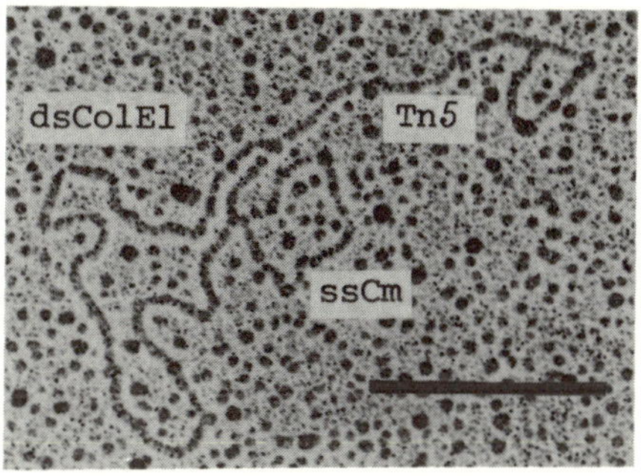

Fig. I. Electron micrograph of heteroduplex between pMKI::Tn335I and ColEI DNA. Bar, 0.5 um. Abbreviations:ss, single-stranded DNA; ds, double-stranded DNA.

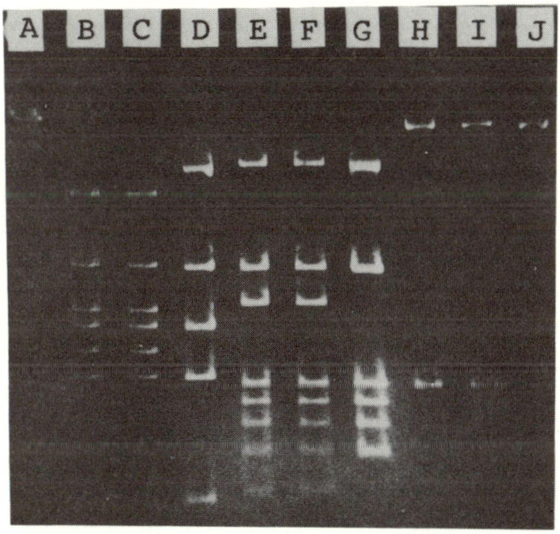

Fig. 2. Restriction cleavage patterns of pMKI::Tn335I, pMKI::Tn3352, and pMKI. Gel electrophoresis was conducted in 1.5% agarose at 100V for 6h. The DNA and the restriction enzyme for cutting are as follows:(A) HindIII(lambda);(B-D)PvuII(pMKI::Tn335I,pMKI::Tn3352, pMKI);(E-G)PstI(pMKI::Tn335I,pMKI::Tn3352,pMKI);(H-J) EcoRI(pMKI::Tn335I,pMKI::Tn3352,pMKI).

DISCUSSION

The transposability of the gene encoding Cm resistance from one replicon to another was first reported by Kondo and Mitsuhashi (9). They found that *E. coli* phage P1 could receive the Cm-resistant gene from an R plasmid, pMS14. Further studies revealed that such transposition of the Cm-resistant gene can occurr independently of the host *recA* gene function (4), and this transposable Cm-resistant gene is now called Tn*9* (11). Other transposable Cm-resistant genes and transposable multiple resistance genes with Cm resistance have been so far demonstrated on R plasmids originating from various bacteria, including *Shigella* (3,5,15), *Salmonella ordonez* (12), *Haemophillus influenzae* (7), and *Pseudomonas aeruginosa* (6,13).

Our results also indicate the existence of such transposable Cm-resistant genes among R plasmids of naturally occurring *S. naestved* and *E. coli* of domestic animal origin. This finding is further evidence for the wide distribution of the transposable Cm-resistant gene in nature. Two Cm transposons (Tn*3351* and Tn*3352*) described herein were indistinguishable in their restriction enzyme cleavage patterns, indicating that they are mostly identical in DNA sequences despite their difference in origin. This result suggests common ancestry for these two Cm transposons. It is not unreasonable, therefore, to speculate that the reciprocal transposition of the Cm-resistant gene might have occurred between these two species in nature.

As stated above, recent epidemiological studies have revealed a high incidence of R plasmids conferring Cm resistance among *Salmonella* strains of domestic animal origin (14,16,17). It seems likely that the transposable nature of the Cm-resistant gene may account for such increased occurrence of these R plasmids among *Salmonella* strains. Frequent use of drugs, especially Cm in treating animals will undoubtedly give a selective advantage to such events in nature.

Tn*3351* and Tn*3352* are quite similar in both size (about 1.7 Mdal) and a structural feature (no inverted repeats) to Tn*9* and other Tn*9*-like Cm transposons originally recognized on pMS 14 and RNR1 (1,3,10,15,19). In addition, restriction enzyme cleavage analysis showed that these two Cm transposons are very similar to Tn*9*. For instance, the number of cutting sites with *Eco*R1 (one site) and *Pst*1 (two sites) on both Tn*3351* and Tn*3352* was found to be consistent with that of Tn*9* (1), although a small difference was observed with *Pvu*ll digention, e.g., two sites on both, whereas one on Tn*9* (1). These results indicate that both have DNA sequences in common with Tn*9*. Considering that Tn*9* and other Tn*9*-like Cm transposons originated from naturally occurring *Shigella* strains of human origin, it is suggested that such Tn*9*-like Cm transposons might be widely distributed among various enteric bacteria of human and animal origin.

REFERENCES

1. Alton,N.K. & Vepnek,D. (1979) Nature(London) 282,864-869
2. Calos,M.P. & Miller,J.H. (1980) Cell 20,579-595
3. Chandler,M.,Tour,E.B.T.,Willems,D. & Caro,L. (1979) Mol. Gen. Genet. 176,221-231
4. Gottesman,M.M. & Rosner,J.L. (1975) Proc. Natl. Acad. Sci. U.S.A. 72,5041-5045
5. Iida,S. & Arber,W. (1977) Mol. Gen. Genet. 153,259-269
6. Iyobe,S.,Sagai,H. & Mitsuhashi,S. (1981) J. Bacteriol. 146, 141-148
7. Jahn,G.,Laufs,R.,Kaulfers,P.M. & Kolenda,H. (1979) J. Bacteriol. 138,584-597
8. Kingsbury,D.T. & Helinski,D.R. (1970) Res. Commun. 41,1538 -1544
9. Kondo,E. & Mitsuhashi,S. (1964) J. Bacteriol. 88,1266-1276
10. MacHattie,L.A. & Jackowski,J.B. (1977) in DNA insertion elements, plasmids, and episomes (Bukhari,I.,Shapiro,J.A. & Adhya,S.L.,ed.) pp.219-228, Cold Spring Harbor Laboratory , Cold Spring Harbor, New York
11. Rosner,J.L. & Gottesman,M.M. (1977) in DNA insertion elements, plasmids, and episomes (Bukhari,I.,Shapiro,J.A. & Adhya,S.L.,ed.) pp.213-218, Cold Spring Harbor Laboratory, Cold Spring Harbor, New York
12. Roussel,A.,Carlier,C.,Gerbaud,G. & Chabbert,Y.A. (1979) Mol. Gen. Genet. 169,13-25
13. Rubens,C.E.,McNeill,W.F. & Farrar,W.E. (1979) J. Bacteriol. 139,877-882
14. Sato,G.,Furuta,Y.,Kodama,H.,Iwao,T. & Oka,M. (1975) J. Am. Vet. Res. 36,839-841
15. Suzuki,K.,Mise,K. & Nakaya,R. (1980) Microbiol. Immunol. 24,309-320
16. Terakado,N.,Ohya,T.,Ueda,H.,Isayama,Y. & Ohmae,K. (1980) Jpn. J. Vet. Sci. 42,543-550
17. Timoney,J.F. (1978) Infect. Dis. 137,67-73
18. Yamamoto,T. & Yokota,T. (1980) Mol. Gen. Genet. 178,77-83
19. Yun,T. & Vepnek,D. (1977) Virology 77,376-385

TRANSPOSITION OF CARBENICILLIN- AND OXACILLIN-HYDROLYZING β-LACTAMASE GENES CARRIED BY PLASMIDS FROM GRAM-NEGATIVE BACTERIA

H. NAKAZAWA, K. KATSU, AND S. MITSUHASHI

Department of Microbiology
School of Medicine, Gunma University
Maebashi, Japan

I. INTRODUCTION

After the first investigation of transposition of ampicillin resistance (Ap^r) gene on RP4 plasmid(10), a number of antibiotic resistance gene on plasmid have been shown to be located in transposable DNA units termed transposons. These descret DNA segments are capable of *rec*A-independent transposition from one location to another on the same or a different molecule and are structurelly defined by repeated DNA sequence at their termini (5). Many Ap transposons have been reported, i.e., Tn*1* (10), *2* (11), *3* (15), *401* (3), *801* (2), *901* (8), *902* (25), *1701* (21), *2601*, *2602* (22), and *2603* (23,24) collectively termed TnA, and these Ap resistance are due to the formation of penicillin β-lactamase (PCase). Mitsuhashi and Inoue classified PCases into four types by their enzymological and immunological properties (18). All the TnA except for Tn*2603*, encode the type I (TEM type) PCase which is the most common enzyme and is most often produced by plasmids of a veriety of incompatibility groups (17). The type II (oxacillin-hydrolyzing, OXA-1), type III (OXA-2, OXA-3), and the type IV (carbenicillin-hydrolyzing. PSE) PCases are less frequently seen in clinical isolates carrying plasmids. TnA encoded the type II, type III, or type IV PCase genes have not yet been reported except for Tn*2603* which carry the type II PCase gene. We attempt to prove the transposability of Ap^r genes carried by Rms213 or Rms433 (13) plasmid which encoded the formation of the type II or type IV PCase, respectively.

II. TRANSPOSITION OF Ap^r GENES ON Rms213 AND Rms433

E. coli ML4903 carrying either the Rms213 or the Rms433

plasmid and strain ML4902 carrying the pCR1 plasmid (1) were mixed. Either the Rms213 or Rms433 plasmid was transfered to ML4902, and the transconjugants were spread on plates containing a high concentration of Ap. Using this procedure, highly Ap resistant clones were selected at a frequency of 10^{-5}. It is well known that degree of resistance to Ap is changed in propotion to the number of copies of the β-lactamase gene on the plasmid genome (19). In order to analyze the highly Ap resistant clones, we isolated the plasmid DNA and transformed the plasmid to the *pol*A214 strain (6). All the transformants could grow on the plates containing Ap at 30°C. In contrast, they were unable to grow on the Ap plates at 42°C. It was possible to demonstrate that the Ap resistance gene on the Rms213 or the Rms433 was transposed to the pCR1 plasmid, because the pCR1 plasmid requires the *pol*A gene product for its DNA replication. To isolate a *pol*A-independent strain carrying Ap resistance, *pol*A214 strain carrying pCR1::Ap was incubated at 42°C. Clones which grown on the Ap plate at 42°C were obtained at a frequency of about 10^{-4} or 10^{-5}. We further transposed the Ap^r gene from *pol*A214 chromosome to another plasmid. Plasmid pACYC184 (4) was transformed into strain χ2206::Ap. Highly Ap-resistant clones were isolated at a frequency of 10^{-4} or 10^{-5} in the presence of a high concentration of Ap. In addition, we examined whether the transposition of the type II or type IV Ap^r gene was *rec*-independent. The recombinant plasmid pACYC184::Ap was transformed to *rec*A^- or *rec*A^+ strain and then plasmid TP114 (9) was transffered to *rec*A^- or *rec*A^+ strain carrying pACYC184::Ap. These strains were used as the donors for conjugation experiments. The plasmid pACYC184 is not mobilized by conjugative plasmids such as TP114. Selection with the resistance marker of TP114 showed that TP114 was transfered at a frequency of 10^{-2}, and transconjugants which aquired Ap resistance were obtained at a frequency of 10^{-7}. Transconjugants had acquired the resistance marker of TP114, but not the pACYC184 plasmid such as tetracycline (Tc) and chloramphenicol (Cm). These results indicate that the transposition of Ap^r gene can occur independently of the *rec*A function of the host strain. Transposition frequencies of the Ap transposons encoding for the type II and the type IV PCase genes are summarized in Table 1. Transposition frequency of Ap^r gene from plasmid to plasmid was about 10^{-5}, and from plasmid to chromosome was about 10^{-5} to 10^{-4}. Their transposition frequency from chromosome to plasmid occured approximately the same frequency. The type I PCase gene on R1 plasmid was transposed from plasmid to plasmid at a frequency of 10^{-2}.

The Ap resistance levels, reletive PCase activities, and resistance markers of the replicons are summerized in Table 2. There were large differences in the degree of resistance to Ap

Table 1. *Transposition frequency of the type II and the type IV PCase genes*

Donor DNA	Recipient DNA	Host function	Frequency Type II	Type IV
Rms213 or Rms433	*pCR1*	$recA^{+}$	2×10^{-5}	5×10^{-5}
pCR1	*Chromosome*	$polA^{ts}$	9×10^{-4}	1×10^{-5}
Chromosome	*pACYC184*	$polA^{ts}$	2×10^{-4}	3×10^{-5}
pACYC184	*TP114*	$recA^{-}$	8×10^{-5}	5×10^{-5}
R1	*pCR1*	$recA^{-}$	10^{-2}	

depending on whether the Ap^{r} gene was located on a plasmid, the chromosome or the number of plasmid copies. For example, strains encoding for chromosomal resistance to the type II enzyme were 8 times more susceptible to Ap than strains carrying multiple copies of plasmid pCR1 and pACYC184 but had equivalent resistance to strains carrying rather less copies of plasmid Rms213 and TP114. Also both types of transposon, II and IV, are accompanied by streptomycin (Sm), sulfonamide (Su), and mercury (Hg) resistance.

Table 2. *Resistance levels, PCase activity, and resistance determinants*

PCase type	Replicon	Resistance level of Ap (μg/ml)	Relative PCase activity[a]	Resistance pattern
II	*Rms213*	200	1.0	*Ap Sm Su Hg Tc Cm*
	pCR1	1600	8.5	*Ap Sm Su Hg Km*
	Chromosome	200	0.6	*Ap Sm Su Hg*
	pACYC184	1600	7.9	*Ap Sm Su Hg Tc Cm*
	TP114	200	1.2	*Ap Sm Su Hg Km*
IV	*Rms433*	400	1.0	*Ap Sm Su Hg*
	pCR1	>3200	11.1	*Ap Sm Su Hg Km*
	Chromosome	200	0.5	*Ap Sm Su Hg*
	pACYC184	>3200	10.8	*Ap Sm Su Hg Tc Cm*
	TP114	400	1.0	*Ap Sm Su Hg Km*

[a]*PCase activities were assayed by the method of Waley (20).*

III. DETERMINATION OF THE MOLECULAR SIZE OF TRANSPOSONS

Three types of mutants were obtained when the type II or

the type IV PCase gene was transposed to the pACYC184 plasmid and these plasmid DNA were transoformed to the *rec*A$^-$ strain. Those transformants were resistant to (Ap, Sm, Su, Tc, Cm), (Ap, Sm, Su, Tc), and to (Ap, Sm, Su, Cm). For the molecular studies of three types plasmids, the plasmid DNA were isolated (12, 16). The three types of plasmid DNAs were digested by *Eco*RI endonuclease. The enzyme treatment of the plasmids encoding the type II PCase genes generated seven DNA fragments, and the five in the seven fragments were common size and remaining two fragments had the different size in each plasmids.

Table 3. *EcoRI digestion products of the type II and the type IV Ap transposons*

Plasmid	*Type II*			*Type IV*			*pACYC*
	(1)	*(2)*	*(3)*	*(1)*	*(2)*	*(3)*	*184*
Fragment size(Md)	*4.09*	*4.09*	*4.09*	*4.66*	*4.42*	*5.53*	*2.65*
	3.54	*3.70*	*3.58*	*3.52*	*3.52*	*3.52*	
	2.68	*2.68*	*2.68*	*2.26*	*2.46*		
	2.01	*1.83*	*1.93*	*1.68*	*1.68*	*1.68*	
	1.12	*1.12*	*1.12*			*1.39*	
	0.95	*0.95*	*0.95*				
	0.78	*0.78*	*0.78*				
Total (Md)	*15.17*	*15.15*	*15.13*	*12.12*	*12.08*	*12.09*	*2.65*

In the case of pACYC184 encoding the type IV PCase gene, four fragments were generated by *Eco*RI digestion. And also two fragments in the four were common and other two fragments were different sizes. Fragment sizes are shown in Table 3. *Eco*RI cleaved the pACYC184 at the only one site, therfore the different size fragments include the pACYC184 DNA and the other common fragments are derived from the transposons. Difference in fragment length is probably caused by the site and orientation of the insertion of the transposon. Since the molecular size of the pACYC184 is 2.65 Md, the actual molecular weight of the two transposons are 12.5 and 9.5 Md, respectively.

IV. ELECTRON MICROSCOPY OBSERVATION

Heteroduplex molecules were formed between pACYC184 and pACYC184 encoding the type II or the type IV Ap transposon (7).

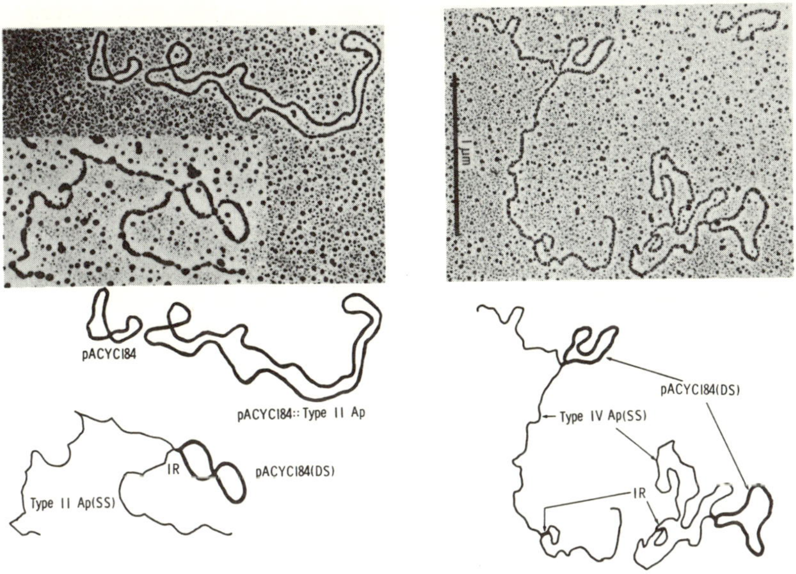

Fig. 1. Heteroduplex between pACYC184 and pACYC184::Ap

Electron microscopy revealed three kinds of molecules. Two of them were double-stranded DNAs, and the third consisted of heteroduprex molecules. The heteroduplex has a single stranded and double stranded region, which corresponded to the transposon DNA and pACYC184 DNA, respectively. The type II Ap transposon contains one stem-like structure indicating inverted repeat DNA sequence, but the heteroduplex morecule of the type IV Ap transposon has two inverted repeat sequences. Heteroduplex analysis also indicated that the transposition of the type II or the type IV Ap^r genes into vector did not cause any significant deletion.

V. CONCLUSION

We have isolated new transposons which encode the formation of the type II (oxacillin-hydrolyzing) or the type IV (carbenicillin-hydrolyzing) PCase together with resistance to Sm, Su, and mercury. They are capable of transposition among various replicons such as plasmids and host chromosome, and to various sites of the replicon independently of the normal *rec*A function of the host cell. The transposition frequencies of the type II and the type IV Ap transposon are 10^{-5} to 10^{-4}. A number of Ap transposons have been reported. These Ap transposons except *Tn2603* encoded the type I (TEM type) PCase production and consists of approximately a 4500 base pairs (b.p.) segment flanked by inverted repeats of about 40 b.p. sequence. Two our transposons are different from these Ap transposons in its type of β-lactamase, molecular size, and their resistance determinants. Tn*2603* encodes the formation of the type II PCase from the R plasmid collected by this laboratory. But our type II Ap transposon is also different from Tn*2603* in its molecular size and a number of *Eco*RI cuting sites. Reports on another multiple resistance transposon, Tn*4* (15), have indicated that it encodes for the same resistance as our transposons, but Tn*4* is a composite transposon consisting of Tn*3* inserted into Tn*21* (14). The Ap^r element in Tn*4* is capable of transposition without accompanying the other resistance determinants. Our two transposons are always transposed to the replicons as a unit of all resistance markers. Transposition frequencies of our transposons are lower than the frequencies reported for other Ap transposons which encoded for the type I PCase.

REFERENCES

1. Armstrong, K.A., Hershfield, V. & Helinski, D.R. (1977) Science *196*, 172-174.
2. Benedick, M., Fennewald, M., & Shapiro, J. (1977) J. Bacteriol. *129*, 809-814.
3. Bennett, P.M. & Richmond, M.H. (1976) J. Bacteriol. *126*, 1-6.
4. Chang, A.C.Y. & Cohen, S.N. (1978) J. Bacteriol. *134*,1141-1156.
5. Cohen, S.N. (1976) Nature (London) *263*, 731-738
6. Curtiss III, R., Inoue, M., Pereia, D., Hsu, J.C., Alexander, L. & Rock, L. (1977) in Molecular cloning of recombinant DNA (Schott, A. & Werner, R., ed.) pp.99-114, Academic press, New York
7. Davis, R.W., Simon, M. & Davidson, N. (1971) Methods Enzymol. *21*, 413-428.
8. Embden, J.D.A., Veltkamp, E., Stuitje, T., Andreoli, P.M. & Nijkamp, H.J.J. (1978) Plasmid *1*, 204-217.
9. Grindley, N.D.F., Grindley, J.N. & Anderson, E.S. (1972) Mol. Gen. Genet. *119*, 287-297.
10. Hedges, R.W. & Jacob, A.E. (1974) Mol. Gen. Genet. *132*, 31-40.
11. Heffron, F., Sublett, R., Hedges, R.W. Jacob, A. & Falkow, S. (1975) J. Bacteriol. *122*, 250-256.
12. Ike, Y., Hashimoto, H., Motohashi, K. & Mitsuhashi, S. (1980) J. Bacteriol. *141*, 577-583.
13. Katsu, K., Inoue, M. & Mitsuhashi, S. (1981) J. Antibiot. *34*, 1504-1506.

14. Kopecko, D.J., Brevet, J. & Cohen, S.N. (1976) J. Mol. Biol. *108*, 333-360.
15. Kopecko, D.J. & Cohen, S.N. (1975) Proc. Natl. Acad. Sci. U.S.A. *72*, 1373-1377.
16. Kupersztoch-Portnoy, Y.M., Lovett, M.A. & Helinski, D.R. (1974) Biochemistry *13*, 5484-5490.
17. Matthew, M. & Hedges, R.W. (1976) J. Bacteriol. *125*, 713-718.
18. Mitsuhashi, S. & Inoue, M. (1981) in Beta-lactam antibiotics (Mitsuhashi, S., ed.) pp.41-56 Japan Scientific Societies Press and Springer-Verlag
19. Uhlin, B.E. & Nordstrom, K. (1977) plasmid *1*, 1-7.
20. Waley, S.G. (1974) Biochem. J. *139*, 780-789.
21. Yamada, Y., Calame, K.L., Grindley, J.N. & Nakada, D. (1979) *137*, 990-999.
22. Yamamoto, T., Katoh, R., Shimazu, A. & Yamagishi, S. (1980) Microbiol. Immunol. *24*, 479-494.
23. Yamamoto, T., Tanaka, M., Nohara, C., Fukunaga, Y. & Yamagishi, S. (1981) *145*, 808-813.
24. Yamamoto, T., Tanaka, M., Baba, R. & Yamagishi, S. (1981) Mol. Gen. Genet. *181*, 464-469.
25. Yun, T. & Vapnek, D. (1977) in DNA insertion elements, plasmids and episomes (Bukhari, A.I., Shapiro, J.A. & Adhya, S.L., ed.) pp.229-234 Cold Spring Harbor, New York

MOLECULAR PROPERTIES OF TN*2603*, A TRANSPOSON ENCODING AMPICILLIN, STREPTOMYCIN, SULFONAMIDE AND MERCURY RESISTANCE

TOMOKO YAMAMOTO, MICHIYASU TANAKA, AND TETSUO SAWAI

Division of Microbial Chemistry, Faculty of Pharmaceutical Sciences, Chiba University, Chiba, Japan

I. TRANSPOSONS IN THE SPREAD OF THE OXACILLIN-HYDROLYSING β-LACTAMASE GENE

Plasmid-mediated resistance to penicillins and cephalosporins is mostly associated with the elaboration of β-lactamase and many of the β-lactamases have been well characterized. In order to define the genetic basis of the spread of β-lactamase among microorganisms, we have investigated the potential transposability of various β-lactamase genes classified on the basis of their enzymological and immunochemical properties.

As previously reported (1, 2), we demonstrated the evidence that the oxacillin-hydrolysing β-lactamase (type II or OXA-1) gene specified by R plasmid RGN238 is capable of transposition and we called this transposable element Tn*2603*. Tn*2603* has a molecular size of 20 kilobase pairs long and is flanked by small inverted repeat sequences of about 0.2 kilobase pairs long. Tn*2603* is a single transposable unit encoding a multiple resistance to ampicillin (Ap), streptomycin (Sm), sulfonamide (Su), and mercury (Hg). In most cases, the type II enzyme gene has been found together with Sm, Su, and Hg resistances on naturally occurring plasmids. We are interested in finding out whether Tn*2603* is involved in the distribution of the type II enzyme gene and have recognized the transposons indistinguishable from Tn*2603* on R plasmids of R455 (3) and R656a (4). These evidences suggest that Tn*2603* plays a major role in generation and distribution of R plasmids encoding the type II β-lactamase gene among microorganisms.

II. PHYSICAL AND GENETICAL STRUCTURE OF TN*2603*

A restriction endnuclease cleavage map of Tn*2603* was constructed by an analysis of restriction cleavage patterns of plasmid pMK1::Tn*2603* DNA and its deletion derivative DNAs (Fig. 1). Each region necessary for expression of resistance to Ap, Sm, Su, or Hg was located on the cleavage map. Furthermore, the functional promoter region and the direction of transcription from each promoter were determined by using plasmid pMC81 which contains: 1) the *lac* genes of *E. coli* under control of *ara* operon promoter, and 2) an intervening DNA fragment containing a *Hin*d III cleavage site. Insertion of a fragment carrying a promoter region at this site results in *lac* expression in the absence of arabinose induction. Insertion of DNA fragment having a termination signal prevents arabinose-induced expression of *lac*. *Hin*dIII-generated fragment of H4, H5, or H7 encoding resistance to Ap, Su ,or Sm, respectively from pMK1::Tn*2603* was introduced into the *Hin*dIII cleavage site on pMC81 and then the expressions of *lac* and resistance were examined (Fig. 2). The results indicate that the fragments of H4 and H7 contain a functional promoter for the expression of Ap resistance and Sm resistance, respectively. The direction of transcription from each promoter was decided by the orientation of the fragment introduced which results in an arabinose-independent expression of *lac*. The Su resistance from pMC81-H5 was expressed only in the presense of arabinose and the potential arabinose-induced expression of *lac* was prevented, suggesting that the H5 fragment contains no functional promoter but a termination

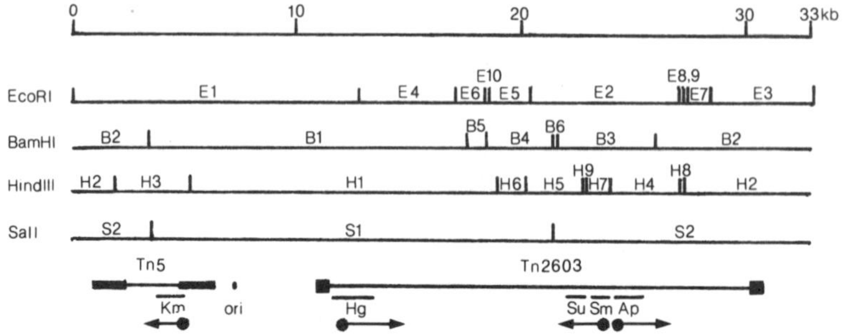

Fig. 1. Functional map of pMK1::Tn2603#1 plasmid. Heavy lines represent the inverted repeat sequences. Solid circles indicate the promoter. Arrows indicate the direction of transcription from each promoter.

signal. The transcription in the H5 fragment was determined to proceed in the same direction from the H7 fragment containing a promoter for the expression of Sm resistance. The H9 fragment did not affect the expression of Su gene. These results suggest the possibility that the Sm and Su genes comprise a single transcriptional unit. The results are summarized in Fig. 1.

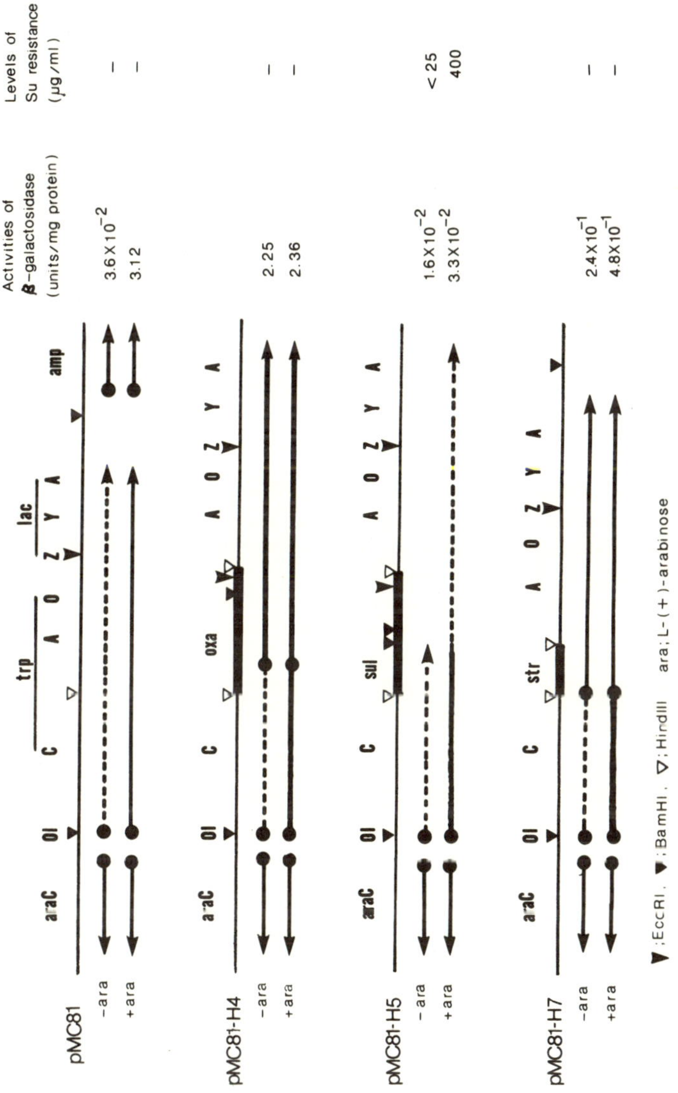

Fig. 2. Expression of *lac* of *pMC81* derivatives containing *HindIII*-digested fragment from *pMK1::Tn2603* in the presence or absence of arabinose induction.

III. GENES REQUIRED FOR TRANSPOSITION OF TN*2603*

To define the genes required for transposition of Tn*2603*, various deleiton mutants were isolated from pMK1::Tn*2603* DNA by partial digestion with *Eco*RI and their transposability were examined. The deletion of each fragment of E4, E5, E6, and E10 did not affect its transposability. The deletion affecting the transposition was that of E3 fragment containing one of the inverted repeat sequences at both ends or that of E7 fragment. The results indicate that the gene(s) encoded in the region covering the E7 fragment and the inverted repeat segments at the ends are required for transposition of Tn*2603*.

Recently, we have found that Tn*2603*, Tn*21* (Sm, Su, Hg) and Tn*501*(Hg) are closely related by a restriction endonuclease digestion analysis and an electronmicroscopic analysis of heteroduplex molecules. It would be speculated that these transposons were generated from common ancestor though these encode for diverse phenotypic character.

REFERENCES

1. Yamamoto, T., Tanaka, M., Nohara, C., Fukunaga, Y. & Yamagishi, S.(1981) J. Bacteriol.*145*, 808-813
2. Yamamoto, T., Tanaka, M., Baba, R. & Yamagishi, S.(1981) Mol. Gen. Genet. *181*, 464-469
3. Hedges, R.W., Datta, N., Coetzee, J.N. & Dennison, S. (1973) J. Gen. Microbiol. *77*, 249-259
4. Hedges, R.W., Datta, N., Kontomichalou, P. & Smith, J.T. (1974) J. Bacteriol. *117*, 56-62

HOST FUNCTIONS REQUIRED FOR TRANSPOSITION OF Tn5 FROM λb221cI857rex::Tn5 : THE ROLE OF lon+ GENE PRODUCT

Y.Uno and M.Yoshikawa

Department of Microbiology
The Institute of Medical Science
The University of Tokyo, Tokyo, Japan

INTRODUCTION

Transposable element moves to various sites without homologous recombination(3,11,14). Although the mechanism of transposition is not yet known, several models for transposition have been proposed(3,8,14). The following two characteristics of transposition have been recognized: i) Transposition results in the duplication of a few nucleotides of the target sequence. ii) Transposable element duplicates itself and leaves one of them at its original site. Transposon itself encodes several functions necessary for transposition. On the other hand, it has been reported that several host functions such as DNA polymerase I(5,13,18) and DNA topoisomerase I(15) also play important roles. We show here that the lon+ gene product is also an important determinant of the frequency of Tn5 transposition from λ::Tn5.

RESULTS

I. Lon-Deficient Mutation Decreased the Frequency of Tn5 Transposition from λ::Tn5 to the Chromosome

Two pairs of lon mutants and their respective parents were examined for their transposition frequency from λ::Tn5 to the chromosome by comparing the number of kanamycin resistant colonies per viable cell. After 2 days incubation, the frequency of appearence of kanamycin resistant colonies per viable cell was always lower with lon mutants than with their parents (Table 1). Introduction of F'13 or F'254, both bearing the lon^+ locus, recovers the transposition frequency in Lon-deficient cells.

Table 1. Tn5 Transposition from λ::Tn5 to the Chromosome

Strain code	lon	Transposition Frequencies($\times 10^{-2}$)
AB1157	+	3.9 ± 1.2
AB1899	lon1	0.6 ± 0.3
AB1899NM	lon1 (non-mucoid)	0.2 ± 0.1
MC100	+	1.9
T206	−	0.09

The method described by Sasakawa *et al*.(13) was followed.

In order to exclude the possibility that the rate with which cells express Km-resistance phenotype is affected by Lon-deficiency, the time of shaking at 30°C for phenotypic expression after λ::Tn5 infection and before plating on selective kanamycin plates was altered. The number of kanamycin resistant colonies per viable cell was always larger in Lon-proficient cells than in Lon-deficient cells. After prolonged incubation of selective plates, the number of kanamycin resistant colonies increased both in Lon-proficient and deficient cells. Nevertheless, the number was always higher in Lon-proficient cells than in Lon-deficient cells. We also determined the minimal inhibitory concentration(MIC) of kanamycin of both Lon-proficient and deficient cells with or without Tn5. Without Tn5, both Lon-proficient and deficient cells were inhibited by kanamycin at the same concentration. Although each clone with Tn5 exhibited different MIC, the MIC was not always higher in Lon-proficient cells with Tn5 than in Lon-deficient cells with Tn5. In order to rule out the possibility that λ multiplication rather than transposition is affected by the lon^{+} gene product, adsorption rate, relative efficiency of plating, host cell killing and burst size as determined by one step growth experiments by the use of λ::Tn5 at 30°C were carried out. All these parameters were shown not to affect the conclusion on the role of the lon^{+} gene product in Tn5 transposition. More than 1,000 Km-resistant transductants of both AB1157 and AB1899NM were examined for additional nutritional requirements due to insertional inactivation by Tn5. The results indicated that Tn5 located at various chromosomal sites among these colonies.

II. DL-pantoyllactone Recovers Normal Transposition Frequency in AB1899NM

DL-pantoyllactone is known to diminish various effects of a lon mutation(1). The transposition frequency in AB1899NM was recovered almost to the same level of that in AB1157 when agar plates contained DL-pantoyllactone at 0.04M. The effect of DL-pantoyllactone in the growth medium and in the reaction mixture was also observed.

Table 2. Effects of lon Mutation in Three Established Experimental Systems of Tn5 Transposition

Expt.	Transposon Donor Replicon	Target Replicon	lon	Transposition Frequencies
I	R388::Tn5	λ bb	+	6.7×10^{-8}
			_	4.9×10^{-8}
II	λ bb::Tn5	R100-1	+	1.2×10^{-4}
			_	8.8×10^{-5}
III	pSC101::Tn5	R100-1	+	1.1×10^{-5}
			_	0.9×10^{-5}

These three established experiments were described in (13).

III. In Established Transposition Experiments, the Effect of lon^+ Gene Product on Transposition Was Unclear

As described in the previous papers(13,18), established transposition experiments were defined as experiments in which the transposon donor replicon had existed long before transposition occurred. We used the following three established experiments: i) Tn5 transposition from a conjugative plasmid to exogenously infected λ b515b519cI857Sam7(abbreviated as λbb) ii) Tn5 transposition from a prophage λ bb::Tn5 to R100-1 iii) Tn5 transposition from pSC101::Tn5 to R100-1. As shown in Table 2, little difference was observed between Lon-proficient and deficient cells in these experiments.

DISCUSSION

The lon^+ gene product seems to be an important determinant in transposition of Tn5 from λ::Tn5 to the chromosome. This is supported by the facts that the low frequency of Tn5 transposition in Lon-deficient cells as compared to that in Lon-proficient cells is recovered by adding DL-pantoyllactone in selective agar plates, or in the growth medium and in the reaction mixture. In established transposition experiments, however, the difference between Lon-proficient and deficient cells could not clearly be detected. The reason may be explained as follows: In established transposition experiments, the transposase operon is under the repressed state, and thus transposase rather than partially defective host functions may become rate-limiting for overall transposition process.

What is then the role of lon^+ gene product? We consider the multiphenotypes of a lon mutation(1,6,7,9,10,12,16,19) should be ascribed to a single exzymatic activity. Recently, Charette *et al*. identified the lon^+ gene product as ATP dependent protease which can bind to double stranded as well as to single stranded DNA(4). We may suppose that the lon^+ gene product is required to bind to DNA so that the physical form of DNA is altered conformationally to be suitable for transposition. Alternatively, ATP dependent protease activity may be essential for regulation of gene expression related to transposition. Transposase may be activated only after proteolytic cleavage by ATP dependent protease coded by the lon^+ gene. The report by Trinks *et al*. seems to make this possibility highly probable in the case of IS4(17).

This study was supported by the Ministry of Education, Science and Culture, the Japanese Government (No.56480128).

REFERENCES

1. Adler,H.I. and Hatdigree,A.A.(1964) J.Bacteriol.87,720-726
2. Berg,D.E.(1977) in:"DNA insertion elements, plasmids and episomes," (Bukhari,A.I., Shapiro,J.A., and Adhya,S.L.,ed.) pp.205-212, Cold Spring Harbor Laboratory, Cold Spring Harbor
3. Calos,M.P., and Miller,J.H.(1980) Cell 20, 579-595
4. Charette,M.F., Henderson,G.W., Doane,L.L. and Markovitz,A. (1981) Abstract of 81st Annual Meeting of ASM-1981
5. Clements,M.B., and Syvanen,M.(1981) Cold Spring Harbor Symp.Quant.Biol. pp.201-204
6. Gayda,R.C., and Markovitz,A.(1978) Mol.Gen.Genet. 159,1-11
7. Gottesman,S., and Gottesman,M.(1981) Cell 24,225-233
8. Hershey,R.M., and Bukhari,A.I.(1981) Proc.Natl.Acad.Sci. U.S.A. 78,1090-1094
9. Howard=Flanders,P., Simson,E., and Theriot,L.(1964) Genetics 49,237-246
10. Kantor,G.J., and Deering,R.A.(1968) J.Bacteriol.95,520-530
11. Kleckner,N.(1977) Cell 11,11-23
12. Leighton,P.M., and Donachie,W.(1970) J.Bacteriol.102,810-

814
13. Sasakawa,C., Uno,Y., and Yoshikawa,M.(1981) Mol.Gen.Genet. 182,19-24
14. Starlinger,P.(1980) Plasmid 3,241-259
15. Sternglanz,R., Dinardo,S. Vockel,K.A., Nishimura,Y., Hirota,Y., Beckerer,K., Zunstein,L., and Wang,J.C.(1981) Proc.Natl.Acad.Sci.U.S.A. 78,2737-2751
16. Takano,T.(1971) Proc.Natl.Acad.Sci.U.S.A. 68,1469-1473
17. Trinks,K., Habermann,P., Beureuther,K., Starlinger,P., and Ehring,R.(1981) Mol.Gen.Genet. 182,183-188
18. Yoshikawa,M., Sasakawa,C. and Uno,Y.(1981) (Levy,S. _et al_. ed.) pp.391-400. Plenum publishing Co., New York
19. Walker,J.R., Ussery,C.L., and Allen,J.R.(1973) J.Bacteriol. 113,1326-1332

USE OF TRANSPOSONS TO IDENTIFY AND MANIPULATE *SHIGELLA* VIRULENCE PLASMIDS

Dennis J. Kopecko, Philippe J. Sansonetti,*[1]
Samuel B. Formal*, and Louis S. Baron

*Departments of Bacterial Immunology and Bacterial Diseases**
Walter Reed Army Institute of Research
Washington, DC, U.S.A

I. INTRODUCTION

During the past decade, many bacterial genetic determinants (e.g., antibiotic resistance genes) have been found to exist on discrete DNA units, termed transposons, that are transposable inter- and intra-chromosomally and that can promote a variety of macro-evolutinary chromosomal genetic rearrangements (e.g., deletions, inversions) (reviewed in ref. 1). The expression of easily identifiable phenotypic properties by transposons and their transposability make them very useful genetic experimental tools (2). It is important to emphasize that in the following studies, transposons were indispensable to the identification and characterization of plasmid-borne virulence determinants of shigellae. The major objective of this report will be to review our current knowledge of plasmid-mediated virulence properties in *Shigella*. In addition, this knowledge has recently been used to construct a bivalent oral vaccine strain aimed at protecting against typhoid fever and shigellosis due to *Shigella sonnei*. This vaccine study will be described briefly.

Bacterial diseases of the gastrointestinal tract usually occur by one of three overall mechanisms. The first mechanism , termed "intoxication", occurs by bacterial secretion of an exotoxin that oftentimes is preformed in food prior to ingestion by the host. This process is exemplified by staphylococcal or clostridial food poisoning. In contrast, the remaining two processes require living and multiplying disease agents. In the "enterotoxigenic" mechanism, bacteria colonize the small intestine, usually in the jejunum or duodenum. These bacteria multiply on the

[1]*Present address: Laboratoire des Enterobacteries, Institut Pasteur, Paris, France.*

intestinal surface and elaborate an enterotoxin that stimulates excessive fluid and electrolyte efflux resulting in a watery diarrhea. Enterotoxigenic *Escherichia coli* and *Vibrio cholera* serve as typical examples. In contrast to these two toxigenic mechanisms that directly involve only the small intestine, the third group of organisms, termed "invasive," also affect the large intestine. In the case of *Shigella*, the colon is the primary site of involvement. Invasive enteric bacteria actually penetrate the epithelial mucosa of the intestine. Subsequently, these organisms multiply intracellularly and disseminate within or through the mucosa. This latter mechanism, classically typified by *Shigella* and *Salmonella*, is now thought to be used by invasive strains of *E. coli*, *Yersinia*, *Vibrio parahemolyticus* and, possibly, *Campylobacter* and *Aeromonas hydrophilia*. Unlike other invasive bacterial diseases such as salmonellosis in which the invading bacteria are disseminated throughout the host, shigellosis is a disease in which the bacteria are normally confined to the intestinal lining. Toxigenic organisms generally require a large dose of organisms to cause disease but previous studies have shown that as few as ten virulent cells of *Shigella* can cause disease in humans. Thus, these features distinguish the toxigenic from the invasive mechanisms of intestinal disease (see reviews 3, 4).

Two common and essential features of virulent *Shigella* are their ability to penetrate and to multiply within the epithelial cells of the colon(3,4). Chromosomal mutants of *Shigella* strains that fail to penetrate or that penetrate but cannot multiply intracellularly have been isolated. Both types of mutants are avirulent. The process of invasion has thus far been characterized in microscopic, but not in biochemical detail. The first visible alteration in the host intestinal epithelium is a localized destruction of the microvilli, the outermost structure of the intestinal lining. The invading bacteria are then engulfed by means of an invagination of the intestinal cell membrane and are contained intracellularly within vaculoes. Subsequently, the microvilli are reestablished and intracellular bacterial multiplication occurs. These bacteria then destroy the vacuole and disseminate to adjacent cells, causing necrosis and resulting in acute inflammation and focal ulceration of the epithelium. The resulting dysentery is characterized by a painful, bloody and mucous diarrhea normally of relatively small volume.

Genetic studies of *Shigella flexneri* have previously resulted in the conclusion that virulence is multideterminant, with at least two widely separated bacterial chromosomal regions being required for invasion (3,4). Furthermore, these studies have shown that not only is a smooth lipopolysaccharide bacterial cell surface necessary for intestinal invasion, but also that only certain O-repeat unit polymers are effective in this process; this is true for both shigellae and invasive *E. coli*. Until recently, plasmids did not appear to play a role in the invasion process or in the virulence of *Shigella*. Recent evidence amassed over the past four years, however, demonstrates that plasmids of *Shigella* are involved in the invasion process (5-7, 11-13).

II. RESULTS

A. *Virulence Plasmids of* Shigella sonnei

Shigellosis is still an important disease worldwide, with approximately 15,000 cases reported in the U.S. during 1980. Of the 4 species of *Shigella, S. sonnei* is currently responsible for greater than two-thirds of all shigellosis cases in the U.S. and Europe. Because of its importance, this species was chosen as the initial focus of our studies. Unlike the other *Shigella* species, all *S. sonnei* strains fall into a single serotype. This serotype is due to a somatic antigen, termed form I, that is required for epithelial cell invasion. Chemical studies have revealed that the form I antigen is the O-side chain (8).

1. *Colonial Morphology Transition of* S. Sonnei. Upon restreaking on agar medium, smooth even-edged form I colonies generate at a relatively high frequency rough uneven-edged colonies, termed form II. Form II colonies appear in different strains at frequencies varying from 1 to 50%. Further study has shown that these rough colonies have irreversibly lost the form I antigen and are always avirulent due to the inability to invade epithelial cells (5,6). The ability to penetrate epithelial cells can be monitored in several laboratory systems. In the Sereny test, virulent *Shigella* that can penetrate into and multiply within corneal epithelial cells will elicit a keratoconjunctivitis within 72 hours following inoculation of a rabbit or guinea pig eye with a bacterial suspension (9). In addition to the Sereny test, invasiveness in the following studies was verified using cultured Hela cells (10).

2. *Plasmid Analyses of Form I and II Strains.* The high frequency and irreversible nature of the form I to II transition, which always resulted in the loss of virulence, suggested the involvement of a plasmid in this phenomenon. Thus, the plasmid DNA's of various *S. sonnei* strains, obtained from different parts of the world, were examined (5). Plasmid

DNA's from the form I strains contained a large plasmid which is estimated, for most strains, to be 120 megadaltons (Mdal) in size (Fig.1A, C. E, G). This large plasmid is missing in all form II derivatives (Fig.1B, D, F, H). This observation has been independently confirmed for many other *S. sonnei* isolates (6, 11).

3. *Conjugal Transfer Studies.* Direct proof that this large plasmid is involved in form I antigen synthesis and virulence can only be obtained by reintroduction of this

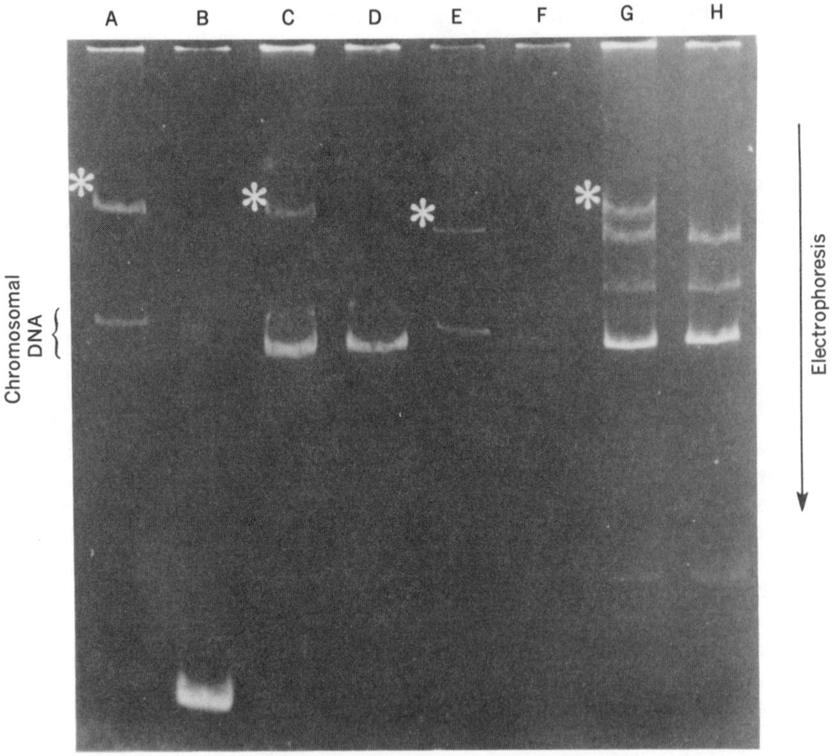

Fig. 1. Agarose gel electrophoretic profiles of circular plasmid DNA obtained from sets of isogenic form I and II S. sonnei *strains. Plasmid profile of: (A) strain 53G form I; (B) 53G form II: (C) 50E form I: (D) 50E form II: (E) 9774 form I; (F) 9774 form II: (G) MBI form I; and (H) MBI form II. The asterisks mark the large plasmids in the form I strains that are lost in form II derivatives. The gel position expected for fragmented chromosomal DNA is indicated. DNA isolation and gel electrophoresis procedures are described elsewhere (5).*

plasmid into form II recipient cells with concomitant reestablishment of these properties. However, neither the form I antigen nor virulence phenotypes are useful as selective markers to monitor plasmid transfer. Therefore, we attempted to identify any marker of selective value expressed by the form I plasmid. To date, about 175 biochemical and antibiotic resistance characters have been tested for, but we have been unable to detect any other trait encoded on this large plasmid. In addition, the results of further studies indicate that neither bacteriocin production nor iron-chelating systems are encoded by this form I plasmid (11). To circumvent this problem, the form I plasmid was phenotypically tagged with the ampicillin resistance transposon, Tn*3*, or with transposons Tn*5* or Tn*10*. These transposons were introduced into the appropriate strains on a carrier F'_{ts}*lac* replicon that is temperature sensitive for replication (5). Strains in which the form I plasmid had been tagged, expressed the appropriate transposon-encoded antibiotic resistance; and, this resistance was always lost during the transition to form II cells.

Attempts to detect conjugal self-transfer of these tagged form I plasmids, using antibiotic selective pressure, were unsuccessful, indicating that these large plasmids are not self-transmissible. However, using various conjugative plasmids of ten different incompatibility (Inc.) groups, several systems have been developed to mobilize the form I plasmid to recipient cells (5, 11). In one system, the Inc. FI plasmid R386 was found capable of mobilizing the form I plasmid. The plasmid DNA profiles of donor, recipient, and transconjugant *S. sonnei* strains from this mobilizing system are shown in Fig. 2. Some transconjugants received only the form I plasmid (Fig. 2E), while others also inherited the R386 plasmid (Fig. 2F). As discussed in more detail below, only transconjugants that had received the form I plasmid and that did not contain the R386 plasmid were virulent. Using a similar mobilization system, the form I plasmid and concomitant form I antigen synthesis has been introduced into *Shigella flexneri*, *Escherichia coli*, *Proteus mirabilis*, *Salmonella typhi*, *Salmonella typhimurium*, and *Serratia marcescens*, none of which typically encode a form I antigen (5, 11, 12). These data establish that form I antigen synthesizing ability and virulence are encoded by the 120 Mdal form I plasmid of *S. sonnei* (5, 11). Furthermore, cell invasion assays indicate that the form I plasmid is required for the invasive ability of *S. sonnei*. Transposon insertions into various form I plasmids have occasionally resulted in loss of virulence without affecting form I antigen synthesis, indicating that these plasmids encode a virulence determinant(s) other than form I antigen (11). At present, specific involvement of the form I antigen or other plasmid-mediated properties in the invasive process has not been determined and awaits clarification following the cloning of these determinants.

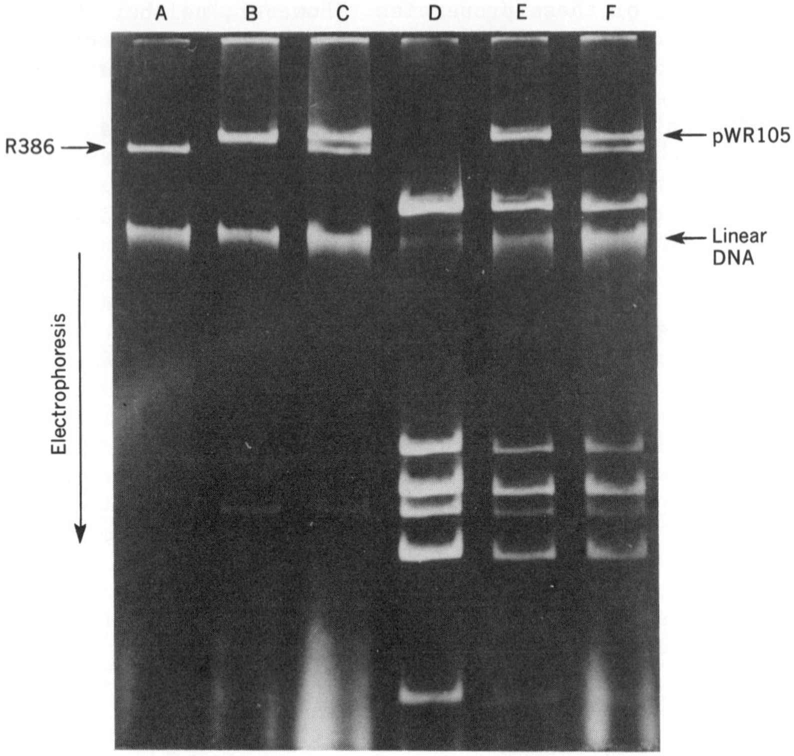

Fig. 2. Mobilization of the form I plasmid by R386. The agarose gel electrophoretic profiles of circular plasmid DNA obtained from donor, recipient, and transconjugant strains: (A) E. coli *J53 carrying R386; (B)* S. sonnei *482-79 carrying pWR105, a Tn5-tagged form I plasmid; (C) donor 482-79 with pWR105 and R386; (D) recipient form II* S. sonnei *Rudy; (E) Rudy transconjugant carrying pWR105; (F) Rudy transconjugant carrying pWR105 and R386. Experimental details are described elsewhere (11).*

4. *Sereny Test Inhibition by Inc. FI Plasmids.* During these studies, we have observed that all Inc. FI plasmids examined, when present intracellularly in *Shigella* or invasive *E. coli* strains, will prevent these organisms from invading the guinea pig corneal epithelium, resulting in a negative Sereny test. As noted above, and described elsewhere (5, 12) this phenomenon presented an experimental barrier to our attempts to mobilize the form I plasmid and reestablish virulence in a form II *S. sonnei* recipient strain. Only transconjugants that did not receive the R386 plasmid

were found to have regained virulence. Further study of this phenomenon has revealed that the intracellular presence of Inc. FI plasmids in invasive *E. coli* or *Shigella,* though affecting the Sereny test response, does not affect their ability to penetrate Hela cell monolayers. Although limited, these data suggest that Inc. FI plasmids specifically interfere with the Sereny reaction and do not actually inhibit bacterial virulence (Sansonetti, Kopecko, and Formal, unpublished data).

5. *Incompatibility of the Form I Plasmid.* Next, an attempt was made to identify the Inc. group of the form I plasmid. Various reference plasmids of 10 different Inc. groups were conjugally transferred to an *S. sonnei* strain containing a Tn*5*-tagged form I plasmid. The resulting strains, purified on antibiotic selective media and each carrying the form I plasmid and a reference plasmid, were streaked onto MacConkey lactose agar. The stability of the form I colony type was then monitored. None of the reference plasmids, except R386, significantly affected the normal form I to II transition as compared to the wild-type *S. sonnei* strain (11). Control studies showed that all of the reference plasmids are stably maintained in the isogenic form II *S. sonnei* derivative strains. Virtually identical results were obtained when two different form I plasmids were tested for incompatibility. Although these experiments are hampered by the natural instability and nonselftransferability of the form I plasmid, these data suggest that the form I plasmid is of the FI Inc. group (11).

B. *Virulence Plasmids of* Shigella flexneri

S. flexneri is responsible for a significant proportion of bacillary dysentery throughout the world. Earlier genetic analyses had established that virulence in *S. flexneri* is associated with several chromosomal loci (3, 4) but plasmid involvement in virulence was not previously recognized. Unlike the single serotype of *S. sonnei, S. flexneri* strains fall into six serotypes which differ significantly in their surface antigens. However, some *S. flexneri* strains have been observed classically to undergo an irreversible, spontaneous transition from translucent (T) to opaque (O) colonial morphology, when examined on agar medium, that results in loss of virulence. In 1979, we reported that this colonial morphology transition did not appear to involve a noticeable change in bacterial plasmid DNA profiles as observed by agarose gel electrophoresis (14). However, in light of the

S. sonnei findings, we decided to reexamine the contribution of plasmids to *S. flexneri* virulence.

1. *Plasmid Content of* S. flexneri. Initially, representative virulent strains of the six serotypes of *S. flexneri*

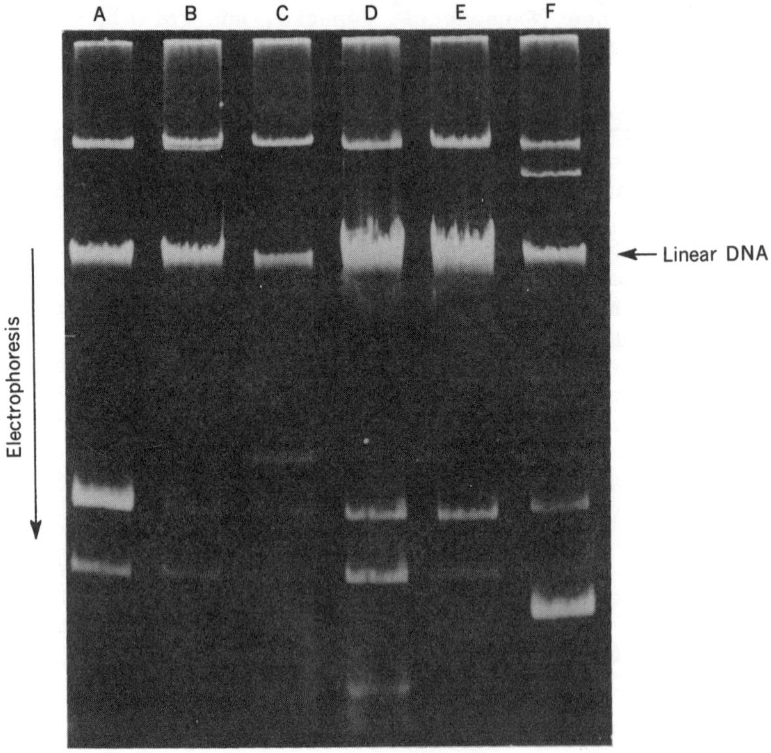

Fig. 3. Agarose gel electrophoretic profiles of plasmid DNA obtained from virulent strains representing the six serotypes of Shigella flexneri. *(A) Strain M25-8, serotype 1b. (B) Strain M4243, 2a. (C) Strain J17B, 3a. (D) Strain M7639, 4b. (E) Strain M90T, 5. (F) Strain CCH060, 6. Molecular sizes of the two large plasmids carried by M4243 (B) have been reported previously (14) to be 140.1 ± 3.0 and 106.4 ± 4.0 megadaltons (Mdal). These plasmids served as internal size standards. The gel position of fragmented plasmid or chromosomal DNA, generated during DNA isolation, has been labelled "linear DNA." All DNA bands, except the band labelled "linear DNA", represent covalently-closed circular plasmid species, as determined by controlled gamma irradiation studies (Unpublished data). Specific details are published elsewhere (13).*

were examined for plasmid content. Fig. 3 shows the plasmid profiles obtained following plasmid isolation and analysis by agarose gel electrophoresis of DNA obtained from 6 *S. flexneri* strains. A large plasmid of approximately 140 Mdal in size was observed in all virulent *S. flexneri* strains, regardless of serotype. Additionally, some strains contained other plasmid species (12, 13).

2. *Isolation and Characterization of Avirulent* S. flexneri *Variants.* Normally, on agar medium *S. flexneri* colonies appear smooth and translucent when illuminated by oblique lighting. An attempt was made to detect colonies that had spontaneously lost virulence but still retained the smooth, translucent appearance of wild type colonies. Several colonies of each *S. flexneri* strain, obtained by restreaking cells from old agar slants, were screened for their ability to elicit a keratoconjunctivitis in guinea pigs. Smooth, avirulent derivative colonies were obtained at a frequency of approximately 5-10% from virulent strains of serotypes 1b, 2a, and 5 (i.e. strains M25-8, M4243, and M90T, respectively). These smooth, avirulent derivatives were agglutinated by specific type and group antisera and did not appear to be altered in their surface characteristics. The plasmid DNA profiles of the parental strains and these avirulent derivatives were examined next. As shown in Fig. 4, most had lost the 140 Mdal plasmid species. In several avirulent M90T isolates, this plasmid appeared to have undergone a large deletion (Fig. 4F). These findings gave indirect evidence that some segments of the 140 Mdal plasmid are required for virulence (13).

3. *Search for Easily Scorable Phenotypic Traits on the Large* S. flexneri *Plasmid.* Direct proof that the large 140 Mdal plasmids are necessary for *S. flexneri* virulence requires their transfer to avirulent derivatives lacking these plasmids, with concomitant reestablishment of virulence. Since virulence is not an easily selectable marker with which to monitor plasmid transfer, an attempt was made to detect any marker of selective value expressed by the 140 Mdal pladmid. However, isogenic strains, irrespective of the presence of the 140 Mdal plasmid, displayed identical antibiotic susceptibility patterns. Both the parental and isogenic derivative strains grew equally well in glucose minimal medium containing nicotinic acid, in minimal medium containing $FeCl_3$, or in minimal medium containing the iron chelator α,α'-dipyridyl. No bacteriocins were detected even though several different colicin-sensitive detection strains were employed and mitomycin-C was used for colicin induction. In addition, none of the six *S. flexneri* strains studied was found to be immune

to any of the common colicins including colicin E_1. In summary, attempts to detect an easily scorable phenotypic trait encoded by the 140 Mdal plasmid have thus far been unsuccessful (13).

4. Transposon-tagging of the Large Plasmid in an S. flexneri *Serotype 5 Strain.* The kanamycin resistance transposon, Tn*5*, was used to label the large plasmid in

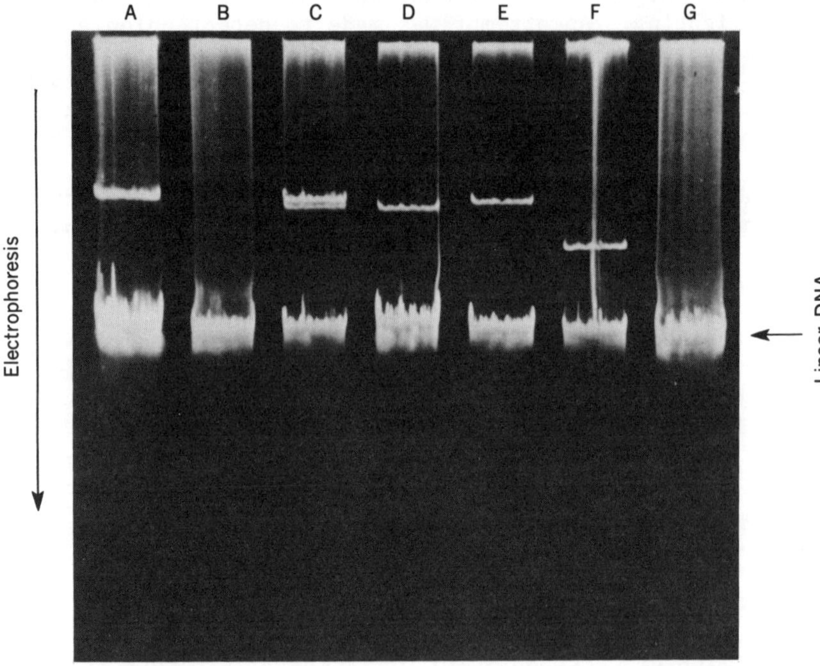

Fig. 4. Agarose gel electrophoretic profiles of plasmid DNA obtained from three serotypes of virulent S. flexneri *strains and their respective avirulent derivatives. (A) Strain M25-8, serotype 1b, virulent (vir*$^+$*). (B) Strain M25-8A, 1b, avirulent (vir*$^-$*). (C) Strain M4243, 2a, vir*$^+$*. (D) Strain M4243A, 2a, vir*$^-$*. (E) Strain M90T, 5, vir*$^+$*. (F) Strain M90TA*$_1$*, 5, vir*$^-$*. (G) Strain M90TA*$_2$*, 5, vir*$^-$*. The gel position of fragmented plasmid or chromosomal DNA generated during DNA isolation, has been labelled "linear DNA". All plasmid species shown above represent covalently closed circular DNA, as discussed in the legend to Fig. 3 (13).*

S. flexneri strain M90T. As mentioned earlier, the F'ts114 *lac*::Tn*5* plasmid served as the Tn*5* donor. M90T clones which had received Tn*5* but had lost the $F'_{ts}114$*lac* plasmid were selected following growth at 42°C. Six of 84 independently isolated, kanamycin-resistant M90T clones were observed to generate kanamycin sensitive colonies during growth under conditions selected to enhance plasmid curing; i.e. 42°C in a subinhibitory concentration of rifampicin. Plasmid analyses of these strains revealed that the 140 Mdal plasmid was tagged with Tn*5* in all six isolates. Of these six Kan^r isolates, three were avirulent, indicating that Tn*5* insertion into the 140 Mdal plasmid had resulted in the loss of virulence. One of the remaining 3 virulent, Tn*5*-tagged plasmid derivatives, termed pWR110, was chosen for subsequent studies (13).

5. *Conjugal Transfer of a Tn5-tagged* S. flexneri *Plasmid*. Attempts were made to conjugally transfer the Tn*5*-tagged plasmid pWR110 from the serotype 5 strain M90T to recipients that had lost the 140 Mdal plasmid (i.e. strain M25-8ANal^r, nalidixic acid resistant mutant of serotype 1b; and strain M4243A$_1$Nal^r, serotype 2a). Although separate conjugal matings were conducted at 20,32,37, and 42°C, no self-transfer of pWR110 was observed following broth or surface mating procedures. Next, we attempted to mobilize pWR110 with three conjugative plasmids; R386 (inc. group F_I), R64*drd*11 (Inc. Iα), and R16 (Inc. O), all of which encode resistance to tetracycline (Tet) but not to kanamycin (Kan). Each of these plasmids was conjugally transferred to strain M90T containing pWR110. The resulting strains were used as donors in conjugal mobilization experiments with the two serotypically heterologous recipient strains. All three conjugative plasmids were self-transferable at a frequency of 10^{-2} to 10^{-3}. More importantly, each conjugative plasmid mobilized pWR110 to both recipient strains, though the frequency of mobilization varied 20 fold. Transconjugants for pWR110 regained virulence with the following exception. Transconjugants which harbored both pWR110 and R386 were avirulent due to the presence of the R386 plasmid as discussed previously. However, as expected, tetracycline sensitive (Tet^s) segregants of the latter transconjugants, which had lost R386, regained virulence. In fact, these Tet^s segregants were observed at a frequency of 50%, suggesting that the 140 Mdal plasmid of *S. flexneri* and the R386 plasmid are incompatible.

Plasmid DNA profiles of strains from mobilization experiments involving R64*drd*11 and pWR110 are shown in Fig. 5. The donor strain carrying R64*drd*11 (Fig. 5C) displayed an extra large plasmid band (i.e. R64*drd*11) which was absent in the M90T parent strain carrying only pWR110 (Fig. 5A). Neither of the avirulent recipient strains harbored the large 140 Mdal plasmid (Fig. 5B, G), and all of the transconjugant strains had acquired one or more large plasmids. One M25-8ANal^r transconjugant (Fig. 5D) had acquired a large cointegrate molecule apparently formed between pWR110 and R64*drd*11. This transconjugant was virulent and could retransfer resistance to Kan, streptomycin, and Tet "en bloc" at a high frequency (5×10^{-3}) to an *E. coli* J53 recipient. When retransferred back from *E. coli* to an avirulent M25-8ANal^r strain, this cointegrate molecule reestablished virulence (data not shown). Two other types of M25-8ANal^r transconjugants were observed;

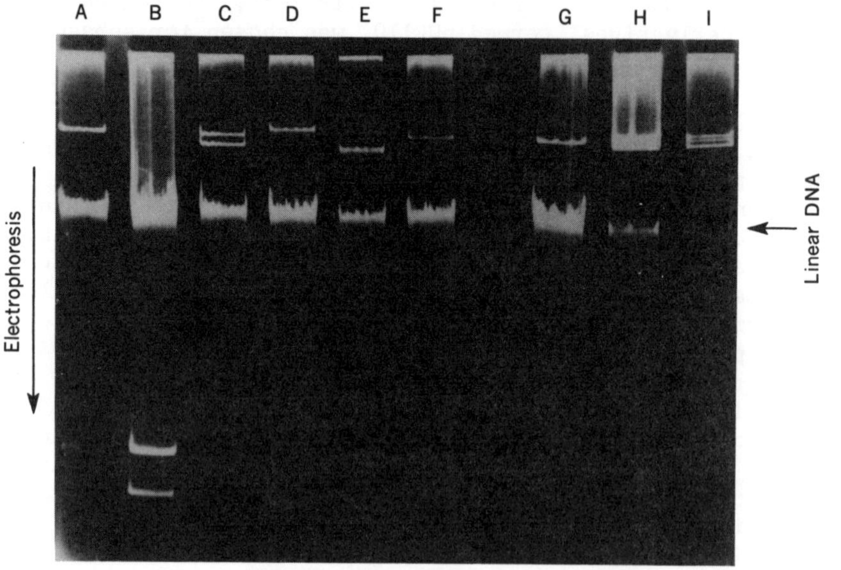

Fig. 5. Demonstration of the physical transfer of the virulence-plasmid pWR110, to avirulent S. flexneri *variants, by comparison of plasmid DNA profiles of parental and transconjugant strains. For each DNA preparation described below, the strain, its serotype and its virulence (vir) are listed in order. (A) Parent strain M90T (pWR110), serotype 5, Vir$^+$. (B) Recipient strain M25-8ANal^r, 1b, Vir$^-$. (C) Donor strain M90T (pWR110 + R64*drd*11*), 5, Vir$^+$. (D) Transconjugant strain M25-8ANal^r (pWR110-R64*drd*11*), 1b, Vir$^+$. (E) Transconjugant strain M25-8ANal^r(R64*drd*11*), 1b, Vir$^-$. (F) Transconjugant strain M25-8ANal^r (pWR110), 1b, Vir$^+$. (G) Recipient strain M4243A$_1$Nal^r, 2a, Vir$^-$. (H,I) Transconjugant strains M4243A$_1$ Nal^r (pWR110 + R64*drd*11*), 2a, Vir$^+$. The gel position of fragmented plasmid or chromosomal DNA, generated during DNA isolation, has been labelled "linear DNA". Specific procedures are detailed elsewhere (13).*

one type received only the mobilizing plasmid and remained avirulent (Fig. 5E); the other type, which occurred infrequently, received only pWR110 and had reacquired virulence but could not retransfer it in further matings (Fig. 5F). Plasmid profiles of DNA from transconjugants of the serotype 2a recipient strain M4243A_1 Nal^r are shown in Fig. 5H and I. These transconjugants, which had reacquired virulence, carried 3 large plasmids: the 105 Mdal plasmid which was originally present in the M4243$A_1$$Nal^r$ recipient (Fig. 5G), and the cotransferred pWR110 and R64*drd*11 species.

The virulence of *S. flexneri* strains was routinely monitored by the guinea pig keratoconjunctivitis assay, and strains which lack the 140 Mdal plasmid were uniformly avirulent. Furthermore, there was a complete correlation between avirulence in the Sereny test and the inability to invade Hela cells (13). Thus, these data directly demonstrate that these large 140 Mdal *S. flexneri* plasmids encode or regulate the expression of some function(s) required for epithelial cell penetration. Additionally, virulence was restored in serotypes 1 and 2 strains by the serotype 5 plasmid, demonstrating that these 140 Mdal plasmids are functionally interchangeable. Unlike the *S. sonnei* form I plasmid, the *S. flexneri* virulence plasmid does not appear to encode any lipopolysaccharide surface component.

C. *Construction and Testing of Bifunctional Oral Vaccine*

1. *Vaccine Strain Construction.* Parenterally administered *Shigella* vaccines have not been successful, probably because shigellosis is an infection limited to the superficial layer of the colonic mucosa. Therefore, circulatory antibodies do not appear to be protective against shigellosis. On the other hand, attenuated shigellae vaccines administered orally have been effective in protecting against this disease, suggesting that the local intestinal immune response is induced by the living oral vaccine (16). However, attenuated *Shigella* vaccines have not been widely used because of difficulties in isolating safe (i.e., nonreverting) and effective strains. Recently, Germanier and Furer (17) have reported on the isolation and characterization of a galactose-epimeraseless (*galE*) mutant of *Salmonella typhi*, the typhoid bacillus. This attenuated strain has been tested in more than 15,000 volunteers and has been shown to be a safe and highly effective oral vaccine(18). We considered the possibility that this strain might be modified so as to be protective also against shigellosis due to *S. sonnei*. Therefore, the plasmid responsible for *S. sonnei* form I antigen synthesis was mobilized, as described earlier, into the *galE S. typhi*

strain. The resulting derivative *S. typhi* was shown to contain the form I plasmid. Furthermore, serological studies demonstrated that this derivative strain expresses not only the somatic antigens of the *S. typhi* parent, but also the *S. sonnei* form I antigen. It appeared that this derivative strain would be a good vaccine candidate (7).

2. *Mouse Protection Tests.* To test the effectiveness of this vaccine, preliminary animal tests were conducted (7). Groups of mice were inoculated with one of several vaccines or with a saline control. Four weeks post-immunization, all mice were challenged with virulent strains of *S. typhi* or *S. sonnei* and deaths were recorded 72 hours after challenge. The living *S. typhi galE* typhoid vaccine protected against the homologous challenge strain, but not against the heterologous (i.e., *S. sonnei*) challenge. Similarly, a living *S. sonnei* vaccine protected against challenge by *S. sonnei*, but not by *S. typhi*. However, the form I *galE S. typhi* derivative vaccine protected against both challenge organisms. These preliminary studies indicate that the form I-expressing *galE S. typhi* strain is an effective immunizing agent in mice for protection against *S. typhi* and *S. sonnei* (7). Volunteer studies are currently underway.

REFERENCES

1. Kopecko, D.J. (1980) *In*: Progress in Molecular and Subcellular Biology (Hahn, F., ed.) Vol. 7, pp. 135-234, Springer-Verlag Press, Heidelberg
2. Kleckner, N., Roth, J., & Botstein, D. (1977) J. Mol. Biol. *181*, 125-129
3. Formal, S.B., Gemski, P., Giannella, R.A., & Takeuchi, A. (1976) *In*: Acute Diarrhoea in Childhood-Ciba Symposium, Vol. 42, pp. 27-43, Elsevier/North Holland.
4. Gemski, P. & Formal, S.B. (1975) *In*: Microbiology-1975 (Schlessinger, D., ed.), pp. 165-169, ASM, Washington, DC
5. Kopecko, D.J., Washington, O., & Formal, S.B. (1980) Infect. Immun. *29*, 207-214
6. Sansonetti, P., David, M., & Toucas, M. (1980) C.R. Acad. Sci *290D*, 879-882
7. Formal, S.B., Baron, L.S., Kopecko, D.J., Washington, O., Powell, C., & Life, C.A. (1981) Infect. Immun. *34*, (*in press*)
8. Kenne, L., Lindberg, B., Petersson, K., Katzenellenbogen, E., & Romanowska, E. (1980) Carbohydrate Res. *78*, 119-126
9. Sereny, B. (1955) Acta. Microbiol. Acad. Sci. Hung. *2*, 293-296
10. Hale, T.L., & Formal, S.B. (1981) Infect. Immun. *32*, 137-144

11. Sansonetti, P.J., Kopecko, D.J., & Formal, S.B. (1981) Infect. Immun. *34*, 75-83
12. Kopecko, D.J., Sansonetti, P.J., Baron, L.S., & Formal, S.B. (1981) *In*: Molecular Biology, Pathogenicity, and Ecology of Bacterial Plasmids (Levy, S.B., Clowes, R.C., Koenig, E.L., eds.), pp. 111-121, Plenum Press, N.Y.
13. Sansonetti, P.J., Kopecko, D.J., & Formal, S.B. (1981) Infect. Immun. *34*, (*in press*)
14. Kopecko, D.J., Holcombe, J. & Formal, S.B. (1979) Infect. Immun. *24*, 580-582
15. Johnson, D.A. & Willetts, N.S. (1980) J. Bacteriol. *143*, 1171-1178
16. Mel, D.M., Terzin, A.L., & Vuksic, L. (1965) Bull. Wld. Hlth. Org. *32*, 647-655
17. Germanier, R. & Furer, E. (1975) J. Infect. Dis. *131*, 553-558
18. Wahdan, M.H., Serie, C., Germanier, R., Lackany, A., Cerisier, Y., Guerin, N., Sallam, S., Geoffroy, P., El Tantaivi, A.S., & Guesry, P. (1980) Bull. Wld. Hlth. Org. *58*, 469-474
19. Casse, F., Boucher, C., Julliot, J.S., Michel, M., & Denarie, J. (1979) J. Gen. Microbiol. *113*, 229-242

STUDIES ON Tn*3* TRANSPOSITION

J. Miyoshi, S. Ishii, K. Shimada[1] and Y. Takagi

Department of Biochemistry
Kyushu University School of Medicine
Fukuoka, Japan

I. INTRODUCTION

The bacterial transposon Tn*3* encodes β-lactamase and other functions required for the transposition process: *tnpA* gene product, a transposase, and *tnpR* gene product which represses the synthesis of the transposase as well as itself (1, 2). Tn*3* transposition from one replicon to another is considered to be a two step process (3,4). In the first step, Tn*3* mediated replicon fusion produces a cointegrated structure containing two directly repeated copies of Tn*3* at the junction between two replicons. In the second step, cointegrates resolve into two replicons each of which contains a copy of Tn*3* by site-specific recombination between the IRS of duplicated Tn*3*. Recent studies demonstrated that the second step is independent of *tnpA* but is dependent on the product of *tnpR*, "transposon resolvase" (5-7), although there is no definite evidence supporting that cointegrates do actually exist in normal Tn*3* transposition. We investigated these mechanisms using the cosmid-phage λ system coupled with density dependent fractionation.

II. RESULTS AND DISCUSSION

A. The Cosmid-Phage λ System

We studied the mechanism of Tn*3* transposition between two replicons by applying the cosmid-phage λ system (8). Packaged cosmids of different sizes ranging from 34 kb to 51 kb can be readily isolated with the combined use of density gradient centrifugation.

This system allows for detection of cosmids carrying an

[1]Present address: Kumamoto University, Kumamoto, Japan.

additional DNA sequence as the packaging efficiency is enhanced. It also has the advantage that these cosmids can be detected not only at Tn*3* transposition but also in other DNA rearrangements such as replicon fusions, separately and quantitatively.

We characterized DNA structures and genetic properties of recombinant cosmids after transducing each fraction onto KS2008 cells[a].

B. Detection of Cosmids Carrying Tn3

The ColE1-derived cosmid pKY96 and the compatible pKY1::Tn*3* plasmids were marked differentially, one with *guaA*$^+$ and others with *cea*$^+$ *amp*^{r} genes, respectively. A deletion mutant in *tnpA* was constructed by eliminating the *Pst*I fragment, and a *tnpR*$^-$ mutant by repair ligation at the *Bam*HI site. Heat-induced lysates of KS2127 cells[b] harboring pKY96 and one of the Tn*3* donors were tested for these transducing abilities (Table 1).

Table 1. *Transducing Abilities of Recombinant Cosmids*

Conditions of Tn3 donors	*Transducing ability per ml*		
	GuaA$^+$	*GuaA*$^+$ *and Amp*^{r}	*Cea*$^+$ *(%)*^{a}
none	5.4×10^5	-	-
wild Tn3	4.6×10^6	4.0×10^6	13
tnpR$^-$	1.0×10^7	9.6×10^6	88
tnpA$^-$	1.2×10^5	2.0×10^3	100

^{a}*Cea*$^+$ *ratio was determined by GuaA*$^+$ *Amp*^{r} *Cea*$^+$*/GuaA*$^+$ *Amp*^{r} *transductants.*

In wild Tn*3* and *tnpR*$^-$ conditions, *amp*^{r} transducing phages were classified into *guaA*$^+$ *amp*^{r} and *guaA*$^+$ *amp*^{r} *cea*$^+$.

In the *tnpA*$^-$ condition, only *guaA*$^+$ *amp*^{r} *cea*$^+$ transducing phages were detected, and the titer was lower than 0.1 %, compared with findings under different conditions. Simple transposition of Amp resistance was not detected.

The increase in the packaging efficiency of pKY96 probably

[a]KS2008: *recA* HfrHΔ*(gal-att*λ*-bio)*Δ*(guaA-guaB)*
[b]KS2127: *recA* HfrHΔ*(guaA-guaB)(*λBAM*)*
λBAM: λ*c*I857Δ*(int-xis-red-gam-c*III*)*

follows the transposition of Tn*3*, because pKY96::Tn*3*, 80 % length of phage λ, is packaged about 200-fold more efficiently than pKY96, 70 % length of phage λ. We fractionated these transducing phages by CsCl density gradient centrifugation and isolated recombinant cosmids, independently.

C. Density-Dependent Fractionation of Recombinant Cosmids

The *amp*^{r} transducing phages were classified into three fractions. Fr.I transduced *guaA*$^+$ *amp*^{r} genes, Fr.II and Fr.III transduced *guaA*$^+$ *amp*^{r} *cea*$^+$ genes. The average densities of Fr.I, Fr.II and Fr.III were 1.484, 1.499 and 1.509 g/cm^3, respectively, thus corresponding to the DNA sizes of 39, 45 and 49 kb (Fig. 1).

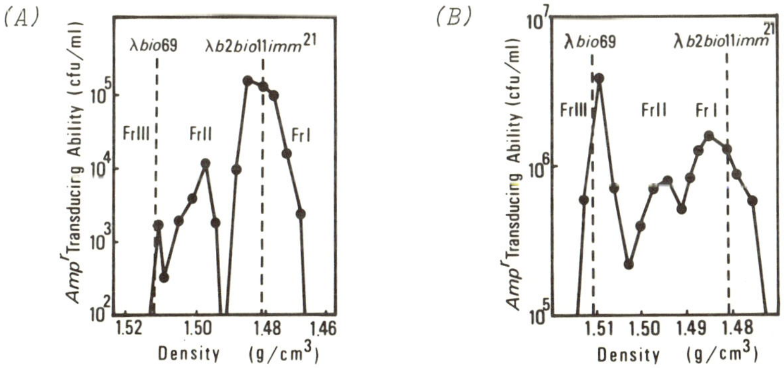

Fig. 1. Density gradient analysis of recombinant cosmids. Donor conditions were wild Tn3 (A) and tnpR$^-$ *(B). λbio69 and λb2bio11imm*^{21} *were density references.*

In the *tnpR*$^-$ condition, three peaks were fundamentally the same as in the wild Tn*3* condition, but the titer of Fr.III was 10^4-fold higher than in the wild condition. In the *tnpA*$^-$ condition, we did not detect Fr.I and Fr.III and found a single peak of the *guaA*$^+$ *amp*^{r} *cea*$^+$ transducing phages at a density of 1.500 (data not shown).

These findings suggest that Fr.I and Fr.III were dependent on *tnpA* gene function and Fr.III was enhanced by a *tnpR*$^-$ mutation.

D. DNA Structures of Recombinant Cosmids

An aliquot from each fraction was placed onto KS2008 cells and cosmid DNAs were isolated from the Amp^r transductants.

DNA molecules extracted from Fr.I transductants were pKY96::Tn*3*, the final products in this assay system. Simple Tn*3* insertions occurred at many sites on pKY96.

The DNA molecules from Fr.II transductants have novel characteristics such as fused replicons of pKY96 and pKY1::Tn3 stable in a *recA* strain and carried deletions of recipient DNA sequences. Their features are summarized as follows (Fig. 2A).

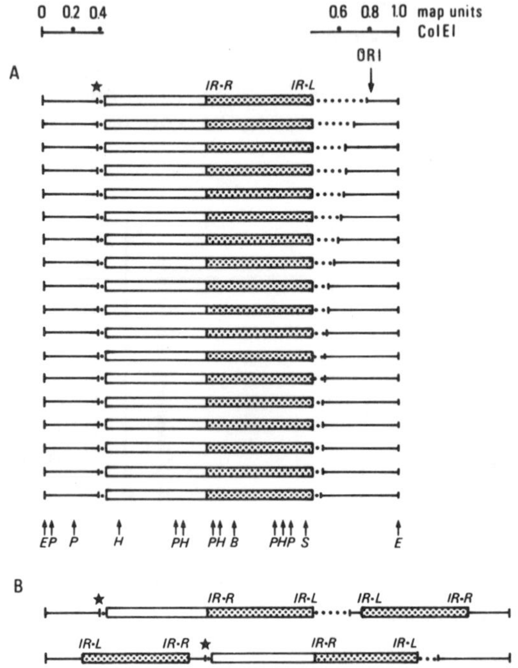

Fig. 2. DNA structures of Fr.II fusions linearized at EcoRI sites on pKY96. Broken lines indicate deletions. Blank areas and dotted areas represent pKY1 and Tn3. IR-L and IR-R indicate the inverted repeats of Tn3. Stars represent 0.37 map units of ColE1. E, EcoRI; P, PstI; H, HincII; B, BamHI; S, SacI.

First, pKY1::Tn3 was always cleaved once at the left inverted repeat terminus of Tn*3* and the entire DNA sequence including pKY1 was integrated into the ColE1 part of pKY96. Second, fused replicons carried a deletion of ColE1 DNA ranging in size from 0.6 to 2.7 kb. These deletions were generated near the integration sites, and originated from one fusion point at 0.37 map units of ColE1 and extended into adjacent regions. Fused replicons carrying another Tn*3* insertion were also detected (Fig. 2B).

Fr.II fusions were dependent on the function of *tnpA*, regardless of that of *tnpR*, because DNA molecules extracted from the *guaA*$^+$ *amp*^{r} *cea*$^+$ transductants under the *tnpA*$^-$ condition were not Fr.II fusions but the original two replicons, pKY96 and pKY1::Tn*3*.

The properties of Fr.III fusions were readily distinguished from Fr.II fusions. Fr.III fusions were highly unstable in the *recA* strain and rapidly segregated into two replicons. Re-packaging of cosmids within Fr.III transductants disclosed the presence of *guaA*$^+$ *amp*^{r} transducing phages which had lost *cea*$^+$ gene, and plasmids extracted were always two replicons, pKY96::Tn*3* and pKY1::Tn*3*. These properties were consistent with the intermediates in normal Tn*3* transposition.

As the DNA within Fr.III phages in wild condition has an extremely low yield, direct characterization is not feasible. However, we did find that *tnpR*$^-$ mutation increased the titer

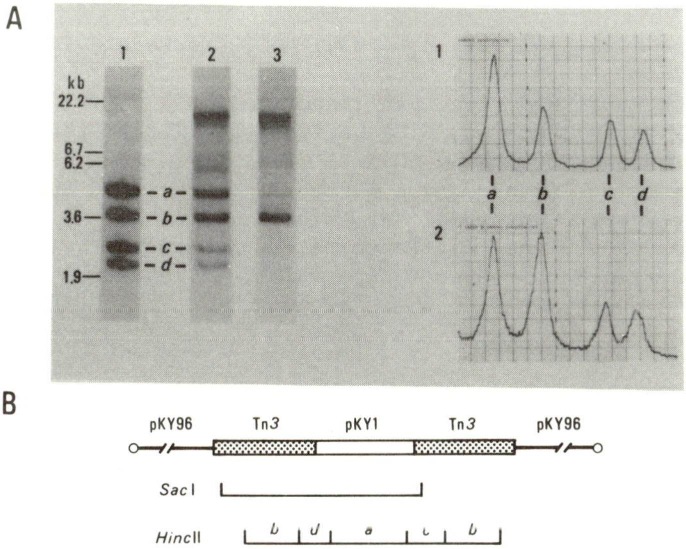

Fig. 3A. HincII digestion of pKY1::Tn3 (lane 1), Fr.III fusions (lane 2) and Fr.I cosmids (lane 3).
3B. Fr.III fusions linearized at cos sites.
Digestion with HincII of Fr.III fusions revealed four bands identical to those generated by the circular donor plasmid, and HincII-b derived from Tn3 was doubly greater in lane 2 than in lane 1 as analyzed by densitometer. All lanes were hybridized to the donor specific probe.

of Fr.III phages and these fusions possessed qualities equal to those of the intermediates. Southern blotting analysis of Fr.III fusions collected on a large scale under *tnpR*$^-$ condition revealed that they were cointegrates containing directly duplicated Tn*3* at the junction of pKY1 and pKY96 (Fig. 3).

ACKNOWLEDGMENT

We thank M. Ohara for comments on the manuscript.

REFERENCES

1. Gill, R.E., Heffron, F. & Falkow, S. (1979) Nature *282*, 797-801
2. Chou, J., Lemaux, P.G., Casadaban, M.J. & Cohen, S.N. (1979) Nature *282*, 801-806
3. Shapiro, J.A. (1979) Proc. Natl. Avad. Sci. USA *76*, 1933-1937
4. Arthur, A. & Sherratt, D. (1979) Mol. Gen.Genet. *175*, 267-274
5. Reed, R.R. (1981) Proc. Natl.Acad. Sci. USA *78*, 3428-3432
6. Kostriken, R., Morita, C. & Heffron, F. (1981) Proc.Natl. Acad. Sci. USA *78*, 4041-4045
7. Sherratt, D., Arthur, A. & Burke, M. (1980) Cold Spring Harbor Symp. Quant. Biol. *45*, 275-282
8. Shimada, K., Umene, K., Nakamura, T. & Takagi, Y. (1979) Cold Spring Harbor Symp. Quant. Biol. *43*, 991-998

A TRANSPOSON-LIKE STRUCTURE CONFERRING UV SENSITIVITY AND KANAMYCIN RESISTANCE ON *ESCHERICHIA COLI* HOST

Y. Terawaki, Y. Itoh, A. Tabuchi
Y. Furuta and Y. Kamio

Department of Bacteriology
Shinshu University School of Medicine
Matsumoto, JAPAN

A kanamycin resistance plasmid, Rts1, is a naturally occurring temperature sensitive plasmid belonging to the T incompatibility group(1-4). Because of the large molecular size of Rts1 DNA(140 megadaltons), it was difficult to treat this plasmid for molecular studies. Accordingly, we have used several of its deletion derivatives for investigating the structure and functions of Rts1.

Recently we constructed a cleavage map of pTW20, which is a deletion derivative(41 Md) of Rts1, and found that the kanamycin resistance gene(*kan*) was located on a *Sal*I fragment (1.5 Md) within the 7.2 Md *Bam*HI fragment of E1(the largest *Eco*RI fragment of pTW20)(5). The pACYC184 chimeric plasmids containing the 7.2 Md *Bam*HI fragment or 1.5 Md *Sal*I fragment were revealed to sensitize the *Escherichia coli* host to UV irradiation. An electron microscopic study of the 7.2 Md *Bam*HI fragment showed the presence of a snap-back structure in the fragment(5). In contrast, Iida *et al.* found a direct repeat flanking the kanamycin resistance gene on the parent plasmid Rts1 and confirmed one of the repeating fragments to have the IS functions(S. Iida, personal communication).

In this report, we describe the character of the kanamycin fragment cloned in pBR322 and determine the location of the genetic loci for kanamycin resistance and UV sensitization in relation to the snap-back structure.

CLONING OF KANAMYCIN FRAGMENT OF pTW20 in pBR322

pTW201-BK(pACYC184 plus 7.2 Md *Bam*HI fragment of pTW20) was digested with *Sal*I endonuclease and ligated to the *Sal*I

This work was supported by grants(to Y. T.) from the Ministry of Education, Japan and from the Yakult Foundation.

digested pBR322. By selecting kanamycin(Km) resistance transformants of *E. coli* JC1557(F^- *arg met leu his str*), pBR322 chimeric plasmid containing the 1.5 Md *Sal*I fragment was obtained from the transformants. The 7.2 Md *Bam*HI fragment was also cloned in pBR322 using a similar procedure, but digested pTW201-BK and pBR322 with *Bam*HI. These chimeric plasmids, pTW201-BSK(pBR322 plus 1.5 Md *Sal*I fragment) and pTW201-BBK (pBR322 plus 7.2 Md *Bam*HI fragment), were used for studying the UV sensitization effect and molecular structure of the Km fragment(Fig. 1).

UV SENSITIVITY CONFERRED BY CLONED KANAMYCIN FRAGMENT

UV survivals of *E. coli* JC1557 harboring pBR322 or its chimeric plasmid were examined by the method of Mortelmans and

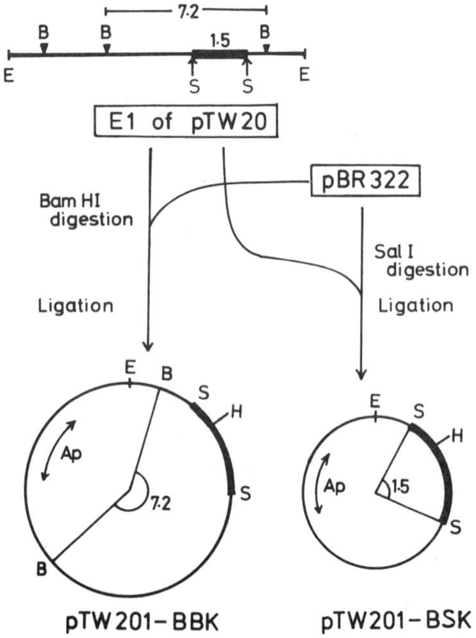

Fig. 1. Construction of pBR322 chimeric plasmids containing kanamycin fragment of pTW20. The largest EcoR1 fragment of pTW20, E1(11.5 Md), contains 7.2 Md BamHI fragment, in which 1.5 Md SalI fragment harboring kanamycin resistance gene is located. The 1.5 Md SalI fragment contains one HindIII cleavage site that is 0.4 and 1.1 Md apart from the SalI cleavage ends. Only relevant cleavage sites are shown. E: EcoRI, B: BamHI, S: SalI, H: HindIII, Ap: ampicillin resistance gene of pBR322.

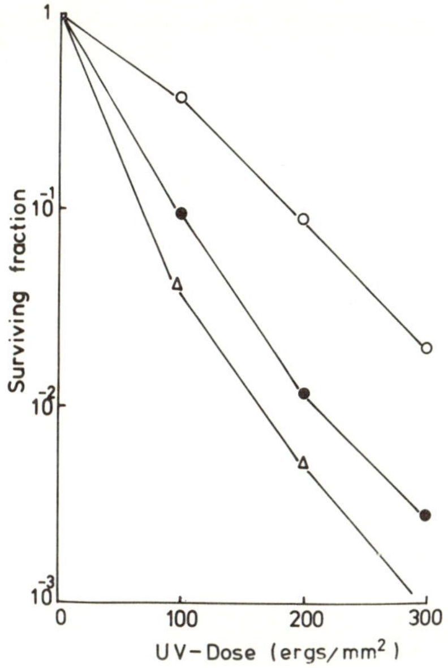

Fig. 2. UV survivals of JC1557 with pBR322 or its chimeric plasmid. UV irradiated cells on PAB agar plate were incubated in dark at 37°C. The number of colonies developed after 18 h of incubation were scored. —○—: *JC1557(pBR322)*, —▲—: *JC1557(pTW201-BSK)*, —●—: *JC1557(pTW201-BBK)*.

Stocker(6). Both pTW201-BSK and pTW201-BBK enhanced the lethal effect of UV irradiation on the JC1557 host(Fig. 2). This UV sensitization effect by the Km fragments was observed in other *E. coli* host W3104, and was also demonstrated with pACYC184 chimeric plasmids containing either of these Km fragments. Thus, we confirmed that the determinants for the UV sensitization effect(*puv*) and *kan* were contained in the 1.5 Md *Sal*I fragment within the 7.2 Md *Bam*HI fragment of pTW20.

ELECTRON MICROSCOPIC FINDINGS OF *kan-puv* FRAGMENT

A snap-back structure consisting of a single stranded loop (0.45 μm) and a double stranded stalk(0.30 μm) was found in the 7.2 Md *Bam*HI fragment(5). To determine whether *kan* and *puv* are located in this transposon-like structure, *Sal*I digested pTW201-BBK(pBR322 plus 7.2 Md *Bam*HI fragment) was denatured, self-annealed and examined in the electron microscope. Essentially the same dimensions of looped structure with short stalks were frequently observed(Fig. 3, left above). The lengths of the single stranded loop and the double stranded stalk were 0.45 μm and 0.095 μm, respectively. This suggests that *Sal*I cleaves at a region approximately two-thirds from the distal end of the stalk of the snap-back structure found in the 7.2 Md *Bam*HI digested molecule. This finding indicates that the entire 1.5 Md *Sal*I fragment harboring *kan* and *puv* is contained in the transposon-like structure within the 7.2 Md *Bam*HI fragment.

KANAMYCIN FRAGMENT FROM OTHER Rts1-DERIVATIVE PLASMIDS AND THEIR UV SENSITIZATION EFFECT

To investigate the origin of the gene *puv*, the Km fragment

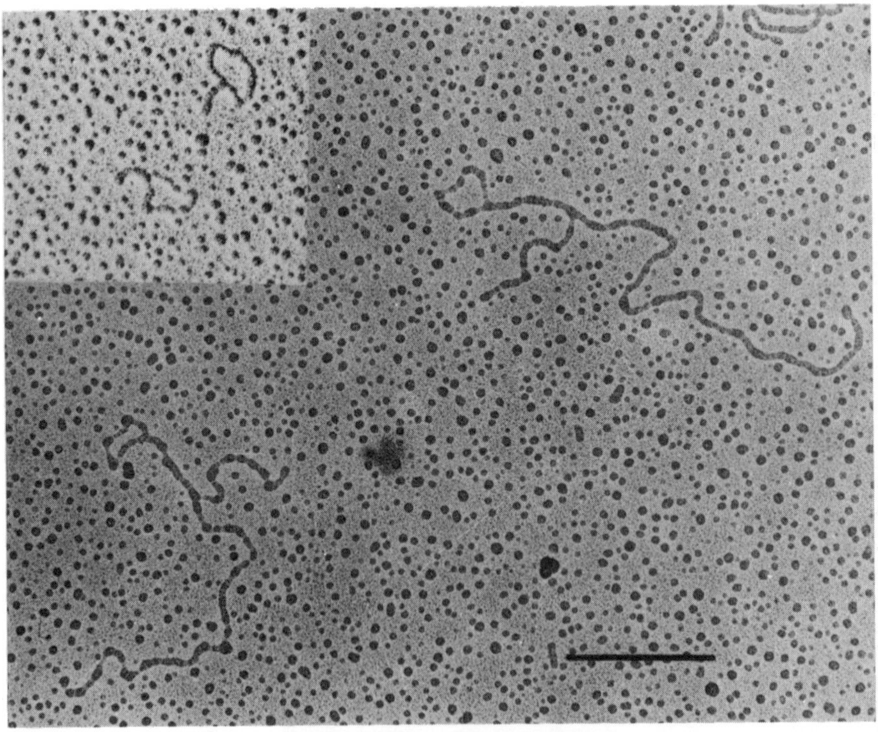

Fig. 3. Electron micrographs of kan-puv fragments. pTW201-BBK digested with SalI was denatured and self-annealed(left above). Two molecules of looped structure with short stalk are seen. On the rest part of the photograph, the 7.2 Md BamHI molecules are observed, in which snap-back structures exist. Two micrographs are shown in one sheet. Bar represents 0.5 μm for both.

of the other Rtsl-derivative plasmids were examined for their UV sensitization effect on the host. For this purpose, pTW3 and pTW10 were used. The former is a spontaneous deletion derivative and the latter is a deletion mutant of Rtsl obtained from the N-methyl-N'-nitro-N-nitrosoguanidine treated cells. Both plasmid DNA cleaved with *Sal*I or *Eco*RI were ligated in pBR322, and Km resistance plasmids were constructed.

The obtained chimeric plasmids, pTW31-BSK(pBR322 plus *Sal*I Km fragment of pTW3) and pTW101-BEK(pBR322 plus *Eco*RI Km fragment of pTW10), enhanced the lethal effects of UV irradiation on the JC1557 host. Thus, these three derivatives of Rtsl conferred UV sensitivity on their host cells, indicating that *puv* must derived from their common parent plasmid, Rtsl.

EFFECT OF *kan-puv* FRAGMENT ON HOST MUTATION

The increase in UV sensitivity of the host by *puv* suggests the possibility that a repair process of the host chromosome is impaired by the determinant. If an excision repair pathway is inhibited, chromosomal mutation of the host would increase. Alternatively, if a post-replication repair is impaired, the host mutation would be suppressed(7). To determine which process is involved, mutation of the host to streptomycin resistance was examined by treating W3104 cells harboring the pBR322 chimeric plasmid with ethyl methanesulfonate. As shown in Table 1, the induced mutation frequency of pTW201-BSK and pTW201-BBK harboring cells were more than 100 times lower than

Table 1. *Effect of kan-puv Fragment on Host Mutation*

Host	*Plasmid*	*Relevant fragment*	*Sm^r-mutation[a] frequency*	*Relative frequency*
W3104	*pBR322*	—	6.2×10^{-5}	*1.0*
	pTW201-BSK	*1.5Md-SalI*	5.3×10^{-7}	*0.008*
	pTW201-BBK	*7.2Md-BamHI*	$< 10^{-7}$	< 0.002

Ethyl methanesulfonate(5 mcg/ml) was added to the exponentially growing cells. After 1 h of incubation at 37°C, cells were harvested, washed with M9 salts and appropriate dilution of the suspension was spread onto PAB agar plate with or without containing streptomycin(10 mcg/ml).

[a]*Number of streptomycin resistant cells per number of total living cells.*

that of pBR322 containing cells. This suggests that *puv* might suppress the chromosomal mutation of the host.

ACKNOWLEDGMENTS

We thank Dr. E. Lederberg(Plasmid Reference Center) for the vector plasmids pBR322 and pACYC184. We also thank H. Katagiri and R. Ichikawa for their technical assistances in electron microscopy.

REFERENCES

1. Terawaki, Y., Takayasu, H. and Akiba, T. (1967) *J. Bacteriol. 94*, 687-690
2. Terawaki, Y. and Rownd, R. (1972) *J. Bacteriol. 109*, 492-498
3. DiJoseph, C. G. and Kaji, A. (1974) *Proc. Nat. Acad. Sci. USA 71*, 2515-2519
4. Coetzee, J. N., Datta, N. and Hedges, W. (1972) *J. Gen. Microbiol. 72*, 543-552
5. Terawaki, Y., Kobayashi, Y., Matsumoto, H. and Kamio, Y. (1981) *Plasmid 6*(2) *in press*
6. Mortelmans, K. E. and Stocker, B. A. D. (1976) *J. Bacteriol. 128*, 271-282
7. Witkin, E. W. (1976) *Bacteriol. Reviews 40*, 869-907

GENERATION OF TRANSPOSABLE MINICIRCULAR DNA OF TN*2001* ORIGIN IN *PSEUDOMONAS AERUGINOSA*

S. Iyobe, T. Kato, and S. Mitsuhashi

Department of Microbiology,
School of Medicine, Gunma University,
Maebashi, Japan

INTRODUCTION

The chloramphenicol (Cm) resistance determinants on many R plasmids in *Pseudomonas aeruginosa* were found to be transposable (1). The Tn*2001* encoding Cm resistance was identified as having a molecular weight of about 1.4 Mdal (2). Later it was found that minicircular DNA appeared after bacterial cells possessing the plasmids with Tn*2001* had been cultivated in broth containing subinhibitory concentrations of Cm. This paper deals with the nature of the minicircular DNA.

PHYSICAL ORIGIN OF THE MINICIRCULAR DNA

Plasmid DNA fractions were isolated from bacterial cells harboring Rlb679::Tn*2001* and it was found that minicircular DNA was also present. By either agarose gel electrophoresis or electron microscopic observation (Fig. 1), its size was estimated to be about 1.5 Mdal. The molecular ratio of minicircular DNA to Rlb679::Tn*2001* DNA was about one after the cells were cultivated in a medium containing Cm. The number of cutting sites in minicircular DNA was examined using several restriction endonucleases and was compared with those of Tn*2001* in Rlb679::Tn*2001*. One cutting site by *Sal*I, *Bam*HI, *Eco*RI or *Pst*I, two by *Sma*I and three by *Sac*II were found. These numbers were the same as the cutting site(s) of Tn*2001* in Rlb679::Tn*2001*, indicating that the minicircular DNA was derived from the Tn*2001* part of this plasmid.

Heteroduplex molecules between minicircular DNA and the plasmid Rlb679::Tn*2001* DNA (Fig. 2) were formed. These consisted of a small double-stranded circle and a larger single-stranded circle. The sizes of these circles showed that the double-stranded was formed between minicircular DNA and the Tn*2001* part of the plasmid Rlb679::Tn*2001*, and the single-stranded circle corresponded to the Rlb679 part of Rlb679::Tn*2001*.

Results obtained so far indicate that minicircular DNA is generated from the plasmid Rlb679::Tn*2001* and identical to Tn*2001*.

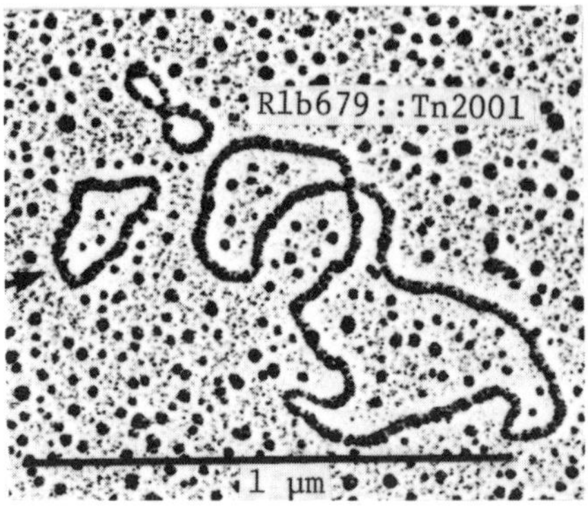

Fig.1. Observation of minicircular DNA.

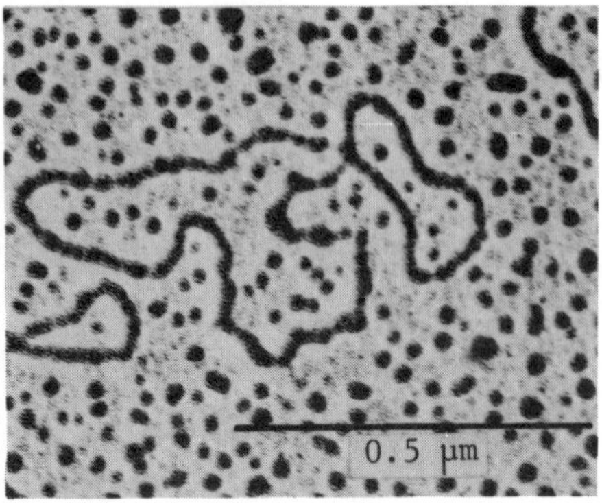

Fig.2. Heteroduplex between Rlb679::Tn2001 and minicircular DNA

TRANSPOSITION OF MINICIRCULAR DNA

It was anticipated that minicircular DNA contained the Cm resistance determinant (Cm^r), because the Tn*2001* encodes Cm resistance. Minicircular DNA was extracted from the agarose gel and three strains were transformed to Cm resistance. One of the recipient strains was PAO2142 (*ilv lys met tyu*), Rp^r Rec^- having no plasmid, and the others were the same strain having RP4 or Rlb679 plasmid. These Cm-resistant transformants were obtained at a frequency of 10^{-5} per number of recipient cells (Table 1). The location of the Cm^r in these transformants was examined as follows.

No minicircular DNA was detected in the extract from the Cm-resistant transformants derived from the PAO2142 Rp^r Rec^- strain. Chromosomal segments were introduced by conjugation from a male strain, PAO4009 (*leu*) $FP5^+$, and recombinants were obtained. About 100 of these recombinants showing either Lys^+ Leu^+ or Met^+ Leu^+ were selected and examined for unselected markers, Cm-sensitive progenies were found among them at a few percent. This suggested that the Cm^r was on the chromosome in the Cm-resistant transformants of the PAO2142 Rp^rRec^- strain after transposition of minicircular DNA on that chromosome.

Table 1. Transformation with minicircular DNA

Plasmid in recipient	*Frequency*	*Location of Cm^r*
–	5×10^{-5}	*Chromosome*
RP4	2×10^{-5}	*Chromosome, RP4*
Rlb679	4×10^{-5}	*Chromosome, Rlb679*

Recipient strain, PA02142 Rp^r Rec^-

Two types of Cm-resistant transformants were obtained from the strain PA02142 $Rp^r Rec^-$ possessing the RP4 plasmid. The Cm^r was conjugally transferred from five of twenty-five transformants and was always accompanied by the resistance determinants of RP4, i.e., Km^r, Tc^r and Cb^r, indicating that the Cm^r was located on the RP4 plasmid. The RP4Cm^r DNAs from five strains were extracted and cleaved by *Pst*I. The patterns of the *Pst*I-cut segments on agarose gel showed that each RP4Cm^r had a new segment instead of one segment derived from the RP4 plasmid. This new segment had a cutting site by *Pst*I, indicating that minicircular DNA was integrated in the segment. Four different cutting patterns were observed in the five RP4Cm^r DNAs. The molecular weight of the transposed Cm^r segments of the RP4 was estimated to be about 1.4 Mdal. These facts indicated that minicircular DNA transposed to the RP4 plasmid in at least four different regions.

Further analysis was done of the Cm-resistant transformants derived from the PA02142 $Rp^r Rec^-$ strain possessing R1b679 plasmid. The plasmid DNAs of 100 transformants were extracted and two were found to harbor plasmid DNA larger than the R1b679 DNA. A transformation experiment with the larger plasmid DNA showed that the plasmid carried Cm^r in addition to the resistance determinants of R1b679, i.e., Sm^r and Su^r. The two R1b679Cm^r plasmids were about 1.4 Mdal larger in molecular weight than the original R1b679 plasmid and showed the same cutting patterns by various restriction enzymes. The number of cutting sites acquired was the same as in minicircular DNA. Therefore, it was concluded that in both R1b679Cm^r plasmids the minicircular DNA had transposed to the same locus in the R1b679. The R1b679Cm^r DNA was shown to be identical to the original plasmid of minicircular DNA, R1b679::Tn*2001*, by agarose gel electrophoresis after digestion with various restriction endonucleases.

CONCLUSION AND DISCUSSION

Minicircular DNA encoding Cm resistance was transposable

and physically almost identical to Tn*2001*. If we presume that there are direct-repeated sequences on the both sides of Tn*2001*, the deletion of Tn*2001* from the plasmid Rlb679::Tn*2001* is easy to explain as it results from the reciprocal recombination between the two sides. However, we have not yet proved the direct-repeated sequences on Tn*2001*. Since the preferential site was identified on the Rlb679 plasmid for the transposition of both Tn*2001* and minicircular DNA and minicircular DNA was generated from that site, a specific target was presumed on the Rlb679 plasmid. The mechanism of the generation of minicircular DNA might be the same as reported by M.Chandler *et al.*(3). They obtained the r-determinant (r-det) as a circular DNA from the R100-1 plasmid, resulting from the recombinational deletion mediated by two IS*1*s directly flanking the r-det. However, they could not obtain any transformant which accepted the r-det. The transposition of circular r-det DNA may not be efficient or the generation of the circular r-det DNA may not be related to the transposition of r-det.

We found that the copy number of minicircular DNA increased when the cells possessing the plasmid Rlb679::Tn*2001* were cultivated in a medium containing Cm, but we could not detect the plasmid DNA corresponding to Rlb679. The amplification and efficient deletion of minicircular DNA from Rlb679::Tn*2001* may occur by the same mechanism as the amplification of the r-det in *Proteus mirabilis* in medium containing Cm, although that mechanism has not been elucidated (4).

REFERENCES

1. Iyobe,S., Sagai,H., Hasuda,K., and Mitsuhashi,S. (1980) in Antibiotic Resistance, Transposition and Othe Mechanisms (Mitsuhashi,S., Rosival,L., and Krcmery,V., ed.) pp43-47, Avicenum, Czechoslovak Medical Press. Prague; Springer Verlag.Berlin.Heidelberg.New York
2. Iyobe,S., Sagai,H., and Mitsuhashi,S. (1981) J. Bacteriol. *146*,141-148
3. Chandler,M., Silver,L., Lane,D., and Caro,L. (1979) Cold Spring Harbor Symp. Quant.Biol. *43*,1223-1231
4. Hashimoto,H., and Rownd,R.H. (1975) J.Bactericl. *123*,56-68

TN*916* (TC) : A CONJUGATIVE NON-PLASMID ELEMENT IN *STREPTOCOCCUS FAECALIS*

M. Cynthia Gawron-Burke
and Don B. Clewell

Depts. of Oral Biology and Microbiology
Schools of Dentistry and Medicine
and The Dental Research Institute
The University of Michigan
Ann Arbor, Michigan U.S.A.

I. INTRODUCTION

Several research groups have recently made the surprising observation that certain drug resistant strains of streptococci are capable of conjugally transferring their resistance traits in the apparent absence of plasmid DNA. Occurring by a DNAse resistant transfer on filter membranes, this phenomenon has now been observed in *S. pneumoniae*(1,2) *S. faecalis*(3,4), *S. pyogenes*(5), *S. agalactiae*(5), as well as in Lancefield groups F and G (5). Shoemaker, Smith and Guild(6) showed that transferrable chloramphenicol and tetracycline(Tc) resistance determinants that were closely linked in *S. pneumoniae* were located on the bacterial chromosome. Here we extend a similar observation made in *S. faecalis* (3,4) where the transferrable resistance determinant is located on a transposon.

S. faecalis strain DS16 harbors a conjugative hemolysin-bacteriocin determining plasmid pAD1(35 Mdal) and a non-conjugative multiple drug resistance plasmid pAD2(16 Mdal) (7). A chromosome- borne tetracycline resistance determinant is located on a 10 Mdal transposon designated Tn*916* and is capable of transposition to several different conjugative hemolysin plasmids (pAD1, pOB1, and pAMγ1) at a frequency of about 10^{-6}(3). Tn*916* is also able to transfer at low frequency (10^{-8}) from plasmid free derivatives of DS16 to plasmid free recipients (JH2-2) in filter matings by

[1]*This work was supported by Public Health Service grants DE02731 and AI0318 from the National Institutes of Health. M.C.G. is a recipient of a Public Health Service post-doctoral fellowship.*

a conjugation-like event requiring direct contact between the donor and recipient. The phenomenon is Rec-independent, and transfer of the mutational markers *str* and *spc* was not observed under these conditions(3).

In this report, we present data obtained from recent DNA filter hybridization experiments which not only confirm the transfer of Tn*916* from plasmid free donors to plasmid free recipients but also indicate that Tn*916* transfer results in the insertion of the transposon into different sites on the recipient chromosome. We also report the isolation of JH2-2 Tc^r transconjugants which exhibit altered transfer frequencies of Tn*916* and address the question of whether transposition involves an excision from its original site.

II. RESULTS

A. *Tn*916 *Transfer*

That Tn*916* is capable of transfer in overnight filter-matings from plasmid free donors is shown in the data of Table I. DS16C3 (a derivative of DS16 cured of both pAD1 and pAD2) is capable of transferring Tc^r at a frequency of about 10^{-8} to the plasmid free recipient strain JH2-2. Transconjugants derived from this mating, such as CG130, are capable of transferring Tc^r in secondary matings to the isogenic recipient strain JH2SS at a similar frequency(Table I). In order to examine this transfer event more closely, chromosomal DNA was isolated from the donor and transconjugant strains listed in Table I, digested with *Hind*III restriction endonuclease, resolved by agarose gel electrophoresis, and blotted on nitrocellulose using the Southern technique (8). These DNAs were then probed with a ^{32}P-labelled *Eco*RI restriction fragment of pAD1::Tn*916*(pAM211) containing the entire transposon. Tn*916* is not cleaved by *Eco*RI, and about 85% of this fragment consists of Tn*916* sequences (3). Since Tn*916* contains a single *Hind*III site, two host-transposon junction fragments should be resolvable after hybridization. An autoradiogram of the Southern blot following hybridization with the ^{32}P-Tn*916* probe is shown in Fig. 1. Gel slot A contained chromosomal DNA from the DS16C3 donor, and the two host-transposon junction fragments (X & Y) are readily detected. Slots B-D contained chromosomal DNA from three different transconjugants derived from the DS16C3 X JH2-2 mating detailed in Table I. In all three transconjugant strains, the hybridization profiles differed from those seen in the donor and, interestingly, also from

Table I. *Transfer of Tn916 from plasmid free donors.*

Donor strain	*Recipient strain(chromosomal markers)*[a]	*Frequency Tc^r transconjugants per recipient*	*Representative transconjugant*
DS16C3	*JH2-2(Rif,Fa)*	*8.5×10^{-9}*	*CG110* *CG130* *CG140*
CG130	*JH2SS(Sm,Sp)*	*1.7×10^{-8}*	*CG131* *CG132* *CG133*
CG110	*JH2SS*	*4.2×10^{-6}*	

[a]*Tc, Rif, Fa, Sm, and Sp indicate tetracycline, rifampin, fusidic acid, streptomycin, and spectinomycin, respectively. JH2-2 and JH2SS strains are described in references 13 and 14.*

[b]*Frequency of spontaneous Tc^r mutation $< 10^{-10}$. Filter matings were carried out using an initial ratio of 1 donor per 10 recipients (0.5 ml recipient, 0.05 ml donor, 4.5 ml AB3 medium). After incubation (18-20 hrs), the mixture was resuspended in 1.0 ml of media and spread on appropriate selective plates (Tc (10 μg/ml, Rif and Fa (25 μg/ml), and Sm (1000 μg/ml). Frequencies are expressed as a function of recipient rather than donor concentration due to the low frequencies encountered and the variation in viable count between donors and recipients after overnight growth(see reference 3).*

each other. Strains such as CG140(gel slot C) gave rise to two host-transposon junction fragments (X and Y), whereas other transconjugant strains, CG110 and CG130, gave rise to at least six and four bands, respectively. Gel slots E-F contained chromosomal DNA from three different transconjugants obtained from the secondary mating experiment shown in Table I that had used strain CG130 (gel slot D) as the donor and JH2SS as the recipient strain. Again the transconjugant patterns differed from the donor and also from each other. Two of the transconjugants (CG131 and CG133) exhibited the simpler banding pattern (fragments X and Y only), whereas

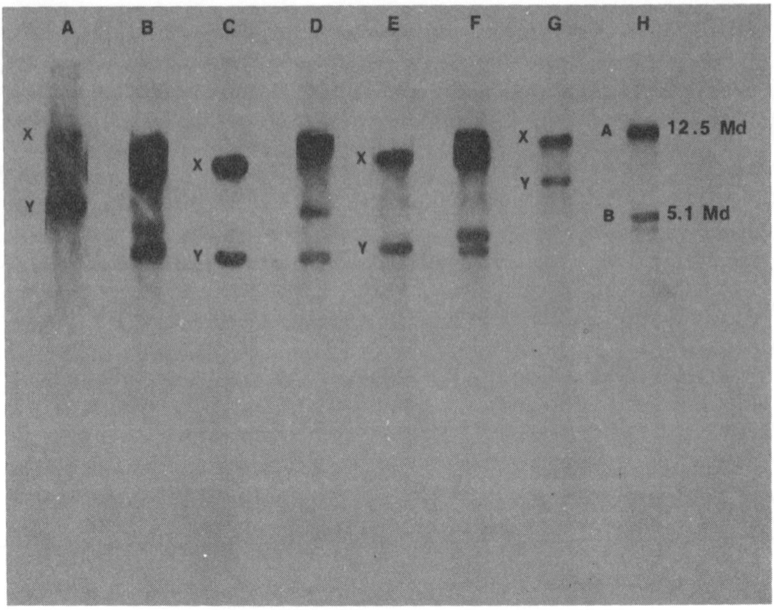

Fig. I. Autoradiogram obtained from Southern blot of Hind*III digested chromosomal DNA following hybridization with* ^{32}P-*Tn*916 *probe DNA.*

Gel Slots B-D contain DNA (1-2 µg) from transconjugants CG110, CG140, and CG130 respectively, which were obtained using DS16C3 (gel slot A) as donor (see Table I). Gel slots E-G contain DNA from transconjugants CG131, CG132, and CG133, respectively, which were obtained using CG130 as donor. Fragments marked X and Y denote presumed host-transposon junction fragments. Gel slot H contains Hind*III digested pAM211 plasmid DNA, and only those fragments containing Tn*916 *sequences (A and B) hybridized to probe DNA. (Isolation of DNA by CsCl-EtBr equilibrium gradient centrifugation and agarose gel electrophoresis was as detailed elsewhere(3). DNA transfer and hybridization were essentially according to the Southern method (8) using a BRL (Bethesda Research Laboratories) Blot Transfer System. Nick translation of probe DNA with either* ^{32}P*-dATP or* ^{32}P*-dCTP was accomplished with a BRL or NEN (New England Nuclear) Nick-translation kit, respectively. Autoradiography employed Kodak X-Omat R film and a Dupont Cronex intensifying screen. Film was exposed 1-4 days at - 70° C).*

the strain CG132 gave rise to multiple bands. Gel slot H contained *Hind*III-digested pAD1::Tn*916* DNA from which the probe had been made.

Since the size of the fragments observed after hybridization varied greatly amongst different transconjugants that had been obtained from matings using the same recipient, Tn*916* appears to insert into different sites on the recipient chromosome. (These results also argue against the possibility that Tn*916* exists on a large plasmid that had previously escaped physical detection; if the latter were true, identical patterns would have been expected).

The multiple bands observed in the case of strains CG110, CG130, and CG132 do not seem to be the result of partial digestion products, since control digests that contained an equivalent amount of JH2-2 chromosomal DNA plus additional pAD1::Tn*916* DNA gave rise only to fragments A and B (i.e. similar to those shown in Fig. 1, gel slot H, for pAD1::Tn*916* plasmid DNA alone). Conceivably, these additional bands are the result of Tn*916* rearrangements occurring as the result of recombination events involving segments within Tn*916* and distal chromosomal sequences. However, transposition of Tn*916* from the chromosome of CG110 to a newly introduced pAD1 gave rise to pAD1::Tn*916* derivatives with typical 10 Mdal insertions of Tn*916*(data not shown). Preliminary experiments involving *Eco*RI digested DNAs probed with ^{32}P-Tn*916* indicate that there may be more than one hybridizable fragment (greater than 10 Mdal) in the chromosome of strains CG110, CG130, and CG132 (data not shown). Since there are no *Eco*RI sites in Tn*916*, the occurrence of multiple copies of the entire element is possible.

B. *Derivatives with Altered Transfer Frequencies*

Certain Tc^r transconjugants of JH2-2 exhibit altered transfer frequencies in secondary matings. Strain CG110 is able to donate Tc-resistance at frequencies of about 10^{-6} (see Table I). We were interested in determining the transposition frequency for Tn*916* in a strain such as CG110, and this was measured in the following way. The conjugative hemolysin plasmid pAD1 was introduced into strains CG110 and CG130, and the frequency of Tn*916* transposition from the chromosome to pAD1 was measured by the frequency of Tc^r hyperhemolytic transconjugants obtained after filter mating these donor strains with JH2SS. The hyperhemolytic phenotype results from an insertion of Tn*916* into or near the hemolysin determinant of pAD1 (3) and is readily detected by an increased diameter of zones of hemolysis on horse blood agar plates. The data presented in Table II indicate that the frequency of Tc^r transconjugants per recipient as well as the frequency of hyperhemolytic Tc^r transconjugants per recipient is more than 100 fold higher when CG110 is the donor strain, compared to when CG130 is the donor strain. Thus both Tn*916* transfer and transposition are increased in strain CG110.

C. *On the Mode of Transfer and Transposition of Tn*916

It is a reasonable possibility that Tn*916* determines functions (at least in part) related to its own transfer; and at 10 Mdal it would seem to be large enough to encode such information. The data indicating both an increase in the transfer and transposition frequencies of Tn*916* in the CG110 strain suggest that these two events may have a common step(s). Thus, the transfer of Tn*916* could represent an elaborate transposition event where the donor and recipient replicons are in different cells.

The notion that both transposition and transfer of Tn*916* may involve an excision of the transposon from the donor chromosome stems from the following observations made while studying derivatives of strain DS16 with increased levels of Tc resistance. Transposition of Tn*916* from the DS16 chromosome to pAD1 has been estimated to occur at a frequency of 10^{-6}. If Tn*916* remained at its original site, a two fold increase in tetracycline resistance should accompany transposition due to the resulting duplication of Tn*916*. (Strains specially constructed to carry Tn*916* on pAD1, as well as on the chromosome, have been shown to have an approximate doubling of the MIC for Tc (i.e. 75μg/ml vs 37μg/ml for the wild type) when compared to DS16. Interestingly, DS16 cells with increased levels of Tc resistance do appear but at a much lower frequency (10^{-8}); and derivatives that appear on selection plates containing elevated concentrations of drug do not contain duplications of the transposon, as ascertained by Southern blot analyses (Gawron-Burke and Clewell, manuscript in preparation). That is, Tn*916* has not been found in different sites on the chromosome nor has it been observed to transpose to pAD1 in any of those derivatives analyzed with increased levels of resistance. This implies that if indeed Tn*916* transposes spontaneously to pAD1 at 10^{-6}, there is not a net increase in transposon copy number. This would be consistent with an excision-insertion mechanism.

One could envision that after transfer the transposon's insertion into the recipient chromosome could be enhanced by a "zygotic induction" of an "integrase" (perhaps the related "transposase"). This supposition would predict that one might be able to detect the excision of Tn*916* from a donor replicon following transfer into a cell devoid of Tn*916*. Such predictions were born out in the following experiments.

Tn*916* was inserted into the erythromycin-resistance

Table II. *Increased transposition frequency in S. faecalis CG110*

Donor strain	*Frequency of Tc^r transconjugants per recipient*[a]	*Frequency of hyperhemolytic Tc^r transconjugants per recipient*
CG110(pAD1)	*9.8×10^{-5}*	*4.9×10^{-6}*
CG130(pAD1)	*3.9×10^{-7}*	*1.8×10^{-8}*

[a]Donor strains filter mated with JH2SS as described in Table I. Transconjugant frequencies are elevated in both matings due to the ability of pAD1 to mobilize chromosomal determinants (15). pAD1 transferred equally well from either host background in control mating experiments (data not shown).

(Em^r)-determining plasmid pAM81(15 Mdal) to generate pAM81::Tn*916* by overnight filter matings between CG110(pAM81) and JH2-2 followed by selection for Em^rTc^r transconjugants capable of transferring both resistance markers at high frequency in secondary matings. [pAM81 originated in a clinical isolate of *S. faecalis*, strain DU81, and is capable of conjugal transfer in filter matings to other *S. faecalis* strains at a frequency of 10^{-5} per recipient (Y. Yagi and L. Dempsey, personal communication)]. When strains containing pAM81::Tn*916* are mated with a JH2SS recipient in overnight filter matings, a high degree of segregation of the Em and Tc resistance determinants occurs in the transconjugants, as shown in the data of Table III. Although both Em and Tc resistance determinants transferred at similar frequency (about 10^{-6} per recipient), a high percentage (43-77%) of the Em selected transconjugants were sensitive to Tc. Tc-selected transconjugants show a higher degree of cotransfer of the Em^r and Tc^r determinants; greater than 90% of the transconjugants were both Em^r and Tc^r. Interestingly, of the Em-selected transconjugants that are also Tc resistant, a varying number represent cells in which the Tc^r and Em^r markers are no longer linked as ascertained by transfer frequencies in secondary matings (Table III).

We were interested in determining the structural integrity of pAM81 DNA sequences in the various transconjugant

Table III. *Segregation of* Em^r *and* Tc^r *in filter matings involving JH2-2(pAM81::Tn916 donors.*

Exp No[a]	Selected on Em[b]		Selected on Tc[b]		Em-selected Em^rTc^r transconjugants with unlinked markers
	Em^rTc^r	Em^rTc^s	Tc^rEm^r	Tc^rEm^s	
1	27/47(57%)	20/47(43%)	41/43(95%)	2/43(5%)	6/12
2	16/48(33%)	32/48(67%)	46/48(96%)	2/48(4%)	1/12
3	11/48(23%)	37/48(77%)	44/48(92%)	4/48(8%)	1/10

[a]*Each donor derived from a single colony isolate and filter-mated with JH2SS (JH2-2 in secondary matings) as described in Table I.*

[b]*Frequency of* Em^r *and* Tc^r *transconjugants ranged from* $2\text{-}5 \times 10^{-6}$ *per recipient* ($0.6 - 1.5 \times 10^{-5}$ *per donor).*

types, and to this end *Hind*III digests of plasmid DNAs were analyzed by agarose gel electrophoresis. Such an analysis is detailed in Fig. 2 where gel slot A contains the original pAM81::Tn*916* donor plasmid, and the arrows denote those fragments which contain Tn*916* sequences. Gel slot B contains pAM81 plasmid DNA and one can readily see that it is fragment G(1.7 kb) that is missing in pAM81::Tn*916*(gel slot A); this fragment must contain the site of Tn*916* insertion. Plasmid DNA from CG184 and CG186, two of the Em^rTc^r transconjugants in which these resistances are no longer linked, are shown in gel slots C and D. The plasmid DNA exhibits a *Hind*III restriction pattern identical to that seen for pAM81; fragment G has returned and the two fragments consisting of Tn*916* sequences are lost. Tn*916* has presumably transposed to the bacterial chromosome, since these strains are Tc-resistant and capable of Tn*916* transfer at low frequency. Plasmid DNA from CG187, representative of the Em^rTc^s trans-conjugant class encountered, also exhibited an identical pattern to that seen for pAM81 (gel slot E). Gel slot F contained plasmid DNA from the Em^rTc^r transconjugant strain CG181, representative of the class in which Em and Tc resistance determinants remain linked. The restriction pattern is identical to that of the original pAM81::Tn*916* donor.

These data are consistent with the notion of an excision step in transfer and transposition, whether this step occurs in the donor prior to transfer or in the recipient after conjugal transfer. It is significant to note that excision from pAM81::Tn*916* appears to be precise within the limits of detection by agarose gel analyses. It will be important to examine plasmid DNA from additional representatives of the various transconjugant types to determine if transposition might in some cases (perhaps less frequently?) leave the element on the donor replicon. Such studies are currently underway.

These data have allowed us to develop a working hypothesis for Tn*916* transfer and transposition as shown in Fig. 3. In this model we envision the transposon to be capable of excision from the donor replicon after which it might then undergo one of three options: 1) reinsertion into the chromosome (perhaps at a different location) 2) insertion into a resident plasmid or; 3) conjugal transfer into another cell. After transfer into the recipient cell the transposon could then insert into the chromosome or perhaps into a plasmid in the recipient. (In this regard, we have observed Tn*916* to insert into pAD1 when recipient cells harboring the plasmid are mated with pAM81::Tn*916* donor strains).

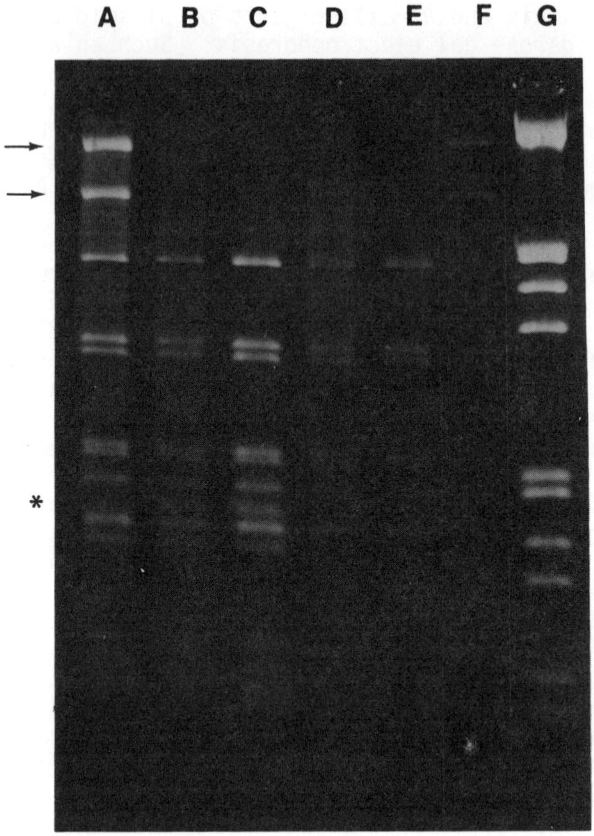

Fig. 2. Hind*III digests of transconjugants derived from JH2-2 (pAM81::Tn*916 *donor.*

Gel slots C-F contain plasmid DNA from the Em^r *transconjugant strains CG184, CG186, CG187, and CG181, respectively. Gel slot A contained plasmid DNA from the JH2-2(pAM81::Tn*916 *donor, and slot B contained plasmid DNA from strain CG130 (pAM81). Arrows mark fragments gained and * marks fragment lost as a result of Tn*916 *insertion into pAM81. Gel slot G contained lambda DNA cut with* Eco*RI and* Hind*III to yield fragments A-L with sizes of 21.8,5.24,5.05,4.21,3.41,1.98,1.9, 1.57,1.32,0.93,0.84, and 0.58 kb respectively(16). Agarose gel electrophoresis utilized a horizontal apparatus, a 1% gel and Tris-borate buffer (17).*

III. CONCLUSION

We have shown by Southern blot hybridization that conjugal

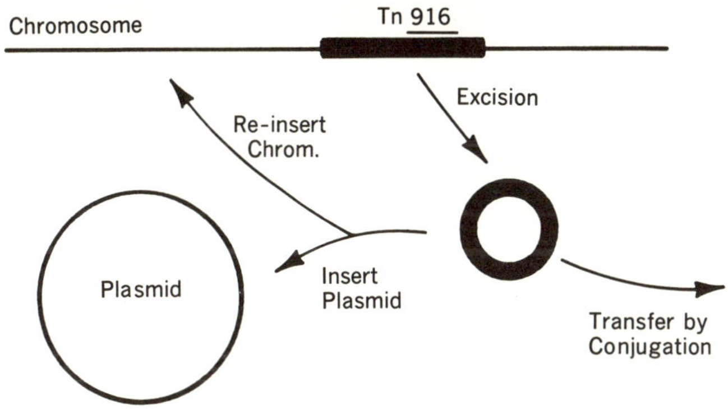

*Fig. 3. Model for Tn*916 *transfer and transposition.*

transfer of Tn*916* results in the insertion of the transposon into different sites on the recipient chromosome. Certain Tc^r transconjugants of JH2-2 such as strain CG110 can donate Tc^r at elevated frequencies, and transposition of Tn*916* from the chromosome to a newly introduced pAD1 is also increased. The basis for the elevation in transfer frequency remains obscure. Conceivably it is related to the apparent redundancy of Tn*916* sequences. [Despite the suggestion of more than one copy of Tn*916* in CG110, it is puzzling that the level of Tc resistance in this strain does not appear to be elevated (Unpublished data)]. Finally, we have begun to examine the mechanism of transfer and transposition and have observed an excision of Tn*916* from donor plasmid molecules.

Chromosome-borne resistance determinants which transfer in the absence of conjugative plasmids appear to be common in the streptococci, and it will be of interest to see to what extent these determinants will represent "conjugative transposons". It is important to note that Guild's group has recently demonstrated homology between Tn*916* and the transferrable tetracycline-resistance element they have observed in *S. pneumoniae*(9). We also note recent reports suggesting that transferrable nonplasmid elements exist in *Clostridium difficile* (10) and *Bacteroides fragilis*(11, 12).

ACKNOWLEDGEMENTS

We thank Arthur Franke, Ronald Hart, and Yoshihiko Yagi for helpful discussions.

REFERENCES

1. Buu-Hoi,A. & Horodniceanu,T.(1980) J. Bacteriol. *143*, 313-320
2. Shoemaker,N.B., Smith,M.D., & Guild,W.R.(1980) Plasmid *3*, 80-87
3. Franke,A. & Clewell,D.B.(1981) J. Bacteriol. *145*, 494-502
4. Franke,A. & Clewell,D.B.(1980) Cold Spr. Harb. Symp. Quant. Biol. *45*, 77-80
5. Horodniceanu,T., Bougueleret,L., & Beith,G.(1981) Plasmid *5*, 127-137
6. Shoemaker,N.B., Smith,M.D., & Guild,W.(1979) J. Bacteriol *139*, 432-441
7. Tomich,P.K., An,F.Y., Damle,S.P. & Clewell,D.B.(1979) Antimicrob. Angents. Chemother. *15*, 828-830
8. Southern,E.M.(1975) J. Mol. Biol. *98*, 503-517
9. Smith,M.D., Hazum,S., & Guild,W.(1981) J. Bacteriol. (in press)
10. Smith,C.M., Markowitz,S.M., & Macrina,F.(1981) Antimicrob. Agents. Chemotherap. *19*, 997-1003
11. Mays,T.D., Macrina,F.L., Welch,R.A., & Smith,C.J.(1981) in Molecular Biology, Pathogenicity, and Ecology of Bacterial Plasmids (Levy,S.B., Clowes,R.C., and Koenig,E.L. eds.) pp 631 Plenum Press, New York
12. Tally,F.P., Shimmel,M.J., Carson,G.R., & Malamy,M.H. (1981) in Molecular Biology, Pathogenicity, and Ecology of Bacterial Plasmids (Levy,S.B., Clowes,R.C., and Koenig,E.L., eds.) pp 631 Plenum Press, New York
13. Jacob,A.E. & Hobbs,S.J.(1974) J. Bacteriol. *117*, 360-372
14. Tomich,P.K., An,F.Y., and Clewell,D.B.(1980) J. Bacteriol. *141*, 1366-1374
15. Franke,A.E., Dunny,G.M., Brown,B., An,F., Oliver,D., Damle,S.P. and Clewell,D.B.(1978) in Microbiology 1978 (Schlessinger,D., ed.) pp 45-47, American Society for Microbiology, Washington, D.C.
16. Phillipsen,P., Kramer,R., and Davis,R.W.(1978) J. Mol. Biol. *123*, 371-386
17. Meyers,J.A., Sanchez,D., Elwell,L.P., and Falkow,S. (1976) J. Bacteriol. *127*, 1529-1538

GENETICS AND MOLECULAR BIOLOGY : VECTOR AND R PLASMID

COMPARATIVE STUDY OF INC N R PLASMIDS

T. Arai and T. Ando

Department of Microbiology
Keio University School of Medicine
Shinjuku-ku, Tokyo, Japan

IncN R plasmids are very popular in a variety of bacteria. This wide distribution suggested that this group of plasmids has wide host ranges. These plasmids are also known to have unique genes for a group specific restriction-modification system and a repair system (1-5). Because of these unique characters, IncN R plasmids have been studied in various labo-

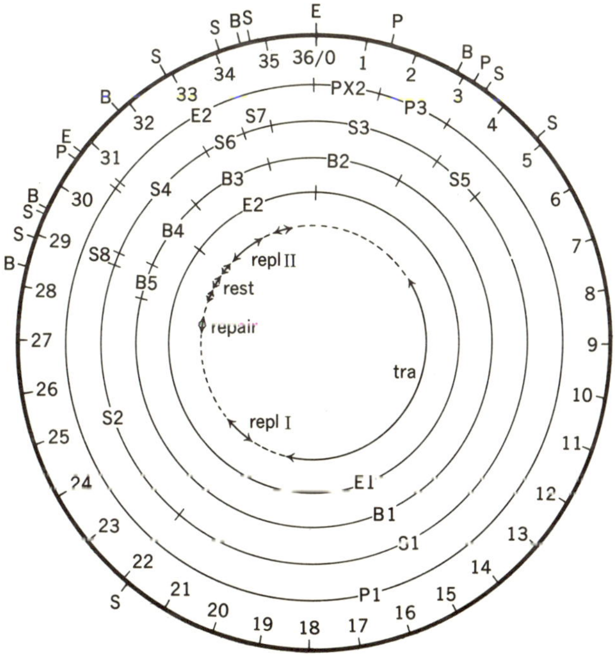

Fig.I. Cleavage and genetic map of N-3. Cleavage sites of Eco-RI, BamHI, SalI, and PstI are abbreviated as E, B, S, *and* P, *respectively. Genes for replication, repair, restriction, transfer, resistances to sulfonamide, streptomycin, and tetracycline are abbreviated as* repl, repair, rest, tra, Su, S, *and* T, *respectively.*

ratories. We have been working on the oldest isolate of this group R plasmids, N-3. The genetic map of N-3 is shown in Fig. 1. It was found that this plasmid had a large portion uncleaved by most of the six base-recognizing restriction endonucleases, and that this large fragment contained all essential and unique genes for this plasmid, although this had no drug resistance genes (6). This finding was supported by the report describing simple EcoRI digestion of various IncN R plasmids (7), and the genetic map of an other IncN plasmid, R46 as well as its derivative, pKM101 (8,9). Then, we could expect the common structure for IncN plasmids in this fragment.

EXPERIMENTAL

The IncN R plasmids studied are shown in Table 1. All methods were as described in the previous paper (6). These plasmid DNAs were digested by EcoRI, BamHI, EcoRI+BamHI, and PstI, and applied on agarose gel electrophoresis. Electrophoretic patterns of EcoRI+BamHI and PstI digested DNAs are shown in Fig.2. Small ampicillin resistant plasmids, R825 and R728 had no EcoRI site and only one BamHI site. pKM101 and pJA4733 had each one site for EcoRI and BamHI. All EcoRI and BamHI fragments from R447a, R454 and R-15 had common sizes, but PstI digestion gave one additional fragment to R454. All plasmids were found to have one large fragment and some small common size fragments.

Table I. IncN R plasmids studied

R plasmid	*Resist. marker**	*Repair*	*Restriction*	*Molec. weight#*	*Origin*	*Reference*
N-3	*SSuT*	+	+	*35.8*	*Shigella*	*I*
R-I5	*SSu*	+	+	*38.0*	*Shigella*	*I*
pJA4733	*SCA*	+	+	*36.0*	*Enterobacter*	*5*
pKMIOI	*A*	+	-	*26.8*	*Salmonella*	*9*
R825	*A*	-	-	*24.4*	*Providence*	*IO*
R728	*A*	-	-	*26.0*	*Providence*	*IO*
R396	*SSuCTA*	+	+	*39.8*	*Proteus rettgeri*	*II*
R447a	*SSuCTA*	+	+	*38.2*	*Proteus morganii*	*I2*
R454	*SSuCTA*	+	+	*35.6*	*Proteus morganii*	*I2*

**Resistances to streptomycin, sulfonamide, tetracycline, chloramphenicol, and ampicillin are abbreviated as S, Su, T, C, and A, respectively. #Molecular weight, M daltons.*

We tried to compare the large fragments of these plasmids.

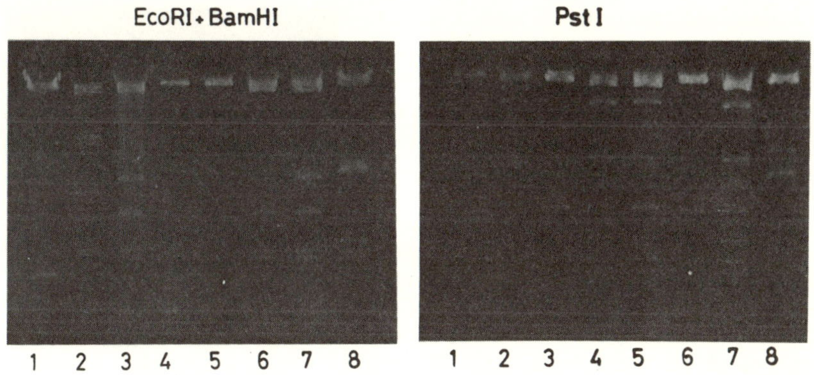

Fig.2. EcoRI and BamHI double and PstI digested products of various IncN R plasmids. Lanes I-8 after EcoRI+BamHI digestion were products of plasmids, pKMI0I, R396, R447a, pJA4733, R825, R454, R-I5,and N-3, respectively. Lanes I-8 after PstI digestion were products of plasmids, R396, pJA4733, R825, R447a, R454, R728, R-I5,and N-3, respectively.

The large fragment of each plasmid was cloned into pBR322 by BamHI. Heteroduplex analysis among these large fragments was carried out by hybridizing the large fragment of N-3 with those of other plasmids after BamHI digestion. It was found that all plasmids tested had long well matched portion on their large fragments with that of N-3. The electronmicrograph of heteroduplex between these fragments of N-3 and R396 is shown in Fig.3. These recombinant plasmids were also subjected to HincII and HaeIII digestions. All digested fragments from the large fragments of R447a and R396 had common sizes, R454 had one additional fragment, and N-3 had some missing and additional fragments. R825 also had some common size fragments after HincII and HaeIII digestions. In contrast, HincII fragments of pJA4733 and pKM101 were quite different from those of other plasmids, although HaeIII digestion of these two plasmids gave some common size fragments with other plasmids (Fig.4).

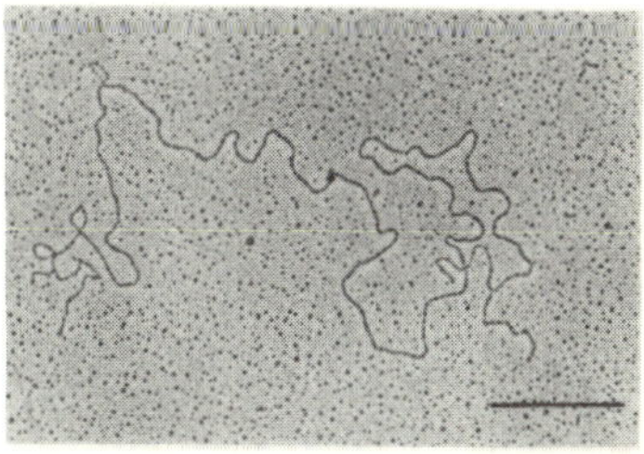

Fig.3. Heteroduplex between the large fragments of N-3 and R396. The line indicated Iμm.

We also cloned sulfonamide and streptomycin resistant fragment of these plasmids into pBR322 by BamHI, and these recombinant plasmids were subjected to EcoRI and BamHI double and HincII digestions. All products after EcoRI+BamHI double digestion had same sizes, but they were divided into two groups after HincII digestion.

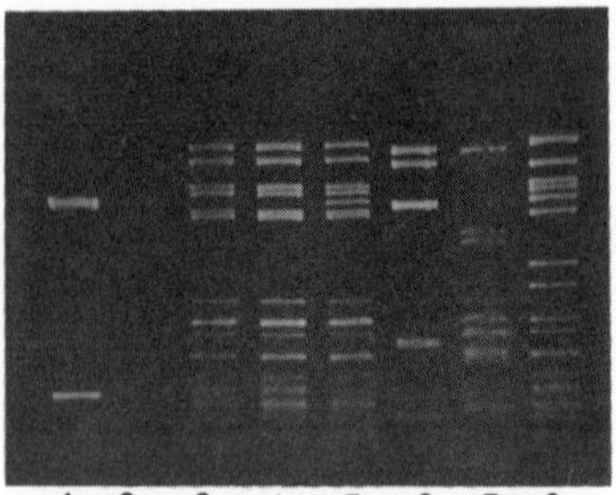

Fig.4. HincII digested products of the recombinant plasmid between pBR322 and the large fragment of each plasmid. Lane I was products of pBR322, and lanes 2-8 were products of the recombinant plasmids from pKMIOI, R447a, R396, R454, pJA4733, R825 and N-3, respectively.

DISCUSSION

All IncN R plasmids so far examined had one large portion uncleaved by most of the six base-recognizing restriction endonucleases, and all genes essential and unique for this group of plasmids were found on this portion. Thus, this portion was expected to be specific for all IncN R plasmids.

We studied how this portion was preserved in all IncN plasmids. That is, we compared the molecular structures of IncN R plasmids detected in various bacteria isolated from different habitats. These R plasmids were classified into two groups by their molecular sizes, production of restriction enzyme and resistances to sulfonamide and streptomycin.

All plasmids were cleaved by the six base-recognizing restriction enzymes. This result confirmed the presences of a large uncleaved portion and some common size small fragments. Then, the large fragments of all plasmids were cloned into pBR322 by BamHI and subjected to heteroduplex analysis and the cleavage by the shorter base-recognizing restriction enzymes. It was found that the large fragments of all plasmids had a rather large well matched portion with some small miss-matchs with that of N-3. This result suggested that the large portion of IncN plasmids had well preserved sequences with some changes. This was also confirmed by the cleavage patterns of these large fragments by smaller base-recognizing restriction endonucleases. That is, some R plasmids had completely common size fragments despite of their different origins, and some R plasmids had some additional or missing fragments other than the common fragments.

We also cloned sulfonamide and streptomycin resitant fragments of these R plamids into pBR322 by BamHI. The sufonamide and streptomycin resistant fragment of N-3 was known to have 2.3 M daltons and single EcoRI site (6). These fragments of all plasmids had the same size and EcoRI cleavage site as that of N-3, although these fragments were divided to two groups by HincII digestion patterns.

CONCLUSION

IncN R plasmids had a large portion uncleaved by most of the six base-recognizing restriction endonucleases, and this portion contained all essential and unique genes for IncN R plasmids. High similarity of the large fragments from sulfonamide and streptomycin resistant R plasmids were proved by both heteroduplex analysis and restriction endonuclease cleavage patterns. The R plasmids which did not have these resistance genes were smaller than others and rather different cleavage patterns, although heteroduplex analysis gave some matchings. This result suggested that the R plasmids which had no cleavable portion had also differences on their uncleaved portions. Similar changes were also observed even on the small sulfonamide and streptomycin resistant fragments.

REFERENCES

1. Watanabe, T., Nishida, H., Ogata, C., Arai, T.,and Sato, S. (1964) J. Bact. *88*,716-726
2. Arai, T.,and Aoki, T. (1977) J. Bact. *130*,529-531
3. Drabble, W.T.,and Stocker, B.A.D. (1968) J. Gen. Microbiol. *53*,109-123
4. Arai, T.,and Ando, T. (1978) Mut. Res. *53*,146-147
5. Arai, T.,and Ando, T. (1979) in Microbial Drug Resistance (Mitsuhashi, S. ed) Vol.2 pp.197-203 Japan. Sci. Soc. Press, Tokyo
6. Ando, T. and Arai, T. (1981) Plasmid *6*, in press
7. Chabbert, Y.A., Roussel, A., Witchitz, J.L., Sanson-LePors, M-J., and Courvalin, P. (1979) in Plasmids of Medical, Environmental and Commercial Importance (Timmis, K.N.,and Pühler, A. eds) pp.183-193 Elsevier/North-Holland, Amsterdam
8. Brown, A.M.C.,and Willetts, N.S. (1981) Plasmid *5*,188-201
9. Langer, P.J., and Walker, G.C. (1981) Mol. Gen. Genet. *182*,268-272
10. Hedges, R. W. (1974) J. Gen. Microbiol. *81*,171-181
11. Coetzee, J.N., Datta, N.,and Hedges, R.W. (1972) J. Gen. Microbiol. *72*,543-552
12. Hedges, R.W., Datta, N., Coetzee, J.N.,and Dennison, S. (1973) J. Gen. Microbiol. *77*,543-552

THE PLEIOTROPIC EFFECTS OF *HIP* MUTATION IN *ESCHERICHIA COLI*

Akihiko Kikuchi

Mitsubishi-Kasei Institute of Life Sciences
Machida, Tokyo, Japan

Integration of phage λ into the host chromosome requires a phage gene product, integrase, and factors encoded by the host, *E.coli* (1). Two mutants of *E.coli* were isolated, one called *hid* (*him*A) and the other, *hip*, both of which were unable to support λ phage integration, *in vivo*. The cell free extracts of these mutants complemented each other in the *in vitro* integration reaction, suggesting that the host provides at least two gene products for λ phage integration.

Hip and *hid* (*him*A) mutants share common pleiotropic phenotypes other than λ phage integration, although these mutations were mapped at different sites in the *E.coli* linkage map; *hip* mutation at 20 min near *aro*A marker, while *hid* (*him*A) at 37.5 min close to *aro*D gene. The isolation of these mutants and their preliminary characterization with respect to the integrative recombination were previously reported (2). Here, I describe briefly other pleiotropic phenotypes caused by *hip* mutation, some of which were observed also in *hid* (*him*A) mutant (3).

A. *The growth of phages*

The *hip* mutant was apparently normal in cell growth and allowed most phages, T4, T5, T7, λ, ϕ80, P2, P1 and St-1, to form plaques. However, a certain phages failed to propagate in *hip* mutant, namely λ phage variants (λ*cin*-1, λ*c*17, λ*Q*80) and Mu phages. Lysogenic phages such as λ, ϕ80 and P2 formed rather tiny and clearish plaques on the lawn of *hip* indicator and their lysogens were hardly obtained. It was suggested that the expression of *c*II region of λ phages was defective in *hip* mutant to cause the failure in establishing the lysogeny and in the growth of some λ variants as well.

Neither Mu phage could grow on *hip* lawns nor Mu lysogens of *hip* mutant was killed by inducing its latent prophages. (The λ lysogens of *hip* mutant was efficiently killed by its induction). Giphart-Gassler, et al. demonstrated that inefficient transcription of early genes of Mu phage was responsible for these phenotypes (7).

The filamentous single-stranded DNA phages, such as fl, fd or M13 could not form plaques on *hip* male strain. After 3 hours of the phage infection, supertwisted DNA of replicative forms were observed in the same quantity in *hip* host as in wild type, though the yield of phage progeny was about 1,000 times lower in *hip* than *hip*$^+$ host. This defect is not general for single-stranded DNA phages, as St-1 phage (spherical like ϕ174) formed plaques with high efficiency and of normal size on *hip* indicator.

The *hip* mutant was not particularly mutagenic, not sensitive to UV irradiation and was proficient in *rec*A-promoted general recombination.

All these phenotypes of *hip* mutant segregated as a unit in P1 phage transduction and the mutation was recessive over wild type allele.

B. *Transposition*

To measure the frequency of transposition, *hip* mutant carrying the conjugative plasmid RTF-*tet*^{R} was lysogenized with phage P1-*cam*^{R}. This *cam*^{R} gene is flanked by one direct repeat of IS1 and transposes as Tn9 to RTF-*tet*^{R} plasmid, which was subsequently transfered into proper recipients to score the occurrence of RTF-*tet*^{R}-*cam*^{R} among RTF-*tet*^{R} (4).

Although the transfer of RTF-*tet*^{R} from *hip* donor cell was somewhat inefficient compared with isogenic *hip*$^+$ strain, the ratio of *cam*^{R}-*tet*^{R} over *tet*^{R} exconjugants roughly indicated the frequency of Tn9 transposition in each strain.

As shown in Table 1, Tn9 transposition was inhibited in *hip* strain by about 100 folds.

Precise excision of Tn10 from *trp* gene (*trp*::Tn10) by

Table 1. *Tn9 transposition*

Donors[a]	*Number of exconjugants*[b]		
	tet^R	*cam*^{R}-*tet*^{R}	*Ratio*
1. *hip*$^+$	3.0×10^7	500	170×10^{-7}
hip	0.8×10^7	0	$< 2 \times 10^{-7}$
2. *hip*$^+$	20×10^7	> 3000	$> 150 \times 10^{-7}$
hip	2.0×10^7	0	1×10^{-7}

[a] *Each donor cells carried RTF-tet*^{R} *and P1 cam*^{R} *plasmids.*

[b] *To select exconjugants, recipient cells were made nal*^{R} *and phage P1 resistant.*

detecting Trp^+ revertant also reduced by at least 100 folds in *hip*$^-$ strain (10^{-8}) compared with that in isogenic *hip*$^+$ strain (10^{-6}). The reversion frequency of *trp*::Tn10 could only be measured in Hfr strains, because the frequency was too low in female strains (less than 10^{-8} Trp^+ in both *hip*$^+$ and *hip*), and no meaningful comparison was made with a female pair.

C. Conjugative plasmids

F factor including Hfr and some R factors such as R100-1, RTF-*tet*^{R} maintained stably in *hip* mutant. However, conjugative transfer of these plasmids was less efficient from *hip* donor strain. For example, at 37°C, transfer of the plasmids was reduced by 1/10 to 1/100, and at 30°C, no transfer was detected. Moreover, adsorption of male specific phages such as f1 (fd, M13) and f2 was poor and again temperature dependent. It was suggested that some functions coded by these plasmids were altered in *hip* mutant.

On the contrary, other functions of these conjugative plasmids, for example, incompatibility, restriction of T7 phage and expression of markers found in F' (*lac*$^+$, *gal*$^+$ etc.) and R (drug resistance determinants) were not affected in *hip* mutant.

D. Multi-copy plasmids

The *hip* mutant carrying plasmids, such as colE1 derivatives, pSC101, λ*gal* N^- could not be constructed as stable clones. These plasmids could be introduced into *hip* host either by transformation (pSC101, RSF2124:colE1-*amp*^{R}), by mobilization with conjugative plasmids (RSF2124, pLC1010:colE1-*lac*$^+$), or by phage infection (λ*gal* N^-). They must be replicated to a limited extent but were not stably maintained. For example the strain LC1010 which harbors both colE1-*lac*$^+$ of Clarke and Carbon's collection and F factor, was mated with *hip* or *hip*$^+$ recipient strain having *lac*Z mutation, and Lac^+ exconjugants were selected on MacConkey lactose agar plates. Approximately the same number of Lac^+ colonies were obtained from both recipients. However, in the case of *hip* recipient, Lac^+ cells were found only in the center of colonies surrounded by pink-white zone of Lac^- cells. Upon purification of these Lac^+ red center by repeated streaking on the same plate without selective pressure for lactose, the *hip* strains segregated Lac^- clones by about 90%, while few Lac were observed from *hip*$^+$ recipient.

A similar result was obtained after infection with λ*gal* N^- phage, which should be maintained in a plasmid state. λ*gal* N^- phage segregated much faster in *hip* than in wild type host.

Table 2. *Influence of hip allele on the maintenance of RSF2124*

A. F'147 *gal*$^{+}$*hip*$^{+}$/RSF2124(*amp*^{R}) × F^{-} *gal*,*hip*,*rps*L

24 Gal^{+} Str^{R}
- 8 Amp^{R} Hip^{+}
- 16 Amp^{S} Hip^{+}

24 Amp^{R} Str^{R}
- 10 Gal^{+} Hip^{+}
- 14 Amp^{S} Gal^{-} Hip^{-}

(2nd cycle purification)

B. F'152 *gal*$^{+}$/RSF2124(*amp*^{R}) × F^{-} *gal*,*hip*,*rps*L

24 Gal^{+} Str^{R}
All Amp^{S}

10^{-6} Amp^{R} Str^{R}

Table 2 summarizes the results of transfer of RSF2124 by mobilization along with F' transfer into *hip* recipients showing that the loss of RSF2124 plasmids was actually related to the *hip* mutation and introduction of F' *hip*$^{+}$ into *hip* cell restored the stable maintenance of the RSF2124 plasmids.

Although these plasmids were very unstable in *hip* mutant a few stable clones which were phenotypically plasmid carriers were obtained, if selective pressure was persistently applied in the presence of ampicilline or tetracycline or using minimal lactose plates in stead of MacConkey lactose plates. They were still *hip* but no plasmids as supertwisted forms were ever recovered out of them.

E. Copy number mutant of phage P1

In contrast to the instability of high copy number plasmids, phage P1 was lysogenized as a low copy number plasmid and maintained stably in *hip* mutant. Sternberg, et al. have isolated a P1 mutant (P1 *cop*-1) which maintains 20-25 copies in stead of one per cell. Unlike wild type P1, this P1 *cop*-1 mutant was not stably maintained in *hip* host (5).

This experiment strongly supported that in general low copy number was associated with stable maintenance of that plasmid in *hip* mutant, while high copy number with instability of plasmids.

F. The role of hip gene product

The two host factors necessary for λ phage integration were purified using in vitro assay system (6). They were

Table 3. *Lysogenization of P1 cop-1 mutant*

	Appearance of non lysogens [a]		*Frequency of P1 lysogen* [b]	
	hip$^+$	*hip*	*hip*$^+$	*hip*
P1 cam^R*,clr100*	10^{-3}	10^{-3}	64/64	66/66
P1 cam^R*,clr100,cop-1*	10^{-3}	1	148/150	0/54

[a] *Fraction of survivors at 42°C in the culture infected by P1 phage at 30°C. (clr100 is ts repressor mutation.)*

[b] *Number of cam*^{R} *cells among the survivors at 30°C.*

rather basic, low molecular weight (9,000-10,000) and heat resistant (no loss of activity at 100°C for 15 min) proteins and found exclusively in the nucleoid fraction prepared in the presence of spermidine. *Hip* protein and *hid* (*him*A) protein were co-purified through all the step and *hip-hid* complex binds supertwisted DNA preferentially. It is suggested that these proteins might be important to form locally the higher order structure of DNA suitable to the substrate for λ phage integration, gene expression or transcription, transposition in some steps and maintenance of certain plasmids.

ACKNOWLEDGEMENT

I thank Bob Weisberg and Howard Nash, with whom the most part of this work was done, Nat Sternberg, Lynn Enquist, Tom Pollack and Chris Gritzmacher for helpful discussions and unpublished communications. I am grateful to Yoshiko Kikuchi, without whose interest and enthusiasm this work would not have been carried out.

REFERENCES

1. Kikuchi, Y., and Nash, H. (1979) Proc. Natl. Acad. Sci. USA *76*, 3760
2. Miller, H., Kikuchi, A., Nash, H., Weisberg, R., and Friedman, D. (1978) Cold Spring Harbor Symposium *43*, 1121
3. Enquist, L., Kikuchi, A., and Weisberg, R. (1978) ibid. *43*, 1115
4. Chandler, M., Silver, L., Lane, D., and Caro, L. (1978) ibid. *43*, 1223
5. Sternberg, N., and Austin, S. (1981) Plasmid *5*, 20
6. Nash, H., and Robertson, C. (1981) J. Biol. Chem. *256*, 9246
7. Giphardt-Gassler, M. and Van de Putte, P. (1979) Gene *7*, 33

INDUCIBLE RESISTANCE TO MACROLIDE, LINCOSAMIDE, AND STREPTOGRAMIN TYPE B ANTIBIOTICS: THE MECHANISM OF INDUCTION IN PLASMID pE194 FROM *Staphylococcus aureus*

Sueharu Horinouchi

Department of Agricultural Chemistry
University of Tokyo
Bunkyo-ku, Tokyo JAPAN

Bernard Weisblum

Department of Pharmacology
University of Wisconsin Medical School
Madison, Wisconsin USA

I. INTRODUCTION

Plasmid pE194, found originally in *Staphylococcus aureus* and described by Iordanescu et al. (10,11) was introduced into *Bacillus subtilis* by Gryczan and Dubnau (4). In view of the small size (2.4 Mdal) of pE194 and the fact that it confers inducible resistance to the macrolide, lincosamide, and streptogramin type B (MLS) antibiotics we have chosen this plasmid as a model system to study in detail induction at the molecular level. From these studies we have obtained useful information on the mechanism by which expression of resistance to the MLS antibiotics is regulated. In addition some of the features of the DNA sequences responsible for replication have been identified and the complete nucleotide sequence of pE194 will be published shortly (9). Other interesting biological properties of pE194 including the peptides which it encodes and mutations involving copy number control have been described (5,15,18).

II. CONTROL OF EXPRESSION OF ERYTHROMYCIN RESISTANCE IN pE194

Erythromycin resistance as it is found in clinical bacterial isolates is mediated by the specific dimethylation

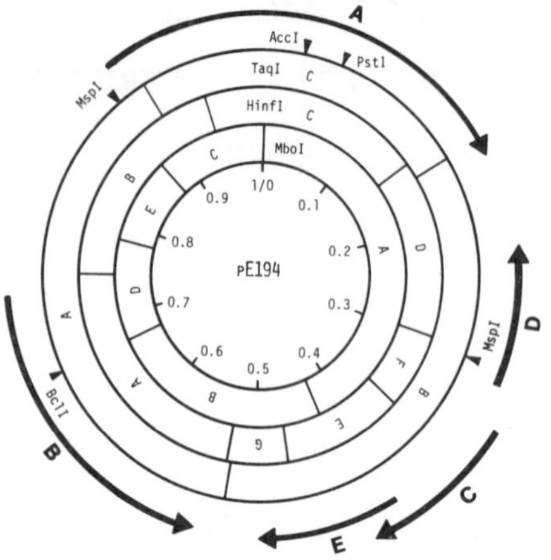

Figure 1. Physical map of pE194.

The physical map of pE194 based on determination of the DNA sequence is shown together with open reading frames identified by computer-aided analysis of the DNA sequence and labeled alphabetically in order of decreasing size. The open reading frame labeled "B" contain the resistance determinant of pE194.

of adenine in 23S ribosomal RNA as a consequence of which the bacterial cell becomes, more generally, co-resistant to all antibiotic inhibitors of 50S ribosome subunit function which belong to the macrolide, lincosamide, and streptogramin type B (MLS) group of antibiotics (12,13). The determination of the coresistance pattern to these three chemically dissimilar classes of antibiotics is easily done by the disc method. MLS resistance specified by plasmid pE194 in S.aureus as well as in B.subtilis to which it can be transferred by transformation (4), is inducible by subinhibitory concentrations of erythromycin and the requirements for induction at the cellular level have been described in detail (17).

The nucleotide sequence of pE194 has been determined and the control region for expression of inducible MLS resistance has been localized between 0.70 and 0.80 map units on the physical map of pE194, shown in Fig.1, by recombinant DNA methods using fragments obtained from the wild type inducible strain and constitutively resistant mutants derived from it (7). The model for regulation of regulation of MLS resistance which we have proposed (7,8) postulates that control is mediated by a translational attenuation mechanism shown in schematic form in Fig.2. A similar mechanism has been proposed by Gryczan et al. (6).

In a formal sense the control mechanism resembles the transcriptional attenuation mechanism by which activity of the operons for biosynthesis of the amino acids is regulated, reviewed by Yanofsky (20). According to the translational attenuation model, the 5' end of the mRNA encoding the inducible MLS resistance determinant contains a set of 4 inverted complementary repeat sequences capable of assuming alternative conformations. The 4 sequences, designated 1, 2, 3, and 4 can associate as 1+2 and 3+4 in view of the complementarity of these respective sequences, or as 2+3, in view of the fact that 2 and 3 (and therefore 1 and 4) are complementary, as well. An additional set of inverted complementary repeat sequences, A and A', bearing no sequence similarity to 1,2,3, or 4, is also present.

We postulate further that the 1+2,3+4 association pattern corresponds to the nascent inactive conformation of the control region. This nascent state of the messenger is inactive since, as shown in Fig. 2, the ribosome loading site for synthesis of the MLS resistance determinant (29K protein)

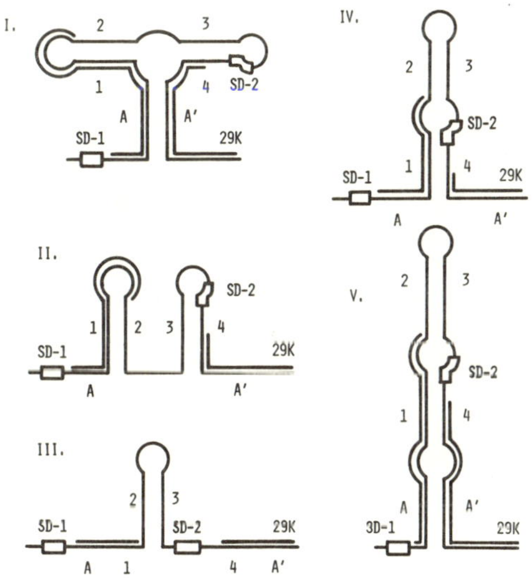

Figure 2. Conformations of 29K control region.

Possible conformations of the MLS resistance control region. Five possible conformations of the mRNA for MLS resistance are shown. Their respective calculated free energies are -30.4, -25.8, -14.2, -29.8, and -34.4 kcal per mole (8). According to our model, activation of the nascent mRNA involves the transition I to II to III, whereas deactivation following removal of the inductive stimulus involves the transitions III to IV to V, rather than III to II to I. The double lines over sequences A,1,4, and A' indicate ability to code for polypeptide.

is sequestered in the loop formed by the association of 3+4. The activation of the mRNA is achieved by a conformational rearrangement of the inverted complementary repeat sequences which results in association of 2+3 thereby unmasking the sequestered ribosome loading site for 29K protein synthesis.

The conformational realignment of the control region results from hindered translation of a 19 amino acid "leader" or "control" peptide encoded by sequences A and 1. By a mechanism similar to that which reorients the conformation of nascent tryptophan operon mRNA (Yanofsky (20)), we postulate that a similar type of hindered translation mediates the induction process, but with two important distinctions: (i) In the case of induced MLS resistance, ribosome function is hindered by the inducer, erythromycin, whereas in the case of the tryptophan operon, the ribosome is hindered by the deficiency of tryptophanyl tRNA (or histidinyl tRNA in the case of the histidine operon etc.), and (ii) Reorientation of the MLS control region results in unmasking the sequestered ribosome loading site for synthesis of the resistance determinant protein, whereas reorientation of the tryptophan operon attenuator results in completion of the tryptophan operon mRNA otherwise prematurely terminated by the fact the association of 3+4 supplies the signal for termination of mRNA synthesis, whereas the form in which 3 and 4 are unassociated does not (i.e. in which 2 and 3 are associated).

With reference to the unmasking of the ribosome loading site, induction can occur either indirectly or directly- i.e. either as: (i) preemptive unmasking in which dissociation of 1+2, allows 2 to preempt 3 freeing up 4 and unmasking the ribosome loading site, and (ii) direct dissociation of 3+4 which can occur owing to mutation in the 3+4 sequence as a consequence of which the association of 3+4 is weakened with spontaneous unmasking of the ribosome loading site with a higher degree of probability than in the wild-type inducible form.

The model which we propose predicts that mutation to constitutive expression of MLS resistance would be most effective for base changes which affect sequences 1,3, and 4, but not 2. Base changes which affect 1 weaken its association with 2 and favor preemptive association of 2 with 3; this situation is not functionally equivalent to one in which stability of 1+2 is reduced owing to a mutation in 2, since base changes in 2 which weaken association with 1 would likewise weaken the association of 2 with 3 thereby reducing the efficiency of the 2+3 preemptive pairing reaction. On the other hand, mutations which affect either 3 or 4 reduce the energy of association of 3 with 4, bypassing the preemptive reaction.

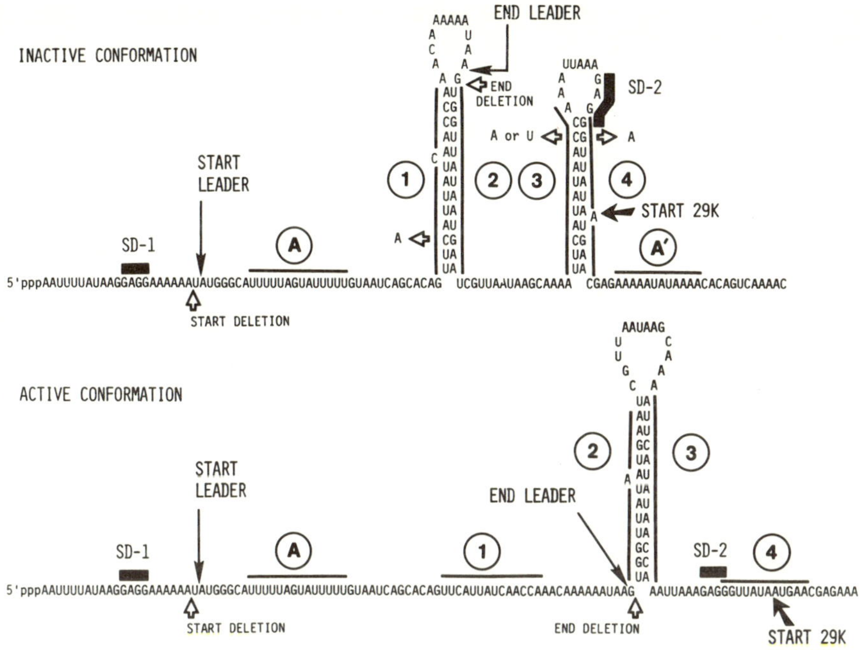

Figure 3. Details of the MLS control region, location of nucleotide changes in 11 mutants to constitutive expression.

The sequence of the mRNA at the control region deduced from the DNA sequence is shown in the inactive and active conformations postulated, and discussed in detail in the text. These correspond to conformations II and III, respectively, shown in Fig.2.

In a test of this model, 11 independent constitutively resistant mutants were selected randomly and the DNA sequence of the MLS control region was determined. In agreement with predictions made on the basis of the model none of these 11 mutants involved sequence 2. Alterations in sequences 3 and 4 were found in the remaining 9 mutants analyzed (5 in sequence 3, and 4 in sequence 4), as shown in Fig.3.

The remaining 2 mutants analyzed were particularly informative. One of these, 3 bases from the 5' end of sequence 1, involved C to A. Insofar as this mutation would have a destabilizing effect on the pairing of the first 2 residues of sequence 1, UU, as well, we infer that a significant level of induction should occur when the control region is disrupted as far as the third of 13 consecutive base pairs in sequence 1. The extent to which the effectiveness of this mutation is mediated by a change in the amino acid sequence of the putative control peptide , His (CAU) to Asn (AAU), has not yet been evaluated. In the other constitutive mutant which involves sequence 1, 62 nucleotides which specify the control peptide (19 amino acids) were excised resulting in a control region containing intact sequences 2, 3, and 4. In the nascent MLS mRNA specified by this mutant, the 2+3 association would form as soon as sequence 3 was synthesized resulting in "instant" induction.

What would happen to the conformation of the control region if the inductive stimulus were removed? Insofar as association of 1+4 is formally possible according to our sequence determination studies, we propose that following removal of the inductive stimulus, the control region reassociates to form a 2+3, 1+4 system which would mask the 29K protein ribosome loading site. This pattern of deactivation would be preferred since it brings the control region to an energetically similar ground state as the nascent mRNA without need for the activation energy required to disrupt 2+3 before realignment to the nascent 1+2,3+4 association pattern.

An additional feature of regulation of MLS resistance was reported by Shivakumar et al (16) who noted hyperinducibility of 29K protein synthesis if the determinant was rendered defective by cleavage, partial resection, and religation at the unique HaeIII site within the 29K structural gene. This suggests that the 29K protein might function as a negative regulatory element by repressing its own synthesis, possibly at the mRNA level, by binding to the control region and preventing the unmasking reaction. This observation suggests the possibility that part of the MLS control region may provide a binding site for the 29K protein acting as an autorepressor. A potential binding site for this reaction might be in the neighborhood of the 3+4 sequence since the loop formed by this association contains the tetranucleotide sequence AAAG shown by Lai et al. (12) to contain the methylatable adenine in S.aureus 23S rRNA and complex formation with 3+4 could stabilize the conformation which sequesters SD-2. A key feature of the model is the fact that induction is only mediated by sensitive ribosomes.

A direct demonstration of the requirement for sensitive ribosomes was reported by Shivakumar et al. (16) who compared inducibility of 29K protein synthesis in a B.subtilis wild type background with synthesis observed in a chromosomal mutant of B.subtilis obtained by selection in the laboratory using oleandomycin. In this latter strain, the mutation was of the type which affects the primary structure of one of the ribosomal of the 50S subunit.

A complementary approach which demonstrated the requirement for an inhibited ribosome in the induction process was based on the work of Pestka et al. (14) who tested a set of erythromycin analog compounds for ability to inhibit ribosome function and to induce MLS resistance. From these studies it was concluded that induction potency was inseparable from ability to inhibit ribosome function. In relation to the hyperinducibility of synthesis of defective 29K protein reported by Shivakumar et al. (16), it may be sufficient that the cells carrying the defective determinant are maximally inducible, since their ribosomes would be maximally sensitive.

III. MLS RESISTANCE IN OTHER ORGANISMS

The inducible MLS resistance determinant from *S.aureus* has served as the model system for our studies of the detailed mechanism of induction. Inducible MLS resistance occurs in natural isolates of a much wider range of organisms such as Bacilli (1), Streptococci (19), and Streptomycetes (2,3). It is in this last group that we have found the most interesting deviations from the *S. aureus* model system (17). In *Streptomyces viridochromogenes* and *Streptomyces lincolnensis*, we have described an inducible adenine monomethylation associated with MLS resistance, whereas a combination of adenine mono- and di-methylation appears to mediate MLS resistance in *Streptomyces fradiae* and *Streptomyces hygroscopicus*. The fact that erythromycin does not induce optimally in *Streptomyces* constitutes additional deviation from the *S.aureus* model system. In *Streptomyces viridochromogenes*, e.g., tylosin effectively induces resistance to erythromycin. By a combination of further recombinant- and DNA sequence studies, we hope to establish some of the rules which govern specificity of induction of these apparently deviant MLS resistance phenotypes which in all probability constitute specific examples of the MLS resistance phenotype in its most general form.

ACKNOWLEDGEMENTS

This work was supported by research grants from The National Science Foundation and from the research foundations of The Upjohn Company and Eli Lilly and Company.

REFERENCES

1. Docherty,A., Grandi,G., Grandi,R., Gryczan,T.J., Shivakumar,A.G., and Dubnau,D.(1981) J.Bacteriol. 145, 129-137.

2. Fujisawa,Y., and Weisblum,B.(1981) J.Bacteriol. 148,621-631.
3. Graham,M.Y., and Weisblum,B.(1979) J.Bacteriol. 137,1464-1467.
4. Gryczan, T.J., and Dubnau, D. (1978) Proc. Natl. Acad. Sci. USA 75,1428-1432.
5. Gryczan,T.J., Shivakumar,A.G., and Dubnau,D. (1980) J.Bacteriol. 141,246-253.
6. Gryczan,T.J., Grandi,V., Hahn,J., and Dubnau,D. (1980) Nucleic Acids Res. 8,6081-6096.
7. Horinouchi, S., and Weisblum, B. (1980) Proc. Natl. Acad. Sci. USA 77,7079-7083.
8. Horinouchi,S., and Weisblum,B. (1981) Mol. Gen. Genet. 182,341-348.
9. Horinouchi,S., and Weisblum,B. (1982) J.Bacteriol. in press.
10. Iordanescu, S., Surdeanu, M., Della Latta, P., and Novick, R. (1978) Plasmid 1,468-479.
11. Iordanescu, S., and Surdeanu, M. (1980) Plasmid 4, 256-260.
12. Lai,C.-J., Dahlberg, J.E., and Weisblum, B. (1973) Biochemistry 12,457-463.
13. Lai,C.-J. and Weisblum, B. (1971) Proc. Natl. Acad. Sci. USA 68,856-860.
14. Pestka,S., Vince,R., LeMahieu,R., Weiss,F., Fern.L., and Unowsky,J. (1976) Antimicrob. Agents Chemother. 9, 128-130.
15. Shivakumar, A.G., Hahn,J., and Dubnau,D. (1979) Plasmid 2,279-289.
16. Shivakumar, A.G., Hahn,J., Grandi,G., Kozlov,Y., and Dubnau,D. (1980) Proc. Natl. Acad. Sci. USA 77,3903-3907.
17. Weisblum, B., Siddhikol,C., Lai, C.-J., and Demohn, V. (1971) J. Bacteriol. 106,835-847.
18. Weisblum, B., Graham, M.Y., Gryczan, T., and Dubnau,D. (1979) J.Bacteriol. 137,635-643.
19. Yagi,Y., McLellan,T.S., Frez,W.A., and Clewell,D.B. (1978) Antimicrob. Agents Chemother. 7,871-873.
20. Yanofsky, C. (1981) Nature 289,751-758.

MOLECULAR AND FUNCTIONAL ANALYSIS OF THE BROAD HOST RANGE PLASMID RSF1010 AND CONSTRUCTION OF VECTORS FOR GENE CLONING IN GRAM-NEGATIVE BACTERIA[1]

M. Bagdasarian, M.M. Bagdasarian, R. Lurz, A. Nordheim[2], J. Frey[3] and K.N. Timmis[4]

Max-Planck-Institut für Molekulare Genetik Berlin, W. Germany

In vitro recombination and gene cloning is especially useful for the analysis of bacteria that have not been extensively characterized genetically and that are not closely related to *Escherichia coli*, such as e.g. soil bacteria. Unfortunately the highly developed host vector systems of *E.coli* K-12 are of limited use in those experiments, in which the cloned genes must be expressed in the new host. Thus, e.g. genes encoding the enzymes of the toluene/xylene degradation pathway are expressed poorly in *E.coli* and complementation studies cannot be carried out, in this host, by following the growth on hydrocarbon substrates that are intermediates of the pathway (1,2). Moreover, the majority of the versatile cloning vectors of *E. coli* are narrow host range plasmids and cannot be introduced into soil bacteria such as e.g. pseudomonads (1,3). On the other hand, some plasmids of gram-negative bacteria exhibit wide host range specificity. They are particularly attractive for the use as replicons in the construction of vectors that should enable gene cloning in many gram-negative bacteria (1).

Plasmids that are appropriate for the construction of cloning vectors are in general small, multicopy replicons, well-characterized structurally and genetically. Indeed, the ex-

[1] *This work was supported by a grant from the Deutsche Forschungsgemeinschaft.*

[2] *Present address: Massachusetts Institute of Technology, Cambridge, Mass./USA.*

[3] *Present address: Dépt. de Biochimie, Université de Genève, Switzerland.*

[4] *Present address: Dépt. de Biochimie Médicale, Université de Genève, Switzerland.*

treme usefulness of the *E.coli* vector pBR322 largely depends on the availability of the nucleotide sequence of the entire plasmid.

In this report we present the structural and functional analysis of the multicopy, broad host range plasmid RSF1010 that is being used as a basis for construction of vectors for gene cloning in gram-negative bacteria.

I. RESTRICTION ENDONUCLEASE CLEAVAGE MAP OF RSF1010

The length of the RSF1010 molecule has been estimated by electron microscopy at 8.5 kb (4,5). Our measurements have given a length of 8.9 kb (6) and this value has been used in the present article.

A restriction endonuclease cleavage map of the plasmid was constructed by agarose gel electrophoresis analysis of the digestion products of purified RSF1010 DNA with a range of individual restriction enzymes and combinations thereof. As shown in Fig.1, RSF1010 molecule has single cleavage sites for *Eco*RI, *Bst*EII, *Hpa*I, *Sst*I and *Pvu*II. It has two *Pst*I sites that are separated by a DNA segment of 0.75 kb and two *Acc*I sites, separated by a segment of 2.0 kb in length. The RSF1010 plasmid contains no cleavage sites for the *Bam*HI, *Bgl*II, *Hin*dIII, *Kpn*I, *Sal*I, *Xba*I, and *Xma*I endonucleases.

II. PLASMID FUNCTIONS ENCODED ON RSF1010

A. Antibiotic Resistance Genes

RSF1010 specifies the resistances to sulphonamide (Su) and streptomycin (Sm). The locations of the antibiotic resistance genes on the genetic map of RSF1010 were determined by Rubens et al. (7) and by our *in vitro* recombination experiments (6). The results of insertional mutagenesis with Tn*3* (7) and of cloning of different DNA fragments between the two *Pst*I sites (6), at the coordinates 7.9 - 8.7 kb (Fig.1), have indicated that the two resistance genes are transcribed as a unit in the direction from Su^r to Sm^r. The promoter for this transcription unit seems to be located at the coordinate of 7.9 kb [Fig.1; see also (6)].

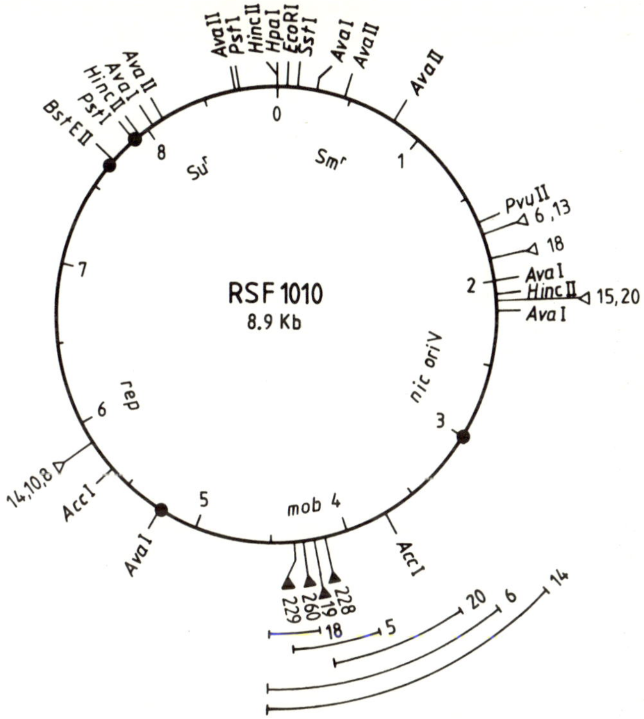

Fig. 1. Physical and genetic map of RSF1010.

Abbreviations: Su^r, Sm^r, resistance to sulphonamide and streptomycin, respectively; mob, nic, oriV, rep, determinants for plasmid mobilization, relaxation nick site, origin of vegetative replication and a positive replication factor, respectively; triangles represent the insertions of Tn3 transposon; filled triangles are the insertions of Tn3 that result in mob^- phenotype, the lines outside of the circle represent the deletions of the mob region of the plasmid; filled circles are RNA polymerase binding sites (6).

B. Plasmid Replication Determinants

Electron microscopic studies of the replicating molecules of RSF1010 have shown that the replication of this plasmid in *E. coli* starts from the origin located at the coordinates of 2.6 kb (Fig.1) and proceeds unidirectionally or bidirectionally with equal frequencies (8).

The attempt to reduce the size of RSF1010 and obtain minimal size replicons from it by using *in vivo* or *in vitro* deletion techniques were unsuccessful (6). This had suggested that essential replication and maintenance determinants of RSF1010, like those of other broad host range plasmids, such as RP4/RK2 (9) and TOL (2), are not clustered in a small segment of the plasmid, but are scattered over a large region of the genome. To find the essential regions of the RSF1010 genome, we have cloned separately different restriction fragments of RSF1010 DNA and tried to complement the functions required to replicate a small circular DNA fragment that contained the origin of RSF1010. The plasmid pMMB2 was constructed by cloning the 5.1 kb *Sst*I fragment from the plasmid pKT228 (RSF1010, carrying a Tn*3* insertion at the 4.2 kb coordinate; Fig.1) into the vector pKT101 (10). The plasmid pMMB2 contains RSF1010 sequences from 4.2 to 8.9 kb and 0.27 kb of the Tn*3* sequence (Fig.1 and 2). The strains of *E.coli* C600 and C600(pMMB2) were then used as recipients for the plasmids obtained by digestion of RSF1010 DNA with the restriction endonuclease *Hae*II and ligation of the fragments in the presence of the *Hae*II DNA fragments carrying ampicillin resistance (Ap^r) or chloramphenicol resistance (Cm^r) genes from the temperature-sensitive plasmid pHSG415 (11). When the selection for Ap^r or Cm^r was performed at 41°C (non-permissive temperature for the pHSG415 replicon) no transformants were obtained with the strain C600. Several transformants have appeared, however, when the strain C600(pMMB2) was used as the recipient. All these transformants contained, in addition to plasmid pMMB2, smaller plasmids. The DNA of the smallest of these plasmids, pMMB4, was purified from an agarose gel and was shown to consist of three *Hae*II fragments, a 2.4 kb Ap^r fragment of pHSG415 and of the two biggest (1.2 kb and 0.8 kb) *Hae*II fragments of RSF1010 (Fig.2). The location of these two contiguous *Hae*II fragments on the genetic map of RSF1010 indicates that they span the origin of replication of the plasmid. The DNA of pMMB4, purified from the agarose gel, could be transformed into the strain C600(pMMB2) but not into the C600 strain. These results indicate that a gene, which is essential for the replication, and codes for a factor, that may be supplied *in trans* to the replication origin, is located between the coordinates 4.2 and 8.9 kb on the genetic map of RSF1010 (Fig.1). It is separated from the origin of replication by a segment of more than 1.6 kb that contains the mobilization determinants, *mob* and *nic* (Fig.1), which are not essential for vegetative replication of the plasmid. The exact location of this replication determinant, that we propose to call *rep* (Fig.1), is not known at present. It is also not known whether one or more trans-acting *rep* factors are encoded between 4.2 and 8.9 kb coordinates of RSF1010. It is clear, however, that the location of replication determinants in RSF1010 follows the same pattern as in other broad host range plasmids. The essential replication genes are scattered in a relatively large portion of the genome and are separated by regions carrying non-essential genes.

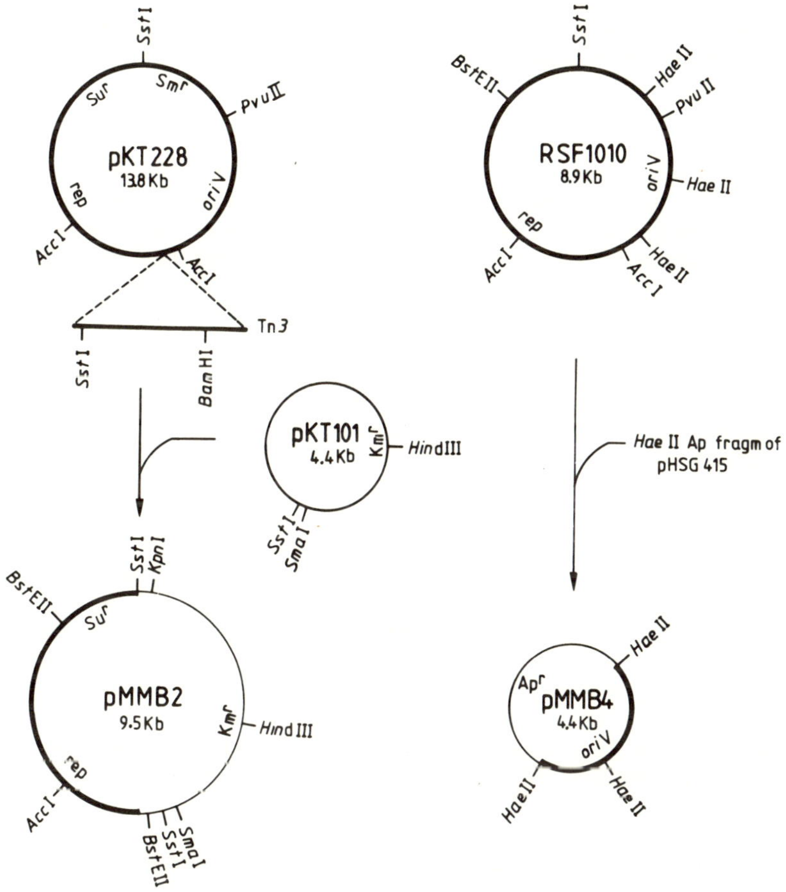

Fig. 2. Diagram showing the construction of plasmids containing oriV and rep of RSF1010. The position of the two biggest HaeII fragments of RSF1010 is shown. The location of the other 15 to 17 HaeII recognition sites on the RSF1010 molecule is not known.

C. Mobilization Functions

RSF1010 is not self-transferable in conjugation. It may be transferred, however, among different bacterial species if fertility functions are provided by a co-existing (conjugative) plasmid, such as the F factor, RP4 or R64 (5,6,12).

In the case of plasmids ColE1 (13) and RK2 (14) it has been

suggested that the origin of conjugal transfer is identical with the "relaxation nick" site, a strand break introduced by protein-denaturing agents at a unique location of the molecule in one specific DNA strand. It is possible to isolate RSF1010 molecules from the host cells in the form of supercoiled DNA-protein "relaxation complexes" that may be converted to open circular, "relaxed", forms by *in vitro* treatment with sodium dodecyl sulphate and proteases (15).

To locate the "relaxation nick" site (*nic*) on the physical map RSF1010, plasmid molecules, relaxed *in vitro*, were labeled by limited polynucleotide chain extension using the free 3'-OH at the *nic* as a primer for DNA polymerase I and [α-^{32}P]dATP as one of the substrate triphosphates. Digestion of the *nic*-labeled plasmid molecules with various restriction endonucleases and separation of the fragments by agarose gel electrophoresis followed by autoradiography showed that the *nic* site was located on the second biggest (0.85 kb) *Hae*II-generated fragment of RSF1010, in the close proximity to the origin of vegetative replication (15; Fig.1 and 2).

Another determinant that is essential for plasmid mobilization, *mob*, has been mapped by Tn*3* transposon mutagenesis. The examination of about 500 mutants, carrying each a Tn*3* insertion in RSF1010, has shown that the insertions clustered between the coordinates 4.1 to 4.3 kb (Fig.1) lower the ability of the plasmid to be mobilized in conjugation by about 3 to 4 orders of magnitude. The *mob* function is either plasmid-specific or *cis*-acting since it cannot be complemented by either RP4 or R64*drd*11. It is also distinct from the *nic* site since, as shown by the results of deletion mapping described below, it is located outside of the 0.85 kb *Hae*II fragment.

The deletions of *mob* region of RSF1010 were obtained in the following way. The DNA of the plasmid pKT260 (RSF1010, carrying a Tn*3* insertion in the *mob* gene at 4.3 kb, oriented opposite to the insertion in pKT228; Fig.2) was linearized by digestion with *Bam*HI at the single *Bam*HI recognition site, located in the Tn*3*. The linear molecules were digested with the exonuclease *Bal*31 for different lengths of time (16). The shortened molecules were circularized by blunt-end ligation and transformed into the *E.coli* strain C600 selecting for Sm^r. The colonies that have lost the resistance to ampicillin were screened for the size of the plasmids they carried by single-colony agarose gel electrophoresis (17). Five plasmids that were smaller than the parental RSF1010 were detected. Their DNA was purified and the extent of deletions estimated by electron microscopic analysis of heteroduplexes, formed with RSF1010 DNA, and by restriction endonuclease digestion. Surprisingly, all five deletions were asymmetrical with respect to the *Bam*HI site of the Tn*3* insertion. None of the deletions extended over the 4.5 kb coordinate of RSF1010 (Fig.1) indicating that a gene essential for replication or maintenance of the plasmid may be located in this region. Deletions 5 and 8 mapped entirely outside of the 0.85 kb *Hae*II fragment while the other extended into this fragment. It must be concluded therefore that the Tn*3* insertions that affect the *mob* gene are also located outside of the 0.85 kb *Hae*II fragment that carries the nick site and that *mob* provides a function distinct from the *nic* function.

III. CLONING VECTOR DERIVATIVES OF RSF1010

As a small, multicopy plasmid that may be easily purified from either *E.coli* or *Pseudomonas* (3,6), RSF1010 is particularly suitable for construction of cloning vectors. Unfortunately, its molecule has few restriction sites that can be used for cloning of DNA segments (Fig.1). To provide RSF1010 with more cloning sites and additional markers that can be used for selection, we have incorporated several different DNA fragments that carry genes specifying the resistance to tetracyline (Tc), chloramphenicol (Cm), kanamycin (Km) or ampicillin (Ap).

The derivatives of RSF1010 containing the Tc^r gene of the plasmid pSC101 (18,19) constructed previously (3) were found to be of limited use as cloning vectors since they were not stably inherited in the absence of selection not only in *Pseudomonas* but also in *E.coli* [M.Bagdasarian, unpublished experiments; see also (20)]. It is not known at present whether this instability resulted from the loss of the Tc^r gene from the plasmid or from plasmid loss from the cell. The reason for the instability is presumably the change in the cell membrane caused by the incorporation of the Tc^r gene protein (see also 21). On the other hand, recombinant plasmids consisting of the RSF 1010 replicon and carrying Cm^r, Km^r or Ap^r genes were stably maintained in *E.coli*, *P.aeruginosa* and *P.putida*. The vector pKT210 was constructed by replacement of the small *Pst*I fragment of RSF1010 by a 3.5 kb *Pst*I fragment of plasmid S-a (22) that carries Cm^r gene that had been previously cloned in the pBR322 vector to form hybrid plasmid pKT205 (3). The plasmid pKT210 does not express resistance to sulphonamide, indicating that at least part of the Su^r determinant is located on the small *Pst*I fragment of RSF1010 (see paragraph on Antibiotic Resistance Genes). Plasmid pKT210 contains single restriction endonuclease sensitive sites for *Eco*RI, *Sst*I and *Hin*dIII. A vector analogous to pKT210 was constructed by insertion into the *Pst*I site of RSF1010 at coordinate 7.8 kb, of a 4.8 kb *Pst*I fragment carrying the Cm^r determinant of the plasmid 621a1a (23). This vector pKT248 contains a single cleavage site for the *Sal*I endonuclease within its Cm^r determinant (Fig.3). The Cm^r determinant of pKT248 is, as far as we know, the only gene, other than the Tc^r gene of the pSC101, in which cloning of DNA fragments into the *Sal*I site results in the insertional inactivation of the gene.

To obtain vectors that enable cloning of the *Hin*dIII, *Xma*I, and *Xho*I generated fragments, the Km^r gene of the Tn*601*/Tn*903* was incorporated in the molecule of RSF1010. The pKT231 vector is a derivative in which the small *Pst*I fragment of RSF1010 has been replaced by two *Pst*I fragments (3.8 kb and 1.0 kb) of a miniplasmid, obtained previously from R6-5 (pKT105; K.N.Timmis, unpublished). The 3.8 kb fragment carries the Km^r gene of Tn*601*/Tn*903*, the 1.0 kb fragment — a promoter that allows the expression of the Sm^r gene [Fig.4; see also (6)]. This vector may be used for cloning of DNA fragments generated by *Hin*dIII, *Xma*I, *Xho*I, *Cla*I, *Eco*RI and *Sst*I. Hybrid plasmids containing insertions in all of these sites, except for *Cla*I, may be detected by insertional inactivation of either Km^r or Sm^r gene (Fig.4; Table 1). A vector similar to pKT231 was obtained by the replacement of the small *Pst*I generated fragment of RSF1010 with the plasmid pACYC177 (24). This vector, pKT230, contains, in addition to most of the cloning sites present in pKT231, a single *Bam*HI cleavage site, but lacks a *Cla*I site (Fig.4). The derivative pKT230 is a double-replicon that in *Pseudomonas* replicates most probably by using RSF1010-specific mechanisms since, as shown previously, neither pACYC177 nor pACYC184 replicon can function in *Pseudomonas* bacteria (1,3). The double-replicon character of pKT230 does not influence its vector properties. It has been used for cloning of DNA fragments of *Pseudomonas* plasmids and active expression of genes, coding for hycrocarbon degradation, was detected in *Pseudomonas* and *E.coli* (2).

For cloning of relatively large (30 - 40 kb) DNA fragments, useful in construction of gene banks of an entire genome, plasmid vectors containing λ*cos* site were used with considerable success. These "cosmid" vectors may be, after ligation with DNA fragments of appropriate size, packaged *in vitro* into λ bacteriophage heads and introduced into recipient bacteria by infection (25,26). Since no *in vitro* packaging system, comparable to that of the bacteriophage λ, exists for any of the *Pseudomonas* phages, it is currently not possible to use cosmid cloning directly in *Pseudomonas*. We have therefore constructed a cosmid derivative of RSF1010 that may be, after packaging into λ heads used to infect *E.coli* strains (Fig.4). These serve as intermediate hosts prior to the transfer of hybrid plasmids into the

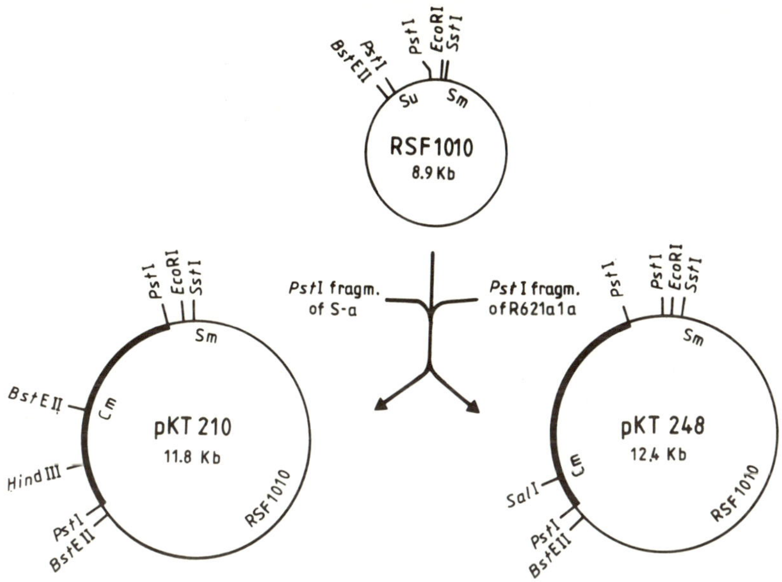

Fig. 3. Schematic representation of the construction of vectors derived from RSF1010 by insertion of chloramphenicol resistance determinant (Cm) (see Table 1).

desired new host system.

The ability of RSF1010 to be mobilized in conjugation by helper plasmids is a distinct advantage in those experiments in which no good transformation method exists for the new host system. Certain cloning experiments, however, are subject to recombinant DNA regulations and must be carried out with certified host:vector systems. One essential feature of the vector component of such system is a low frequency of conjugal transfer or co-transfer (mobilization) with conjugative plasmids. To satisfy these regulatory requirements, we have constructed a series of mobilization-defective vectors, analogous to the RSF 1010 derived vectors, described above, by using the Mob$^-$ deletion plasmid, pKT261 (deletion number 18 of RSF1010; Fig.1; see section on Mobilization Functions). The Mob$^-$ vectors are mobilized by RP4 plasmid with frequencies that are 5 to 6 orders of magnitude lower than those of the corresponding Mob$^+$ derivatives. The mobilization-defective RSF1010 vectors thus exhibit a degree of biological containment that is similar to current EK1 cloning vectors (6). Their basic properties are presented in Fig.3 and 4 and summarized in Table 1.

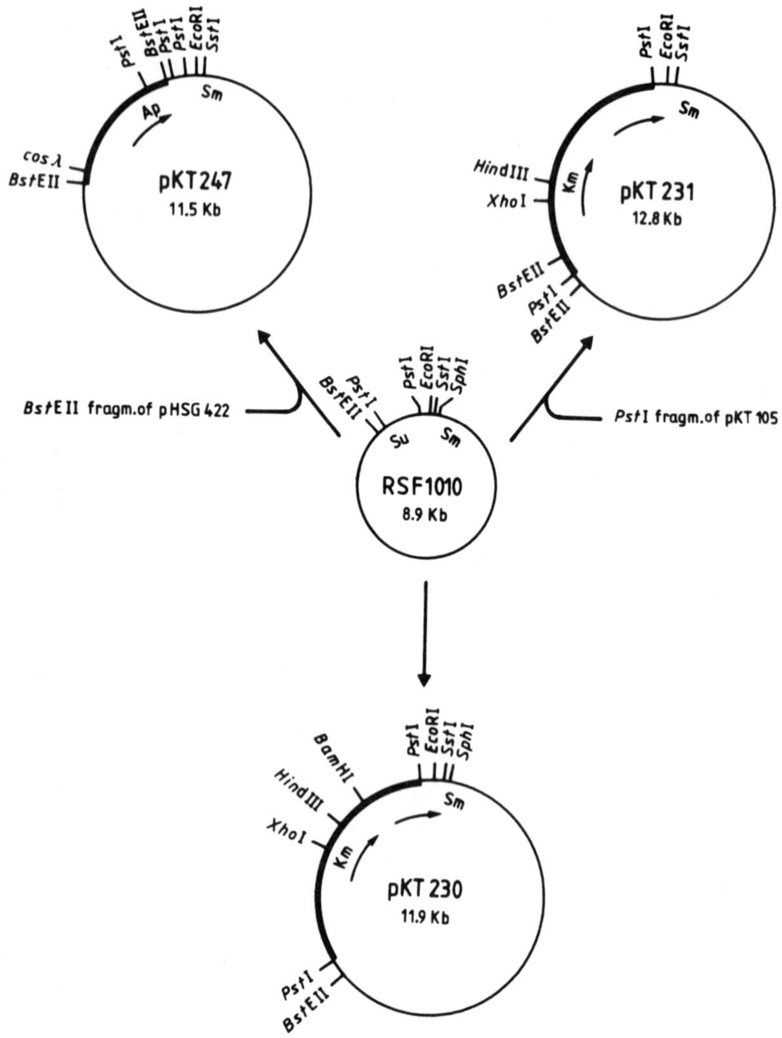

Fig. 4. Schematic representation of the construction of vectors derived from RSF1010 by insertion of kanamycin (Km) and ampicillin (Ap) resistance determinants and of the bacteriophage λcos site. The arrows inside the circles indicate the direction of transcription (see Table 1).

Table 1. Properties of RSF1010 derived vector plasmids for cloning in *Pseudomonas*

Vector	Size (kb)	Copy No.[a]	Cloning sites	Primary selection[b]	Insertional inactivation[b]	Remarks[b]
pKT210	11.8	15-20	*Eco*RI	Cm	Sm	Mob^+
			*Sst*I	Cm	Sm	
			*Hin*dIII	Cm, Sm	—	
pKT248	12.4	15-20	*Eco*RI	Cm	Sm	Mob^+
			*Sst*I	Cm	Sm	
			*Sal*I	Sm	Cm	
pKT230	11.9	15-20	*Eco*RI	Km	Sm	Mob^+
			*Sst*I	Km	Sm	
			*Hin*dIII	Sm	Km	
			*Xho*I	Sm	Km	
			*Xma*I	Sm	Km	
			*Bst*EII	Km, Sm	—	
			*Bam*HI	Km, Sm	—	
pKT231	13.0	15-20	*Eco*RI	Km	Sm	Mob^+
			*Sst*I	Km	Sm	
			*Hin*dIII	Sm	Km	
			*Xho*I	Sm	Km	
			*Xma*I	Sm	Km	
			*Cla*I	Km, Sm	—	
pKT247	11.5	15-20	*Eco*RI	Ap	Sm	Mob^+
			*Sst*I	Ap	Sm	cosmid
pKT262	11.7		deletion mutant of pKT230[c]			Mob^-
pKT263	12.8		deletion mutant of pKT231			Mob^-
pKT264	11.3		deletion mutant of pKT247			Mob^-

IV. CONCLUDING REMARKS

Plasmids that exhibit broad host range specificity have a significant practical as well as theoretical importance. The practical importance lies in the possibility for their use in the construction of vectors that enable gene cloning in several different bacterial species. This provides an *in vitro* gene manipulation system for studying of biological functions that are incompatible with the metabolic activities of *E.coli* (see also 1, 6). RSF1010-derived vectors described in this paper are able to replicate in a particularly wide range of Gram-negative bacteria including *E.coli*, *P.aeruginosa*, *P.putida*, *Azotobacter vinelandii* (27), *Methylophilus methylotrophus* (28), *Rhizobium meliloti* (M.David, personal communication), *Agrobacterium tumefaciens* (M. van Montagu, personal communication), *Acetobacter xylinum* (S.Valla, personal communication), *Alcaligenes eutrophus* (F.C.H.Franklin, personal communication) and *Rhodopseudomonas spheroides* (W.T.Tucker, personal communication). These vectors are useful therefore for both self-cloning experiments in all strains of these bacteria that can be transformed with plasmid DNA and for two-stage cloning, i.e. primary cloning in a convenient intermediate host, such as *E.coli*, followed by transfer of hybrid plasmids to a second host strain that is more appropriate for analyzing of the biochemical activities of the cloned genes.

Although the broad host range vectors currently available for gene cloning in *Pseudomonas* and other soil bacteria have by no means the sophistication of the highly developed *E.coli* vectors, they are nevertheless of considerable utility as evidenced by their use in the analysis of the genes of the toluene/xylene degradation pathway encoded on the TOL plasmid of *P.putida* (2).

Further improvement of the broad host range vector systems depends largely upon the expansion of our basic knowledge about the genetic and physical organization of the plasmids used as the replicons for construction of such cloning vehicles. Recent

---

[a] *Number of copies/genome equivalent*

[b] *Abbreviations: Ap, ampicillin-; Cm, chloramphenicol-; Km, kanamycin-; Sm, streptomycin-resistance; Mob^{+}/Mob^{-}, ability or inability, respectively, to be mobilized by a coexisting conjugative plasmid to other bacteria*

[c] *All Mob^{-} plasmids were derived from a single Mob^{-} deletion of RSF1010 - pKT260*

in vivo and *in vitro* studies have indicated that significant differences exist in the replication mechanisms of different plasmids. Thus plasmids of ColE1 type replicate in the absence of plasmid-coded factors (for discussion see 29, 30). Narrow host range plasmids of the FII incompatibility group require the expression of plasmid genes for their replication and these genes are located in the immediate proximity of the appropriate replication origins (31-33). The broad host range plasmid RK2 requires, for its replication, at least one plasmid-coded function, other than the origin, and the gene for this function is located a considerable distance away from the replication origin (34). It is presumed that this genetic organization may have some significance for the physiology of the broad host range mode of replication of RK2 (9).

It has been shown in this report that the essential replication genes of RSF1010 — a plasmid with extremely wide host range — are also scattered over a large portion of the genome. In the case of a third broad host range plasmid, TOL, it has been found that its minimal replicon cannot be reduced, by simple deletion, to a size smaller than about 50 kb (F.C.H.Franklin and M.Bagdasarian, unpublished). It is possible therefore that the location of essential replication genes in a scattered, as opposed to clustered, organization has indeed certain advantages for the broad host range mode of replication. RSF1010 provides a good model for studying of the molecular mechanisms of replication, inheritance and maintenance of the plasmids in relation to their ability to replicate in different host bacteria. The results of recent experiments that enabled the cloning of essential determinants of RSF1010 on different plasmids and the availability of the *in vitro* systems for the replication of RSF1010 DNA, derived from both *E.coli* and *P.aeruginosa* (35) should be helpful in further elucidation of the regulatory mechanisms involved in the replication of broad host range plasmids.

REFERENCES

1. Bagdasarian, M. & Timmis, K.N. (1981) in Current Topics in Microbiol. and Immunol., Vol.96, pp.47-67, Springer-Verlag, Berlin, Heidelberg, New York
2. Franklin, F.C.H., Bagdasarian, M., Bagdasarian, M.M. & Timmis, K.N. (1981) Proc. Natl. Acad. Sci. USA 78, in press
3. Bagdasarian, M., Bagdasarian, M.M., Coleman, S. & Timmis, K.N. (1979) in Plasmids of Medical, Environmental and Commercial Importance (Timmis,K.N. & Pühler,A., eds.), pp.411-422, Elsevier/North Holland, Amsterdam

4. Guerry, P., van Embden, J. & Falkow, S. (1974) J. Bacteriol. 117, 619-630
5. Barth, P.T. & Grinter, N.J. (1974) J. Bacteriol. 120, 618-630
6. Bagdasarian, M., Lurz, R., Rückert, B., Franklin, F.C.H., Bagdasarian, M.M., Frey, J. & Timmis, K.N. (1981) Gene 16, 237-247
7. Rubens, C., Heffron, F. & Falkow, S. (1976) J. Bacteriol. 128, 425-434
8. De Graaff, J., Crosa, J.H., Heffron, F. & Falkow, S. (1978) J. Bacteriol. 134, 1117-1122
9. Thomas, C.M. (1981) Plasmid 5, 10-19
10. Timmis, K.N., Bagdasarian, M., Brady, G. & Franklin, F.C.H. (1982) manuscript in preparation
11. Hashimoto-Gotoh, T., Franklin, F.C.H., Nordheim, A. & Timmis, K.N. (1981) Gene 16, 227-235
12. Willets, N. & Crowther, C. (1981) Genet. Res. Camb. 37, 311-316
13. Warren, G.J., Twigg, A.J. & Sherratt, D.J. (1978) Nature (London) 274, 259-261
14. Guiney, D.G. & Helinski, D.R. (1979) Molec. Gen. Genet. 176, 183-189
15. Nordheim, A., Hashimoto-Gotoh, T. & Timmis, K.N. (1980) J. Bacteriol. 144, 923-932
16. Legerski, R.J., Hodnett, J.L. & Gray, Jr., H.B. (1978) Nucl. Acids Res. 5, 1445-1464
17. Eckhardt, T. (1978) Plasmid 1, 584-588
18. Cohen, S.N. & Chang, A.C.Y. (1973) Proc. Natl. Acad. Sci. USA 70, 1293-1297
19. Cohen, S.N. & Chang, A.C.Y. (1977) J. Bacteriol. 132, 734-737
20. Wood, D.O., Hollinger, M.F. & Tindol, M.B. (1981) J. Bacteriol. 145, 1448-1451
21. Bochner, B.R., Huang, H.C., Scieven, G.L. & Ames, B.N. (1980) J. Bacteriol. 143, 926-933
22. Watanabe, T., Furuse, C. & Sakaizumi, S. (1968) J. Bacteriol. 96, 1791-1795
23. Hedges, R.W. (1974) J. Gen. Microbiol. 81, 171-181
24. Chang, A.C.Y. & Cohen, S.N. (1978) J. Bacteriol. 134, 1141-1156
25. Collins, J. & Hohn, B. (1978) Proc. Natl. Acad. Sci. USA 75, 4242-4246
26. Hohn, B. & Collins, J. (1980) Gene 11, 291-298
27. David, M., Tronchet, M. & Dénarié, J.L. (1981) J. Bacteriol. 146, 1154-1157
28. Windass, J.D., Worsey, M.J., Pioli, E.M., Pioli, D., Barth, P.T., Atherton, K.T., Dart, E.C., Byrom, D., Powell, K. & Senior, P.J. (1980) Nature 287, 396-401
29. Tomizawa, J. (1978) in DNA Synthesis: Present and Future

(Molineux,J. & Kohiyama,M., eds.) pp.797-826, Plenum Publishing Corp.
30. Staudenbauer, W.L. (1978) in Current Topics in Microbiol. and Immunol., Vol.83, pp.93-156, Springer-Verlag
31. Andres, I., Slocombe, P.M., Cabello, F., Timmis, J.K., Lurz, R., Burkardt, H.J. & Timmis, K.N. (1979) Molec. Gen. Genet. 168, 1-25
32. Miki, T., Easton, A.M. & Rownd, R.H. (1980) J. Bacteriol. 141, 87-99
33. Diaz, R., Nordström, K. & Staudenbauer, W.L. (1981) Nature 289, 326-328
34. Figurski, D.H. & Helinski, D.R. (1979) Proc. Natl. Acad. Sci. USA 76, 1648-1652
35. Diaz, R., Bagdasarian, M. & Staudenbauer, W.L., manuscript in preparation

GENETICS AND MOLECULAR BIOLOGY : PSEUDOMONAS

COMBINED ACTION OF MIDECAMYCIN (MACROLIDE ANTIBIOTIC) AND A CELL WALL-AFFECTING ANTIBIOTIC, CARBENICILLIN, FOSFOMYCIN, DIBEKACIN OR POLYMYXIN B ON *PSEUDOMONAS AERUGINOSA* *IN VITRO* AND *IN VIVO*

J. Yuzuru Homma, Takao Kasai*, Shiro Kanegasaki** and Toshio Tomita**

The Kitasato Institute, Minato-ku, Tokyo, Japan.
**Central Research Laboratories, Meiji Seika Kaisha, Ltd., Kohoku-ku, Yokohama, Japan.*
***Institute of Medical Sciences, University of Tokyo, Minato-ku, Tokyo, Japan.*

Yamamoto *et al.* (1) demonstrated that unstable L-forms of *Pseudomonas aeruginosa* were frequently isolated from clinical specimens of patients with chronic infections due to *P. aeruginosa* being treated with β-lactam antibiotics.

They also reported that antibiotics which are effective against *P. aeruginosa* became far less effective or completely ineffective against the L-forms, while macrolides which are usually ineffective against the parent bacillary forms became remarkably effective against the L-forms (2) and the spheroplasts (3). It is important to note that all antibiotics which are widely used for treatment of pseudomonal infection show loss of activity against the L-forms of *P. aeruginosa*.

Taking the above into consideration the combined action of a macrolide, midecamycin (MDM) or its derivative, 9,3"-di-O-acetylmidecamycin (MOM), and the cell wall-affecting antibiotics, polymyxin (PL) and dibekacin (DKB), and the peptidoglycan synthesis-inhibiting antibiotics, carbenicillin (CBPC) and fosfomycin (FOM), were examined both *in vitro* and *in vivo*.

SYNERGISTIC EFFECT OF A CELL WALL-AFFECTING ANTIBIOTIC AND A MACROLIDE ANTIBIOTIC MDM OR MOM, ON *P. AERUGINOSA IN VITRO*

Cells in logarithmic growth phase in culture medium, growth were received each of the cell wall-affecting antibiotics, PL, DKB, CBPC and FOM, and one or the other of MDM or MOM, at the various concentrations indicated in Fig. 1.

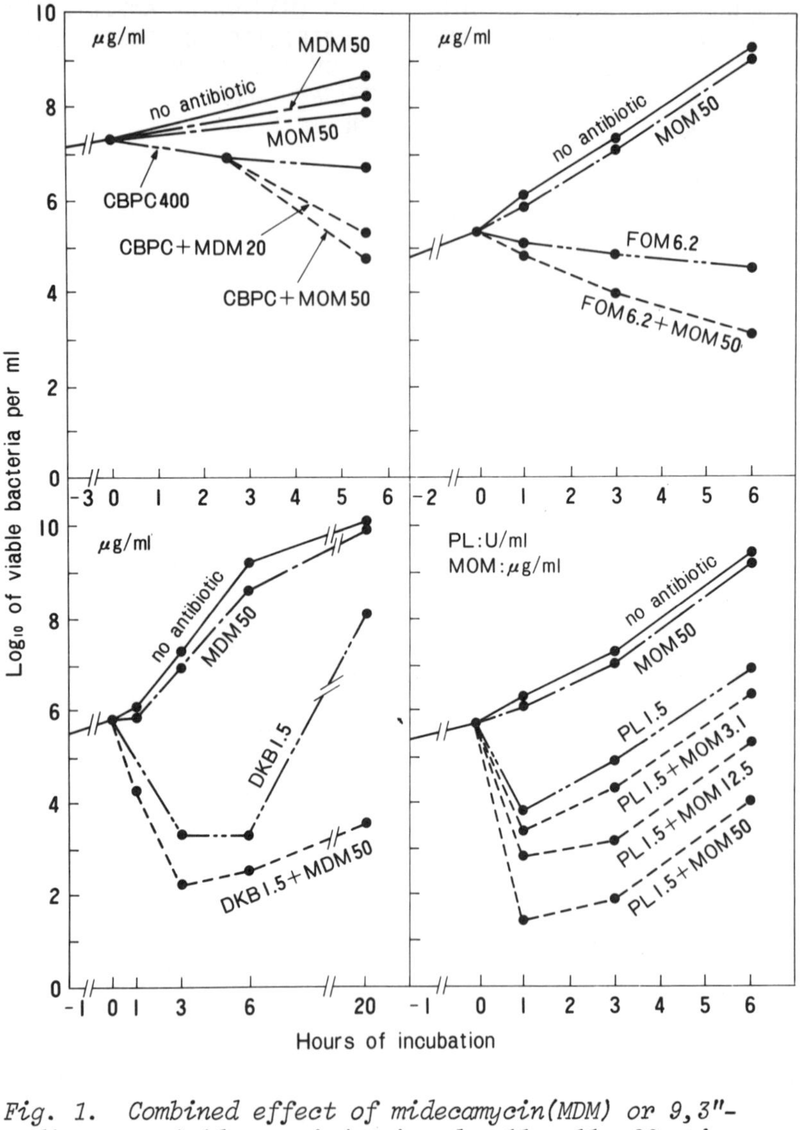

Fig. 1. Combined effect of midecamycin(MDM) or 9,3"-di-O-acetylmidecamycin(MOM) and cell wall-affecting antibiotics on P. aeruginosa

The number of viable cells was determined by estimating the number of colony forming units (CFU) on the L-form medium which provided sufficient osmotic support.

In the case of the macrolide antibiotics alone, no reduction in viable cells was observed, this was similar to the phenomenon observed in the control which had not been treated with any antibiotic. However, the number of viable organisms decreased remarkably upon the addition of a macrolide plus one of the other antibiotics in question. Synergistic bactericidal effect on *P. aeruginosa* was observed for all four antibiotics, PL, DKB, CBPC and FOM.

SYNERGISTIC EFFECT OF A CELL WALL-AFFECTING ANTIBIOTIC AND A MACROLIDE ANTIBIOTIC, MDM OR MOM AGAINST PSEUDOMONAL INFECTION *IN VIVO*

Five-week-old ddY and ICR mice were used. Each mouse infected by *P. aeruginosa* was given the cell wall-affecting antibiotic intramuscularly or subcutaneously and given the macrolide antibiotic, MDM or MOM orally.

In the case of single treatment by each one of the two kinds of antibiotics alone, all mice died within 1 or 2 days. In contrast, combined treatment using a macrolide plus one of the antibiotics, PL, DKB, CBPC or FOM prevented death in a high percentage of the infected mice.

All 17 experiments indicated that the combined treatment was statistically (Fisher's exact method) more effective than single treatment by only one of the two kinds of antibiotics. The doses and P values are indicated in Table 1.

These series of *in vitro* and *in vivo* experiments indicate that *P. aeruginosa* is affected by macrolide antibiotics when the outer membrane is injured by any agent or peptidoglygan synthesis is inhibited.

INCORPORATION OF ^{14}C-MOM INTO *P. AERUGINOSA* CELLS PRETREATED WITH THE CELL WALL-AFFECTING ANTIBIOTIC

Bacterial cells in L-forms medium to which was added one of the antibiotics, PL (50 U/ml), CBPC (500 μg/ml), DKB (3.13 μg/ml) or FOM (1000 μg/ml), were incubated at 37°C for 1 hr and then centrifuged. Under the conditions, no significant decrease of viable cells were observed when they were cultivated on the L-form medium. The precipitated bacterial cells were suspended in the same medium to yield 10-fold higher concentration. Then ^{14}C-MOM in the same medium was added to the concentrated bacterial cells and incubated at 37°C. Samples were taken from the culture at intervals. The bacteria were washed with 0.5 M sucrose solution by centrifugation and radioactivity estimated.

The bacterial cells treated with the antibiotics appeared to incorporate ^{14}C-MOM efficiently. Incorporating the

Table 1. Combined action of a macrolide and a cell wall-affecting antibiotic against P. aeruginosa infection

Expt.	Macrolide (mg/mouse)	A cell wall-affecting antibiotic (U or mg/mouse)		Times administered	P value
1	MOM 10 mg	PL	2000 U	1	0.070
2	MOM 2 mg	PL	500 U	5	0.042
3	MDM 2 mg	PL	500 U	5	0.015
"	MOM 2 mg	PL	500 U	7	0.024
4	MOM 10 mg	CL	4 mg	1	0.010
5	MOM 10 mg	CL	4 mg	1	0.034
6	MOM 4 mg	PL	500 U	4	0.00096
7	MOM 4 mg	PL	500 U	4	0.0027
8	MOM 0.0125 mg	FOM	0.2 mg	6	0.085
9	MOM 0.025 mg	FOM	0.2 mg	7	0.023
10	MOM 0.4 mg	FOM	0.1 mg	9	0.018
11	MDM 2 mg	DKB	0.1 mg	5	0.043
12	MDM 0.2 mg	DKB	0.1 mg	8	0.084
13	MDM 0.2 mg	DKB	0.1 mg	8	0.032
14	MDM 0.05 mg	DKB	0.1 mg	9	0.0006
15	MDM 10 mg	CBPC	40 mg	4	0.028
16	MDM 10 mg	CBPC	40 mg	8	0.0034
"	MOM 10 mg	CBPC	40 mg	8	0.012
17	MDM 10 mg	CBPC	40 mg	8	0.026

radioactivity increased 10-20 min after addition of ^{14}C-MOM and retained the same level for 60 min. As for FOM, incorporation of the radioactivity increased gradually after initial uptake. The bacterial cells which had not been treated with the cell wall-affecting antibiotics did not incorporate any ^{14}C-MOM.

DISTRIBUTION OF ^{14}C-MOM INCORPORATED INTO THE BACTERIAL CELLS

The bacterial cells which had incorporated ^{14}C-MOM were suspended in M/100 tris buffer (pH 8.0) to which an equal volume of lytic buffer (containing Mg DNase Brij 58) was added. The suppernatant of the lysate was subjected to sucrose density gradient centrifugation analysis. When the bacterium was treated with one of the four antibiotics, the radioactivity migrated as a single band, where absorption at 260 nm was maximum. The two peaks comigrated with the 70 S ribosome derived from *E. coli* used as a standard.

These results suggest that ^{14}C-MOM was incorporated into ribosome of *P. aeruginosa*.

To confirm if the incorporated ^{14}C-MOM actually binded to ribosome of *P. aeruginosa*, the following experiment was conducted.

The bacterial cells with ^{14}C-MOM were ground with alumina in a mortar. The supernatant obtained after 12,000 g centrifugation was laid on top of 1.2 M sucrose solution and centrifuged at 200,000 g for 4 hr. The isolated ribosome was divided into two parts, and subjected to two kinds of sucrose density gradient centrifugations: one in 0.4-1.3 M sucrose gradient solution containing 10 mM $MgCl_2$ and the other, in 0.15-0.6 M sucrose gradient solution containing 1 mM $MgCl_2$.

At high concentration of $MgCl_2$, radioactivity was found to correspond with the position of 70 S ribosome, while at low concentration of $MgCl_2$, radioactivity was found to correspond with the position of 50 S ribosome subunit and not with the position of 30 S ribosome subunit.

The results provide evidence that *P. aeruginosa* cells incorporate MDM or MOM when the bacterial cells are exposed to cell wall-affecting antibiotics such as CBPC, FOM, PL and DKB. Protein synthesis is thus inhibited by the macrolides because the macrolides bind 50 S ribosome subunit of *P. aeruginosa*. From these experiments the combined action of a macrolides and a cell-wall-affecting antibiotic has been proved *in vitro* as increasing bactericidal action as well as *in vivo* in increasing survival rates of mice infected with *P. aeruginosa*.

Chemotherapy against chronic infections due to *P. aeruginosa* must take into consideration the ecological fact that conversion of *P. aeruginosa* into L-forms is associated with remarkable alteration in the drug sensitivity of this organism.

REFERENCES

(1) Yamamoto, A. and Homma, J.Y. 1979. Isolation of Unstable L-forms from Clinical Specimens with *Pseudomonas* Infection during Antibiotic Therapy. *Jpn. J. Exp. Med., 49:* 361-364.

(2) Yamamoto, A. and Homma, J.Y. 1978. L-form of *Pseudomonas aeruginosa*. 2. Antibiotic Sensitivity of L-forms and Their Parent Forms. *Jpn. J. Exp. Med., 48:*355-362.

(3) Kawaharajo, K., Kasai, T. and Homma, J.Y. 1979. Combined Action of Semisynthetic Penicillin and Macrolide Antibiotic on *Pseudomonas aeruginosa in vitro* and *in vivo*. *Jpn. J. Exp. Med., 49:* 331-336.

CHROMOSOMAL LOCATION OF THE GENES PARTICIPATING IN THE FORMATION OF β-LACTAMASE IN *Pseudomonas aeruginosa*

Hideki Matsumoto and Yoshiro Terawaki

Department of Bacteriology
Shinshu University School of Medicine
Matsumoto, Japan

Pseudomonas aeruginosa produces an inducible β-lactamase that hydrolyses cephalosporines. The β-lactamase plays an important role, in association with the permeability barrier of the cell wall, on the resistance of this bacterium to β-lactam antibiotics(1, 2). Although the β-lactamase has been belived to be specified by the chromosomal genes, genetics of its synthesis is little known(3). We report here on the chromosomal location of the four *bla*(β-lactamase) genes in *P.aeruginosa* strain PAO.

Mutants producing the β-lactamase constitutively were tetected among mutagenized cells by emulsifying($10^9/ml$) the toluenized cells in 10mM of Tris-HCl buffer(pH 8.0 ; cephazolin, 1% ; phenol red, 0.01%). The constitutive mutants turned color of the solution from red to yellow within 1 min, whereas the inducible cells did not change the color for at least 15 min. Mutants that lost β-lactamase activity were obtained by looking for the clones with increased sensitivity to β-lactam antibiotics. Genetic methods were as previously described (4).

I. BIOCHEMICAL CHARACTERIZATION OF *bla* MUTANTS

Relative activity of the β-lactamase and minimum inhibitory concentration of the *bla* mutants are shown in Table 1.

II. MAPPING OF *bla* GENES

A. *Location of Genes Concerning with Induction System*

As shown in Table 2, existences of the three genes, *blaI*, *blaJ* and *blaK*, were become evident from the crosses between the constitutive and inducible strains. The *blaI* was mapped near *met-9011*, but no cotransduction occurred between the *met-9011* and *blaI* markers. Both *blaJ* and *blaK* were located between *strA* and *proA*.

B. *Location of Gene That Probably Coding for β-Lactamase*

Mutants lacking the β-lactamase activity(*blaP*⁻) were derived from various parents. All of the *blaP*⁻ were mapped between *strA* and *proA*(Table 3). It was impossible to select directly for the *blaP*⁺ transconjugants on the plates containing a β-lactam antibiotic, probably due to the inactivation of the drug by the donor.

Table 1. Relative Activity of β-Lactamase and Minimun Inhibitory Concentration of bla Mutants

Strain	*Mutation(s) in bla gene*[a]	*Relative β-lactamase activity on Cephaloridine*[b] *Uninduced*	*Induced*[c]	*MIC(μg/ml) to Cephaloridine*[d]
PAO1840	* *blaP*⁺	< 1	100	> 6,400
PAO4020	* *blaP*⁺	< 1	100	> 6,400
PAO4098	* *blaP*⁻	.	.	50
PAO4077	*blaI*ᶜ, *blaP*⁺	18	100	> 6,400
PAO4095	*blaI*ᶜ, *blaP*⁺	44	100	> 6,400
PAO4096	*blaI*ᶜ, *blaP*⁻	.	.	6.25
PAO4082	*blaJ*ᶜ, *blaP*⁺	28	100	> 6,400
PAO4089	*blaJ*ᶜ, *blaP*⁻	.	.	6.25
PAO4083	*blaK*ᶜ, *blaP*⁺	7	100	> 6,400

[a] ** indicates no mutation in the induction system ; Small letter c, affixed to blaI, blaJ or blaK indicates constitutive mutant allele ; + or -, affixed to blaP indicates β-lactamase producer of non-producer, respectively.*

[b] *Activity was assayed acidimetrically(5). Sonically disrupted cell extract was used as enzyme. Activity by induced cell was expressed as 100.*

[c] *Cells grown in the presence of 5mg/ml of benzylpenicillin.*

[d] *Tested on nutrient agar plate by inoculating saline suspension(10^8/ml) in exponential phase of growth.*

Table 2. Segregation of blaI, blaJ and blaK Markers in Conjugal Chromosome transfer

Cross and	*Genetic markers tested*[b]							
Selection[a]	*strA*	*blaJ*	*blaK*	*proA*	*hex-9001*	*leu-10*	*met-9011*	*blaI*
PA04077,FP5$^+$	*R*	*i*	*i*	+	+	+	+	*c*
PA01840	*S*	*i*	*i*	+	-	-	-	*i*
hex$^+$	6	.	.	.	100[c]	49[c]	49	39
leu$^+$	14	.	.	.	81	100	94	70
met$^+$	6	.	.	.	79	98	100	89
PA04020,FP5$^+$	*R*	*i*	*i*	+	+	+	+	*i*
PA04095	*S*	*i*	*i*	-	+	+	-	*c*
pro$^+$	31	.	.	100	.	.	2	2
met$^+$	5	.	.	24	.	.	100	92
PA04020,FP5$^+$	*R*	*i*	*i*	+	+	+	+	*i*
PA04082	*S*	*c*	*i*	-	+	+	-	*i*
str^r	100	89	.	75	.	.	5	.
pro$^+$	38	64	.	100	.	.	8	.
met$^+$	6	8	.	11	.	.	100	.
PA04020,FP5$^+$	*R*	*i*	*i*	+	+	+	+	*i*
PA04083	*S*	*i*	*c*	-	+	+	-	*i*
str^r	100	.	95	94	.	.	6	.
pro$^+$	35	.	91	100	.	.	1	.
met$^+$	21	.	27	36	.	.	100	.

[a]*The top was donor and the bottom was recipient ; Contraselected marker was hisI.*

[b]*str, streptomycin ; pro, proline ; hex, hexose utilization ; leu, leucine ; met, methionine ; R or S, resistant or sensitive ; i, inducible, wild type allele ; c, constitutive mutant allele ; +, wild type allele ; -, mutant allele.*

[c]*Number of transconjugants inheriting the donor's allele ; 100 means that selection was made for this marker.*

III. MODE OF THE β-LACTAMASE PRODUCTION BY *blaP*$^+$ RECOMBINANTS

Mode of the β-lactamase production by the *blaP*$^+$ transconjugants and transductants are presented in Tables 3 and 4, respectively. Results in the Tables also indicated the very close linkage of *blaJ* with *blaP*.

Table 3. Segregation of blaP Marker in Conjugal Chromosome Transfer and Mode of β-Lactamase Production

Cross and Selection[a]	Genetic markers tested[b]						Mode of β-lactamase production by $blaP^+$ Transconjugant	
	strA	blaP	blaJ	proA	met-9011	blaI	Cons(%)[d]	Indu(%)[d]
PAO4020,$FP5^+$	R	+	i	+	+	i		
PAO4096	S	-	i	-	-	c		
str^r	100[c]	88[c]	.	83	5	> 5[e]	8	92
pro^+	59	72	.	100	16	>16	24	76
met^+	41	43	.	44	100	>38	91	9
PAO4020,$FP5^+$	R	+	i	+	+	i		
PAO4089	S	-	c	-	-	i		
str^r	100	91	91	84	6	.	0	100
pro^+	48	86	86	100	2	.	0	100
met^+	32	37	37	39	100	.	0	100
PAO4020,$FP5^+$	R	+	i	+	+	i		
PAO4098	S	-	i	-	-	i		
str^r	100	85	.	81	7	.	0	100
pro^+	20	64	.	100	2	.	0	100

[a, b, c] See footnote of Table 2.

[d] Cons, constitutive ; Indu, inducible.

[e] Exact numbers could not be determined, since the transfer of blaI gene was unable to test in $blaP^-$ transconjugants.

IV. LOCATION OF *bla* GENES ON PAO LINKAGE MAP

Map position of the four *bla* genes with respect to their neighboring markers were summarized as follows ;

1) *FP5 — strA — (blaJ, blaP) (blaK) —— proA ——-*

2) *proA ——— hex-9001 — leu-10 — met-9011 — blaI ———*

Table 4. Mode of β-Lactamase Production by blaP+ Transductant

Cross Mutation(s) in bla gene[a] Donor		Recipient		Selected[b] for ;	Mode of β-lactamase production Cons(%)[d]	Induc(%)[d]
*	blaP+	*	blaP−	blaP+	0	100
blaI^c,	blaP+	*	blaP−	"	0	100
blaJ^c,	blaP+	*	blaP−	"	98	2
blaK^c,	blaP+	*	blaP−	"	0	100
*	blaP+	blaI^c,	blaP−	"	0	100
*	blaP+	blaJ^c,	blaP−	"	2	98
blaI^c,	blaP+	blaJ^c,	blaP−	"	4	96
blaJ^c,	blaP+	blaI^c,	blaP−	"	100	0
blaK^c,	blaP+	blaI^c,	blaP−	"	100	0
blaK^c,	blaP+	blaJ^c,	blaP−	"	1	99

[a]*See footnote of Table 1.*

[b]*Selected for resistance to 3,000 μg/ml of benzylpenicillin on nutrient agar plates ; The number of transductants tested ranged from 200 to 400. Phage F116 was used.*

[d]*See footnote of Table 3.*

We presume that the *blaP* is the structural gene for the β-lactamase. The *blaJ*, which is very close to *blaP*, would be the operator gene for the *blaP*. The *blaJ^c* mutation was *cis*-dominat(Matsumoto, unpublished). The *blaI* and *blaK* may be playing some roles on the regulation of the induction system. It would be worthwhile to clarify the function of each of the genes.

REFERENCES

1. Richmond, M.H. & Sykes, R.B. ( ]973) Adv. Microbiol. Physiol. *9, 31-88*
2. Furth, A. (1979) in Beta Lactamase (Hamilton-Miller, J.M.T. & Smith, J.T. eds.) Academic Press, London
3. Holloway, B.W., Krishnapillai, V. & Morgan, A.F. ( ]979) Microbiol. Rev. *43*, 419-443
4. Matsumoto, H., Ohta, S., Kobayashi, R. & Terawaki, Y. (1978) Mol. Gen. Genet. *167*, 165-176
5. Rubin, F.A. & Smith, D.H. (1973) Antimicrob. Ag. Chemother. *3*, 68-73

MAPPING OF PYOCIN GENES ON THE CHROMOSOME OF *PSEUDOMONAS AERUGINOSA* USING PLASMID R68.45

Tomoyuki Shinomiya, Yumiko Sano, Akihiko Kikuchi
and Makoto Kageyama

Mitsubishi-Kasei Institute of Life Sciences
Machida, Tokyo, Japan

Two basic types of pyocin, bacteriocins of *Pseudomonas aeruginosa*, have been described by the authors (1). One is a group of very large particles having structures similar to some bacteriophage tails. The genetic determinant of pyocin R2, a T-even phage tail-like bacteriocin of *P. aeruginosa* strain PAO, has been found to be located on the chromosome linked to *trpC* (2). Pyocins of another group (type S) are distinguished from the above group by their sensitivity to proteases and their relatively low molecular weights. Among them, pyocin S2 of *P. aeruginosa* PAO (3) and pyocin AP41 of *P. aeruginosa* PAF were purified and characterized (4). We have recently found that the genetic determinant for pyocin AP41 is also located at a certain position on the bacterial chromosome. As both pyocin R2 and pyocin AP41 genes were found on the chromosome, we attempted to clone them on a plasmid so that fine analyses of these genes could be done more easily.

The broad host-range plasmid R68.45 conferring resistance to carbenicillin, kanamycin and tetracycline has a chromosome mobilizing ability. Occasionally a fragment of a chromosome is incorporated into the plasmid forming R primes (R'), which can be obtained when recombination-deficient strains are used as recipients. In this way, so-called *in vivo* cloning can be achieved (5). Using this plasmid, chromosomal genes responsible for the production of pyocin R2 and pyocin AP41 have been cloned. Here we report the results of genetic and physical mappings of these pyocin genes.

I. *Fine Genetic Mapping of Pyocin R2 Genes*

The genetic determinant of pyocin R2 has been mapped on the chromosome at 35 min from the origin of transfer by the FP2 plasmid, linked to *trpC* (1,2). Since pyocin R2 is composed of about 20 species of protein subunits, we isolated a number of pyocin R2-less mutants for the analysis of the cistron composition and arrangement. Complementation tests with about 50 pyocin mutants were done by a lacunae counting method combined with phage F116L-mediated transduction. One pyocin mutant was infected with F116 prepared on another pyocin mutant, and mitomycin C was added to induce pyocin production. After 60 min at 37°C, the mixture was plated with a mitomycin-resistant indicator strain on a soft agar plate. In the case of combination of complementary mutations, a considerable number of lacunae were observed after overnight incubation. In this way, 16 cistrons (*prtA* ∿ *P*) have been identified so far.

Linkages of these genes to *trpC* or *trpE* were determined quantitatively by F116L-mediated transduction with crosses $prt^{+}trp^{+}$ × *prt trpC* or *prt* trp^{+} × $prt^{+}trpE$. These results suggest that *prtA* ∿ *N* are located in between *trpC* and *trpE*.

Cloning of pyocin R2 genes on the plasmid R68.45 was carried out as follows. Recombination-deficient (*rec*) *trpC* mutants were mated with appropriate auxotrophic strains carrying R68.45, and exconjugants with the Trp^{+} phenotype were selected. Among them R' plasmids carrying the *trpC* gene were found. A series of R'68.45 $trpC^{+}$ thus obtained was investigated for the occurrence of other chromosomal markers. As expected from the genetic data, some of the R'68.45 $trpC^{+}$ actually carried the pyocin R2 determinant. The arrangement of the chromosomal genes and those cloned on the plasmid is shown in Fig. 1.

Plasmid DNA was extracted from some R' carriers and digested with restriction enzymes HindIII, BamHI or EcoRI, which cleave R68.45 only at one site. Agarose gel electrophoresis of these digests revealed extra DNA bands in the R' plasmids. Partially deleted plasmids were isolated by digestion with the enzyme HindIII followed by ligation. The loci of various markers on the R' plasmids were identified by complementation tests with these deleted plasmids and appropriate mutant bacteria. As shown in Fig. 1a, *argC* and *trpC, D* genes were located on a 24 Kb fragment, pyocin R2 (*prt*) genes *A* to *J* on a 16.2 Kb fragment and *prtK* to *N* and *trpE* on an 8 Kb fragment. Transductional and R' plasmid mapping showed good agreement.

II. Genetic Characterization of the Pyocin AP41 Determinant

The nature of the genetic determinant for S-type (protease-sensitive) pyocins has not yet been studied. We have begun investigating the genetic determinant for pyocin AP41. This pyocin is produced by *P. aeruginosa* PAF and attacks strain PAO. Pyocin AP41 has been purified to homogeneity and characterized (4). It is a complex of two kinds of protein, a large component (MW=90,000) and a small component (MW=6-7,000).

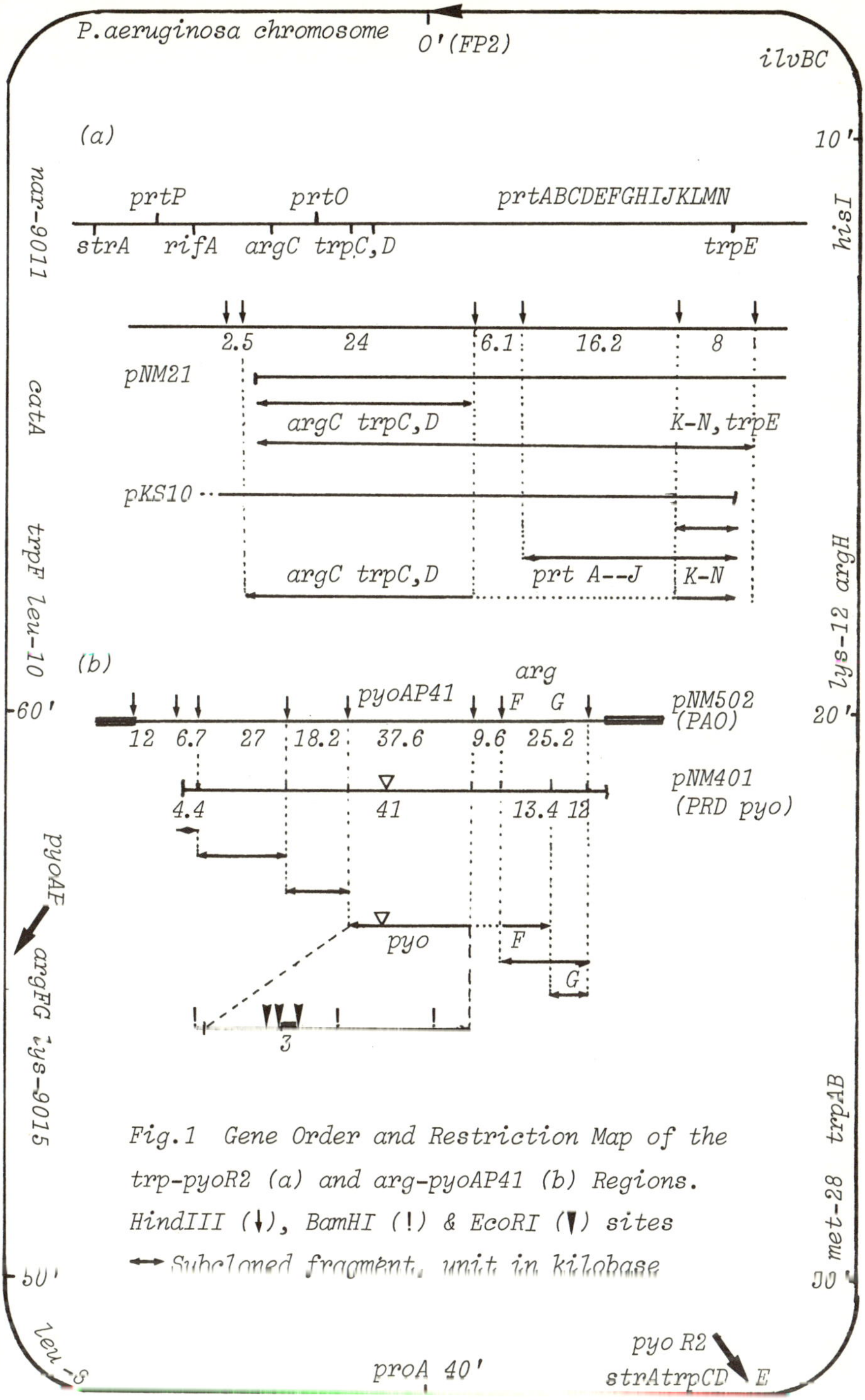

Fig.1 Gene Order and Restriction Map of the trp-pyoR2 (a) and arg-pyoAP41 (b) Regions. HindIII (↓), BamHI (!) & EcoRI (▼) sites

↔ Subcloned fragment, unit in kilobase

The pyocin AP41 complex and its large component showed killing activity. The preferential inhibition of DNA synthesis and the induction of the resident pyocin or phage by this pyocin were observed in the sensitive cells. When the AP41 complex was digested with trypsin, a shortened polypeptide (MW=16,000) was generated from the large component. This tryptic polypeptide had no killing activity but did show DNA endonuclease activity; the small component inhibited this endonuclease activity. In these respects, pyocin AP41 resembles colicin E2, which is coded by a plasmid.

Genetic analyses of pyocin AP41 were done as follows. PAO strains were mated with a PAF derivative carrying R68.45. The recombination frequency of the interstrain cross was very low, but among the recombinants a few were found which inherited the productivity for pyocin AP41. Using these recombinant derivatives (PRD) as donors, R68.45-mediated conjugation gave a good number of recombinants with PAO recipients, enough to perform genetic analysis. Selections were made with various chromosomal markers, and co-inheritance of the pyocin productivity (*pyoAP*$^+$) was investigated. The highest linkage was found with *argF* and *G* markers which were located at about 55 min on the chromosome. The linkage value decreased as the distance between *argFG* and the selected marker increased. No co-transfer with the plasmid markers (drug resistance) was observed. These results suggest that the pyocin AP41 gene, transferred from strain PAF to PAO, is on the chromosome.

To study the nature of this pyocin gene more precisely, *in vivo* cloning of this gene onto R68.45 was attempted taking advantage of its close linkage to *argG*. *Pseudomonas* strains carrying R68.45 were conjugated with *E. coli recA argG* strains, and *arg*$^+$ exconjugants were selected. The transfer frequency of the plasmid by inter-species cross was very low, but several R'*arg*$^+$ plasmids were obtained. When *pyoAP*$^+$ donors, PAF or PRD, were employed, some exconjugants carried R'*arg*$^+$*pyo*$^+$ plasmids. Restriction patterns of these plasmid DNA were compared using HindIII or BamHI. Results are shown in Fig. 2a. Deletions were introduced into the plasmid, R'68.45 *arg*$^+$*pyo*$^+$ (*pNM401*), by partial digestion with HindIII and ligation. The deleted plasmids were purified and characterized. Locations of *argF*, *argG* and *pyoAP* can be deduced from these results (Fig. 2a).

Pyocin AP41 productivity is located on a 41 Kb fragment which is about 3 Kb longer than the corresponding *pyo*$^-$ fragment. Further digestion with EcoRI revealed that a sequence specific for *pyo*$^+$ plasmids (about 3 Kb) is present in this region. This 3 Kb EcoRI fragment was recovered from agarose gels, labeled with [^{32}P]dATP and used as a probe of Southern hybridization. All R'*arg*$^+$*pyo*$^+$ plasmids gave a restriction fragment hybridizable with this probe, whereas those R'*arg*$^+$

obtained from non-pyocinogenic strains showed no hybridization. The chromosomal DNA of 11 independent strains of *P. aeruginosa* (*pyoAP*$^{+}$ or negative) were tested after EcoRI digestion. All pyocin AP41 producers (6 strains) showed a common hybridizable band of about 3 Kb, whereas all non-producers (5 strains) were devoid of it. All *Pseudomonas* strains tested, however, showed 4 hybridizable fragments in the higher molecular weight region. The entity of this homology is now under investigation.

REFERENCES

1. Kageyama, M. (1975) in *Microbial Drug Resistance* (Mitsuhashi, S. & Hashimoto, H., ed.) p.291-305, University of Tokyo Press.
2. Kageyama, M. (1970) *J. Gen. Appl Microbiol. 16*, 523-530 and 531-535.
3. Ohkawa, I., Maruo, B. & Kageyama, M. (1975) *J. Biochem. 78*, 213-223.
4. Sano, Y. & Kageyama, M. (1981) *J. Bacteriol. 146*, 733-739.
5. Holloway, B.W. (1978) *J. Bacteriol. 133*, 1078-1082.
6. Kageyama, M., Shinomiya, T., Aihara, Y. & Kobayashi, M. (1979) *J. Virol. 32*, 951-957.

PLASMID-MEDIATED GENTAMICIN RESISTANCE OF *PSEUDOMONAS AERUGINOSA* AND ITS EXPRESSION IN *ESCHERICHIA COLI*

T. Kato, S. Iyobe, and S. Mitsuhashi

Department of Microbiology
School of Medicine, Gunma University
Maebashi, Japan

I. INTRODUCTION

The isolation frequency of *P. aeruginosa* strains resistant to gentamicin (Gm) has recently increased in Japan and many investigators have isolated plasmids mediating Gm resistance (1, 2). We isolated transferable plasmids mediating Gm resistance at a frequency of 20% from Gm-resistant *P. aeruginosa* strains isolated from clinical specimens in Japan. We also isolated nontransferable Gm-plasmids from 56% of Gm-resistant strains. In this report, we deal with the properties of the nontransferable plasmids mediating Gm resistance and their genetic properties in *E. coli* after being introduced by the transformation method.

II. RESISTANCE PHENOTYPE OF GM RESISTANCE IN *P. AERUGINOSA* AND *E. COLI* STRAINS

A. *Transformation with Nontransferable Plasmids*

Nontransferable plasmids mediating Gm resistance were isolated from Gm-resistant *P. aeruginosa* strains and eight plasmids were randomly selected. Their resistance patterns and molecular weights are shown in Table 1. pMS112 was a Tn1-transposed mutant of the plasmid which mediated resistance to Cm, Km, and Su. pMS113 was a Tn1-transposed mutant of the plasmid which mediated resistance to Gm, Km, Sm, Su, and Cm.

Transformation of *P. aeruginosa* ML4561 (*leu arg ilv his trp rif*) and *E. coli* ML4905 (*lac gal met hsr* Rif^r) was performed with plasmid DNAs which were isolated from *P. aeruginosa* ML5018 Rif^r [a Rec$^-$ mutant of PAO2142 (*lys met ilv*)] by a slightly modified method of Kupersztoch-Portnoy et al. (3). *E. coli* transformants with plasmid DNAs except for pMS110 DNA were obtained on the selection plates containing Cb or Hg, but not on those with Gm, Km, Sm, or Cm. *E.coli* transformants with pMS110 DNA were obtained by Cb, Hg, Sm, or Cm selection, but not by Gm selection. On the other hand, transformants of *P. aeruginosa* ML4561 were selected by all drugs to which resistance was mediated by each of plasmids.

B. Levels of Resistance Mediated by Gm Resistance Plasmids in P. aeruginosa and E. coli Strains

When the levels of resistance were expressed in minimal inhibitory concentration (MIC, μg/ml) in heart-infusion agar, MIC of Gm was markedly decreased in *E. coli* transformants (Table 2). No Gm resistances mediated by 6 plasmids (pMS106-109, 111, and 113) was not detected. The levels of resistance to Km, Sm, or Cm were greatly decreased, whereas those to Cb or Hg were not.

When plasmid DNAs were isolated from either *E. coli* or *P. aeruginosa* transformants and were identified on agarose gel electrophoresis, their molecular weight was found to be the same in the two bacteria. This indicated that no deletion of plasmid DNA occurred in either host even after transformation.

The plasmid DNAs isolated from *E. coli* transformants were reintroduced to *P. aeruginosa* ML4561 by transformation. It was found that the resistance levels of *P. aeruginosa* ML4561

Table 1. *Isolation of Nonconjugative Plasmids from P. aeruginosa Strains*

Plasmid	*Resistance pattern*	*Molecular weight (M daltons)*
pMS106	*Gm Hg Km Sm Su*	22
pMS107	*Gm Hg Km Sm Su*	22
pMS108	*Gm Hg Km Sm Su*	24
pMS109	*Gm Hg Km Sm Su*	23
pMS110	*Cb Cm Gm Hg Km Sm Su*	27
pMS111	*Gm Hg Km Sm Su*	22
pMS112	*Gm Km Su::Tn1(Cb)*	17
pMS113	*Cm Gm Km Sm Su::Tn1(Cb)*	14

abbreviation; Hg (mercury chloride), Km (kanamicin), Sm (streptomycin), Cb (carbenicillin), and Su (sulfanilamide).

Table 2. *Levels of Resistance Mediated by Plasmids in Two Host Bacteria*

Plasmid	*Levels of resistance to*[a]													
	Cb		*Cm*		*Gm*		*Hg*		*Km*		*Sm*		*Su*	
	P.[b]	*E.*[c]	*P.*	*E.*	*P.*	*E.*	*P.*	*E.*	*P.*	*E.*	*P.*	*E.*	*P.*	*E.*
pMS106	*50*	*3.2*	*12.5*	*3.2*	*25*	*0.2*	*100*	*50*	*100*	*3.2*	*>800*	*6.3*	*>6400*	*50*
pMS107	*50*	*3.2*	*25*	*3.2*	*25*	*0.2*	*100*	*50*	*100*	*3.2*	*>800*	*6.3*	*>6400*	*50*
pMS108	*50*	*1.6*	*25*	*6.3*	*50*	*0.2*	*100*	*50*	*200*	*6.3*	*>800*	*6.3*	*>6400*	*100*
pMS109	*12.5*	*3.2*	*200*	*12.5*	*25*	*0.2*	*100*	*50*	*200*	*12.5*	*>800*	*6.3*	*>6400*	*100*
pMS110	*>800*	*>800*	*50*	*12.5*	*100*	*0.8*	*100*	*50*	*200*	*25*	*>800*	*50*	*>6400*	*400*
pMS111	*50*	*3.2*	*12.5*	*3.2*	*25*	*0.2*	*100*	*50*	*800*	*3.2*	*>800*	*6.3*	*>6400*	*100*
pMS112	*>800*	*>800*	*12.5*	*3.2*	*100*	*0.8*	*12.5*	*3.2*	*200*	*100*	*12.5*	*0.8*	*>6400*	*400*
pMS113	*>800*	*>800*	*100*	*3.2*	*25*	*0.1*	*12.5*	*3.2*	*100*	*3.2*	*800*	*0.8*	*>6400*	*3.2*
-	*50*	*3.2*	*12.5*	*3.2*	*1.6*	*0.2*	*12.5*	*3.2*	*25*	*3.2*	*12.5*	*0.8*	*400*	*0.8*

[a] *levels of resistance is expressed in minimal inhibitory concentration (MIC, μg/ml).*
[b] *P. aeruginosa* ML4561.
[c] *E. coli* ML4905.

transformants were restored to those of *P. aeruginosa* ML4561 which had possessed the original plasmids (Table 2).

Accordingly, we could safely conclude that the decrease of Gm or Sm resistance levels mediated by plasmids in *E. coli* host after transformation was not due to mutations occurring in plasmid DNA molecules.

III. MECHANISMS OF GM RESISTANCE MEDIATED BY GM RESISTANCE PLASMIDS

Bioassay of aminoglycoside-modifying enzymes was carried out by a slightly modified of Ozanne et al. (4) using the extracts of plasmid-bearing bacteria. At least two enzymes, one requiring acetyl coenzyme A and the other ATP as a coenzyme, were found in *P. aeruginosa* transformants. Otherwise, no enzyme activity was found in *E. coli* transformants except for those possessing pMS110. It is considered that the transcriptional and/or translational level is decreased in *E. coli* strain possessing plasmids except for pMS110.

The MIC values of Gm were high in all *P. aeruginosa* transformants, but low in the *E. coli* strain possessing pMS110 although enzyme activities were found. This result indicates that synthesis of modifying enzymes is carried out normally in *E. coli* possessing pMS110; however it is not yet known why the MIC value of Gm and enzyme activities are not in proportion.

ACKNOWLEDGMENTS

We wish to acknowledge the technical assistance of Hidehiko Yajima.

REFERENCES

1. Bryan, L.E., Shahrabadi, M.S., & van den Elzen, H.M. (1974) Antimicrob. Agents Chemother. *6*, 191-199
2. Jacoby, G.A. (1974) Antimicrob. Agents Chemother. *6*,239-252
3. Kupersztoch-Portnoy, Y.M., Lovett, M.A., & Helinski, D.R. (1974) Biochemistry *13*, 5484-5490
4. Haas, M.J. & Dowding, J.E. (1975) in Methods in Enzymology (Hash, J.H., ed.) vol.43, pp.611-628, Academic Press, New york

TRANSMISSION OF GENTAMICIN AND AMIKACIN RESISTANCE BY WILD-TYPE PHAGES FROM CLINICAL STRAINS OF *PSEUDOMONAS AERUGINOSA*

H. Knothe

Institute of Hygiene, University of Frankfurt, Main, F.R.G.

S. Mitsuhashi

Department of Microbiology, Gunma University, Maebashi, Japan

V. Krčméry, A. Sečkárová, and M. Antal

Research Institute of Preventive Medicine, Bratislava, Czechoslovakia

F. Výmola

Institute of Hygiene and Epidemiology, Prague, Czechoslovakia

In 1972 we isolated, in Frankfurt University Clinic, a series of gentamicin (GEN)-tobramycin(TOB)-carbenicillin(CAR)-resistant strains of *Pseudomonas aeruginosa (1).* Three of them were identified, in Maebashi Laboratory, to carry R plasmids *(2)* and two of them constitute a new compatibility group *(3).*

During our studies in Maebashi Laboratory, it began to be suspected that wild-type phages, carried by some *P. aeruginosa* strains, including one strain from the above series, could play a role in mediating the transfer of antibiotic-resistance genes. This was confirmed by an independent investigation(G. A. Jacoby, personal communication).

We began to investigate the presence and transmission ability of antibiotic-resistance determinants of wild-type phages isolated from multiresistant–especially GEN-TOB-amikacin(AMI)-resistant *P. aeruginosa* strains isolated in various countries from hospitalized patients. Some strains were typical causal agents of nosocomial infections. In other experiments, we propagated phages F-116 and G-101 in some wild multiresistant *P. aeruginosa* strains/or in some laboratory strains to which corresponding plasmids have been transferred by conjugation/and tested these phage preparations for their ability to transmit resistance determinants. Indeed, it could be shown that phages might play some role in transmission of antibiotic resistance in hospitals. This was documented by the work of other authors, too *(4)*.

MATERIALS AND METHODS

Phages were detected, as a rule, in mixed cultures of wild-type *P. aeruginosa* strain(s) with a recipient strain. Plaques of the phage were then propagated in broth, and Millipore filtered. Titer of the phage preparation was estimated, and, if necessary, the phage was propagated to 10^9 and more p.f.u. by a soft-agar layer method. For transduction experiments, m.o.i. of 0.5 and 0.2 was used. Phenotypic expression for GEN, CAR and/or AMI resistance determinants was extended to 90–120 min (for details see ref. *5* and *6*, where also recipient strains and strains for propagation are characterised. With ML strains No. 4561, 4262 and 4600 we used aminoacid-requirement incompatibility, with strain PAO 1670, generously supplied by Prof. B. Holloway, its rifampicin resistance). Transductants, picked up from single-antibiotic selective plates, were tested for the presence of other resistance determinants with a multiloop applicator technique. At least 50–100 colonies were so tested from each transduction experiment.

RESULTS AND DISCUSSION

Results are summarised, in a chronological schedule, in Table 1. As mentioned previously, first transfers of GEN resistance by conjugation were performed in Frankfurt *(1)* and plasmids were characterised during my stay, and after it, in Maebashi laboratory in collaboration with Drs. Iyobe, Hasuda, Hashimoto and Sagai *(2)*. S. Mitsuhashi was first who encouraged to clarify his suspection that wild-type phage(s) might play a role in low-frequency transfers of some plasmids studies. Plasmids Rms 147 and 149 could be indeed transduced by means of F-116 and G-101 propagation(unpublished results).

Two larger studies in Czechoslovakia, in collaboration with Dr. Húšťavová, indicated that also other properties, *e.g.* "Autoplaque formation," hemolytic toxin production *etc.* might be spread, with or even without resistance determinants, by wild-type phage transmission. Again, Phages F-116 and G-101 could acquire and transmit several resistance determinants.

AMI resistance became actual in 1978 in Frankfurt *(7)*. Among 17 clinical isolates of *P. aeruginosa* highly resistant and transferring GEN, three strains, *i.e.* Ps.70, 78 and 113 were resistant to AMI and all three carried a wild-type phage. After isolation, they were found to transmit AMI resistance, although predominantly by indirect selection. Fertility function, however, was not phage-mediated.

Interestingly, AMI resistance determinant could be transmitted phage AP-113, isolated from the strain Ps. 113, which, although AMI-resistant, did not transfer AMI by conjugation. Dr. Kettner (Slovak Academy of Sciences), studying the aminoglycoside-inactivating enzyme pattern of our as well as other strains, could demonstrate that the strain Ps. 113 did

Table 1. *Chronological Presentation of Phage Transmission Experiments*

Period	*Location (Reference)*	*Designation of plasmids, phages and strains*	*Resistance[a] transmitted by the phage*
1972–1974	*Frankfurt (1)*	*Ps.110,138,142*	*GEN,CAR*
	Maebashi (2)	*Rms147,148,149* *AP-149*	
1975–1977	*Trenčin (CSSR)*	*Pz-3,Pz-5,Pz-6*	*STR,KAN* *hemolytic toxin prod.*
	Prague (5)	*Vy.28,29,34* *AP-29,AP-34*	*GEN,CAR Auto-* *plaque toxin prod.*
		F-116/29,F-116/34	*Via ML-4262*
		G-101/29,G-101/34	*GEN,CAR Auto-plaque[b]*
1978–1980	*Frankfurt (6,7)*	*Ps.70,78,113* *AP-70, -78, -113*	*GEN,TOB,* *AMI[c]*
1980–1981	*Berne (G.Lebek) (8,9)*	*BE 578/19* *AP-19*	*AMI,GEN,* *CAR*
		F-116/BE-11-2 *G-101/BE-11-2*	*GEN,CAR[c]* *AMI[b, c]*
1981	*Brezno (CSSR, P. Sirági)*	*Ps.1,2* *AP-1,-2*	*GEN,CAR*

[a] *Or other property.*
[b] *Conjugal ability was not co-transduced.*
[c] *Predominantly detected by indirect selection.*

not produce inactivating enzyme for AMI and did not posses a DNA band characteristic for plasmid DNA. Thus, impermeability to AMI could be the basis of resistance of this strain. AP-113, thus, could acquire a presumably chromosomal determinant for AMI resistance, by transposition.

In 1980 we investigated a larger collection of AMI resistant strains from Switzerland and three such *P. aeruginosa* strains from Prague hospitals *(8)*, the latter, however, showed no signs of transmissibility. In contrast, the majority of Swiss strains (from G. Lebek) transferred AMI resistance in mixed cultures. Non-transferring strains again did not inactivate AMI enzymatically, showed no band of plasmid DNA and no phage could be isolated from them. Nevertheless, F-116 and G-101 could be propagated on one such strain and phage preparations did mediate the transmission of AMI. Experiments are still in progress, since we suspect chromosomal location of AMI determinant in such non-transferring strains *(9)*.

We isolated altogether three distinct phages from AMI-transferring strains from Switzerland and investigated AP-19 from the strain BE 578. Results are submitted for publication(9) and indicate that this phage transmit AMI, GEN, TOB and CAR resistance directly and few transductants obtained also the fertility function.

Some progress has been also achieved in study of two phages isolated in Czechoslovakia from two *P. aeruginosa* strains, resistant to GEN, infecting succesively the same patient. Both strains and phages are different, however, and we try to correlate the findings with epidemiological data of a nosocomial infection.

Development of GEN and AMI resistance in *P. aeruginosa* in different part of the world offer markers for reliable isolation of colonies, to which resistance determinants are transmitted, in mixed cultures, either by conjugation, or *via* wild-type phages. Thus, we suggest that phages, liberated from hospital multiresistant strains of *P. aeruginosa* by means of physical, chemical or biological stimuli, might contribute to the spread of antibiotic resistance to newer aminoglycosides among clinical strains of *P. aeruginosa.*

ACKNOWLEDGMENT

We thank Professor B. Holloway for his interest and advice during this investigation, particularly in reference to "Autoplaque production" problem.

REFERENCES

1. Knothe, H., Krčméry, V., Sietzen, W., Borst, J. (1973) Chemotherapy *18*, 229–234
2. Sagai, H., Krčméry, V., Hasuda, K., Iyobe, S., Knothe, H., Mitsuhashi, S. (1975) Japan. J. Microbiol. *19*, 427–432
3. Bukhari, A. I., Shapiro, J. A., Adhya, S. L. DNA insertion elements. Cold Spring Harbor Lab. 1977 (see p.650)
4. Iyobe, S. et al., (1979) in Microbial Drug Resistance (Mitsuhashi S., ed.) Vol. 2, pp.187–192, Japan Sci. Soc. Press, Tokyo
5. Krčméry, V., Výmola, F. Mitsuhashi, S. (1977) Zbl. Bakt. A *239*, 361–364
   See also Výmola, F., Krčméry, V., Mitsuhashi, S. (1979) J. Hyg. Epid. Microbiol. *23*, 74–78
6. Knothe, H., Krčméry, V., Mitsuhashi, S. (1980) Zbl. Bakt. A *246*, 373–378
7. Knothe, H., Krčméry, V. (1979). Chemotherapy *25*,. 23–29
8. Lebek, G., Výmola, F., Kettner, M., Krčméry, V., Antal, M., Knothe, H., Mitsuhashi, S. (1981) Zbl. Bakt. A *249*, 557–559
9. Knothe, H., Lebek, G., Krčméry, V., Seginková, Z., Červenka, J., Antal, M., Mitsuhashi, S. (1981) Zbl. Bakt. A (in press)

DRUG AND MERCURY RESISTANCE IN *PSEUDOMONAS AERUGINOSA*

A.M.Boronin and L.A.Anisimova

Institute of Biochemistry and Physiology of Microorganisms, USSR Academy of Sciences, Pushchino, USSR

I. INTRODUCTION

R plasmid often determines drug resistance of hospital isolates of *Pseudomonas aeruginosa*. However some strains are unable to donate resistance in crosses and contain no detectable plasmid DNA (1). The nature of a genetic control of such resistance remains obscure yet.

II. DRUG RESISTANCE PLASMIDS IN *P. AERUGINOSA*

The study on 503 strains of *P. aeruginosa* isolated in hospitals of the Soviet Union showed that 68 (13.5%) strains harbor conjugative or nonconjugative plasmids controlling the resistance to one or several antibiotics, as well as to organic and inorganic mercuric compounds, chromate, borate, tellurite, and ultraviolet irradiation. Plasmid DNA was visualized by agarose gel electrophoresis in all 68 strains. Two or three types of plasmid DNA were seen in 15 of 68 strains. The Table gives a general idea of types of R plasmids detected in strains of *P. aeruginosa* (2). It is worthy of note that 50% of plasmids belong to IncP-2 group.

III. STRAINS OF *P. AERUGINOSA* CARRYING NO DETECTABLE R PLASMIDS

The number of strains resistant to concentrations of antibiotics, exceeding the level of intrinsic resistance in *P. aeruginosa* strains, was far higher than the number of R plasmids found in these strains. Of 503 strains mentioned, 83.6% were resistant to streptomycin, 71% to kanamycin and 70.2% to sulfanilamide. Plasmid DNA was not detected in 83% of strains which are characterized by a high level of resis-

Table. *R Plasmids of Pseudomonas aeruginosa*

Inc group	*Plasmid*	*Properties*	*Molecular weight ($x10^6$)*	*Transfer frequency*
P-1	*pBS52*	*Sm Cb Tra*$^+$	*21*	10^{-3}
P-2	*pBS10*	*Sm Su Hg Tra*$^+$	*260*	10^{-3}
	pBS12	*Sm Cm Hg Mer Ter Uv Tra*$^+$	*280*	10^{-4}
	pBS13	*Sm Tra*$^+$	*280*	10^{-3}
	pBS31	*Sm Tc Cm Su Hg Ter Uv Tra*$^+$	*280*	10^{-4}
	pBS33	*Sm Hg Chr Ter Uv Tra*$^+$	*260*	10^{-3}
	pBS38	*Sm Cm Tc Su Hg Tra*$^+$	*260*	10^{-4}
	pBS48	*Sm Km Ter Uv Tra*$^+$	*280*	10^{-3}
	pBS51	*Sm Chr Ter Uv Tra*$^+$	*260*	10^{-2}
	pBS54	*Sm Km Hg Tra*$^+$	*270*	10^{-5}
	pBS56	*Sm Cm Hg Tra*$^+$	*280*	10^{-2}
	pBS79	*Sm Km Tc Cm Hg Tra*$^+$	*280*	10^{-5}
	pBS91	*Sm Cm Hg Tra*$^+$	*280*	10^{-4}
P-3	*pBS73*	*Sm Cm Tc Km Hg Tra*$^+$	*59*	10^{-5}
P-4	*pBS94*	*Sm Su Tra*$^-$	*5.5*	-
	pBS95	*Sm Su Cb Tra*$^-$	*8.7*	-
P-5	*pBS11*	*Sm Su Hg Chr Mer Tra*$^+$	*131*	10^{-6}
	pBS43	*Sm Hg Bor Tra*$^+$	*83*	10^{-4}
P-7	*pBS14*	*Cm Tra*$^+$	*95*	10^{-5}
?	*pBS59*	*Sm Cm Tc Su Hg*	*42*	10^{-6}
?	*pBS81*	*Sm Hg Tra*$^+$	*42*	10^{-5}
?	*pBS44*	*Sm Tra*$^+$	*65*	10^{-1}
?	*pBS55*	*Hg Tra*$^+$	*147*	10^{-3}
?	*pBS70*	*Sm Cm Tc Km Hg Tra*$^+$	*61*	10^{-4}
?	*pBS74*	*Sm Cm Hg Tra*$^+$	*53*	10^{-5}

tance to more than one antibiotic and fail to be donors of resistance markers. One of such strains, *P.aeruginosa* BS205, is resistant to streptomycin, kanamycin, chloramphenicol, sulfanilamide and mercuric chloride but contains no detectable plasmid DNA. Resistance markers of this strain were transferred to strains *P.aeruginosa* PAO via transduction by phage F116L or mobilization by the RP4 plasmid, but plasmid DNA was not seen in transductants and transconjugants with phenotype Sm Km Cm Su Hg. The study of transconjugants carrying markers of the strain BS205 and RP4 plasmid (Sm Km Cm Su Hg Tc Cb) showed that most of them harbor RP4 plasmid. Some of transconjugants harbor hybrid plasmids 62 megadaltons by molecular weights. Two of such plasmids were named pBS206 and pBS207.

The further investigations showed the decrease in the frequency of transfer of the plasmid RIP64 belonging to the IncP-3 group into strains of transductants and transconjugants carrying markers Sm Km Cm Su Hg. After transfer, this plasmid is eliminated from cells of transconjugants grown in a nonselective liquid medium. Plasmids pBS206 and pBS207 appeared to be incompatible with plasmids of two groups of incompatibility P-1 and P-3 and showed entry exclusion towards RIP64 plasmid belonging to IncP-3 group.

Basing on the results obtained it is assumed that the genes of resistance to streptomycin, kanamycin, chloramphenicol, sulfanilamide and mercury as well as Inc genes specific for IncP-3 group plasmids localized on the DNA segment of 24 megadaltons which could represent a defective (incapable of autonomous existence) plasmid integrated into the bacterial chromosome of the strain *P.aeruginosa* BS205.

IV. TRANSPOSABLE MERCURY RESISTANCE IN *P.AERUGINOSA*

About 62% of all the *P.aeruginosa* strains studied were resistant to mercuric chloride, however only in 19.7% of strains resistant to mercury, the resistance was found to be determined by plasmids.

The use of transconjugants, harboring resistance markers of the same strain *P.aeruginosa* BS205, as donors in experiments on transduction by the phage F116L showed two types of transductants carrying markers of Sm Km Cm Su Hg (frequency $4x10^{-8}$) or Hg marker (frequency $2x10^{-7}$). The study of transductants harboring only Hg marker showed that Hg marker is compatible with plasmids belonging to groups of incompatibility from P-1 to P-11.

A strain of *P.aeruginosa* carrying Hg marker and RP4 plasmid was crossed with the strain *E.coli* HB101. Transconjugants harboring hybrid plasmids of 46 megadaltons appeared at the frequency 10^{-6} on selection by Hg (at 10^{-2} of the RP4 plasmid transfer frequency per a donor cell). We isolated a set of RP4 hybrids carrying the insertion of a DNA segment consisting of *mer* gene(s) into Cb, Km or Tc genes of RP4 plasmid.

One of hybrid plasmids was transferred to the $RecA^-$ strain *E.coli* HB101 harboring nonconjugative R plasmid pBS95 of IncP-4 group. Using this strain as a source of plasmid DNA for transformation, we obtained hybrids of pBS95 with *mer* gene(s) what showed that transposition of a DNA segment coding for the mercury resistance can occur in the absence of a RecA gene product.

The study on hybrid plasmids using restriction endonucleases showed that a *Mer* transposon is a DNA segment of 7 megadaltons which has two EcoRI sites, two HindIII sites and no sites sensitive to SmaI, SalGI, and BamHI enzymes. The transposon found differs in size and restriction sites from the known and evidently identical transposons Tn501 and Tn1861 (3, 4).

V. CONCLUSION

The experiments performed confirm further the prevalence of R plasmids among hospital strains of *P.aeruginosa*. It was shown that the absence of detectable DNA in strains of *Haemophilus influenzae* resistant to antibiotics can be a consequence of using the inappropriate method of plasmid DNA isolation (5). Evidently this holds true also to plasmids of *Pseudomonas* (6). However, the above reported results make it possible to suppose that the lack of plasmid DNA in strains of *P.aeruginosa* resistant to drugs is conditioned in a number of cases by the integration of plasmids into a bacterial chromosome. Evidently, the mercury resistance of *P.aeruginosa* strains samely as resistance to antibiotics can be often controlled by transposable genetic elements.

REFERENCES

1. Jacoby G.A. (1980) in Plasmids and Transposons (Stuttart C., Rozee K., ed.) pp.83-96, Academic Press, New York
2. Anisimova L.A., Boronin A.M. (1981) Antibiotiki *6*, 450-456
3. Bennett P.M., Grinsted J., Choi C.L., Richmond M.H. (1978) *Mol.Gen.Genet. 159*, 101-106
4. Friello D.A., Chakrabarty A.M. (1980) in Plasmids and Transposons (Stuttart C., Rozee K., ed.) pp.249-260, Academic Press, New York
5. Roberts M.C., Smith A.L. (1980) *J.Bacteriol. 144*, 476-479
6. Ingram L., Sykes R.B., Brinsted J., Saunders J.R. (1972) *J.Gen.Microbiol. 72*, 269-279

INTEGRATION OF PLASMIDS INTO THE *PSEUDOMONAS* CHROMOSOME

B.W. Holloway, C. Crowther[2], H. Dean,
J. Hagedorn, N. Holmes[3] and A.F. Morgan

Department of Genetics
Monash University
Clayton, Victoria, Australia

I. INTRODUCTION

The many ways in which a plasmid may contribute to the bacterial phenotype have been extensively documented. There are also many ways in which the plasmid and bacterial genomes may directly interact. In some cases, the entire plasmid may integrate into the bacterial chromosome, as happens in the Hfr forms of *Escherichia coli* K12. Alternatively, specialized segments of either the plasmid or bacterial DNA may be mobilized with a variety of genetic consequences, as happens with transposons or insertion sequences. Of special interest are these genomic interactions which involve plasmids which can freely transfer between unrelated bacteria in view of the potential for novel genotypes.

The IncP-1 group of plasmids have many features of interest including wide bacterial host range and the ability to mobilize bacterial chromosome. R68.45 is a variant of the IncP-1 plasmid R68 which has acquired the ability to mobilize chromosome in many gram negative bacteria (1). It has been recently demonstrated that R68.45 differs from R68 in possessing a duplication of a region of the plasmid genome pre-existing in R68 (2-4). This 2.1 Kb segment located immediately anticlockwise to the kanamycin resistant determinant has been identified by its characteristic pattern of restriction endonuclease sites. Evidence has been presented that this region includes an insertion sequence, IS21, identified *inter alia* by its ability to transpose to an unrelated plasmid, pED815 (2). The present data are not

[1]*Supported by the Australian Research Grants Committee.*
[2]*Present address: Princeton University, Princeton, New Jersey, U.S.A.*
[3]*Present address: Cancer Institute, Melbourne, Victoria*

sufficient to conclude that the 2.1 Kb segment which becomes duplicated and IS21 are identical, but this seems the most likely conclusion. IS21 has been shown not to occur in the *Pseudomonas aeruginosa* strain PAO chromosome (2). It has been proposed that the role of IS21 in the mobilization of chromosome by R68.45 is to mediate the formation of a cointegrate between R68.45 and the bacterial chromosome. In view of the wide range of gram negative bacteria in which R68.45 can mobilize chromosome, IS21 must have a low specificity of sites of transposition to the chromosome of these bacteria.

While this proposed mechanism is in essence similar to that found with the plasmid F and *Escherichia coli* K12, one important difference is that R68.45 does not integrate stably into the *P. aeruginosa* chromosome, nor for that matter, any other bacterial genus in which R68.45 chromosome mediated transfer has been studied (1). R prime plasmids can be derived from R68.45 in various bacterial genera and in these a segment of bacterial chromosome is inserted into the plasmid genome (1). There is some evidence that in such R prime plasmids one copy of IS21 is located to either side of the bacterial chromosome fragment inserted into the R68.45 genome (4; Willetts, personal communication).

We have studied a variety of other interactions between bacterial and plasmid genomes in *Pseudomonas*. For example, inserts of Tn*1* have been identified at a variety of sites in the *P. aeruginosa* PAO chromosome and have been used to create portable regions of homology between plasmid and chromosome which can result in polarized chromosome transfer from a range of chromosomal origins (5).

P. aeruginosa is notable for the number and range of plasmids which can mobilize the chromosome of that species (1, 6,7). Recently a number of plasmids similar to R68.45 have been identified which can mobilize plasmid in *P. putida* (Dean and Morgan, unpublished data). However, none of these plasmids have been shown to integrate into the *Pseudomonas* chromosome to form a donor with properties analogous to the Hfr form of the F plasmid in *E. coli*. In this paper we describe the integration of plasmids into the *P. aeruginosa* and *P. putida* chromosomes and the use that can be made of such integrated plasmids for genetic studies on these bacteria. Recently Haas *et al.* (8) have described the integration of RP1 into the tryptophan synthase gene of *P. aeruginosa* PAO with the formation of an effective donor which can mobilize chromosome from the site of plasmid integration.

II. EXPERIMENTAL

The experimental procedures and media used have been the same as those described in publications from this laboratory as cited in the references. The chromosomal location of the various *P. aeruginosa* PAO markers used are as follows: *argH32* (19'); *his-4* (16'); *leu-8* (50'); *lys-12* (20'); *ser-3* (28'); *trp-6* = *trpC/D* (33') (9).

A. The Isolation of pMO190, a Temperature Sensitive Mutant of R68

Mutants of IncP-1 plasmids temperature sensitive for either replication or maintenance have been previously isolated. In *E. coli*, Danilevich *et al.* (10), Robinson *et al.* (11) and Harayama *et al.* (12), found that by growth at 43°C of strains carrying such temperature sensitive plasmid mutants, derivatives could be isolated in which the plasmid had become integrated into the *E. coli* chromosome, these variants mobilizing chromosome at much greater efficiency than the parent plasmid. Haas *et al.* (8) have obtained similar results with the IncP-1 plasmid RP1 in *P. aeruginosa* PAO.

In our experiments mutants of the IncP-1 plasmid R68 (carbenicillin, tetracycline and kanamycin resistant - Cb^r, Tc^r, Km^r) were sought which either did not replicate or failed to be maintained in *P. aeruginosa* PAO grown at 43°C, but which could replicate at 28°C. Using hydroxylamine as the mutagen, one such mutant was isolated from about 26,000 colonies tested. This plasmid, denoted pMO190, shows a 10^{-4} reduction in colony forming ability on antibiotic supplemented media when grown at 43°C in either *P. aeruginosa* PAO or *E. coli* K12. It has not been determined whether the mutant plasmid is affected at 43°C in replication or maintenance function.

Those colonies of *P. aeruginosa* PAO (pMO190) which do grow on antibiotic containing media at 43°C were examined, and it was shown that the maintenance and replication of pMO190 becomes temperature insensitive following integration into the PAO chromosome. The evidence for integration is provided by the following properties of colonies of *P. aeruginosa* PAO (pMO190) which have grown at 43°C on nutrient agar plus 500 μg/ml carbenicillin.

1. Loss of Transfer Function of Plasmid Antibiotic Resistance Markers. While all the survivors have retained all three antibiotic resistances most have lost the ability to transfer this resistance by conjugations. About 5% of the survivors show a much reduced (10^{-3}) transfer ability of plasmid markers. *P. aeruginosa* PAO (pMO190) grown at 28°C has the same plasmid transfer ability as R68.

2. *Most survivors have acquired bacterial chromosome donor ability.* R68 mobilizes *P. aeruginosa* chromosome very inefficiently, marker transfer being about 10^{-8}/donor parent cell (13). When *P. aeruginosa* PAO5 (pMO190) (*trp-6)* was grown at 43°C on nutrient agar plus 500 μg/ml carbenicillin, of 93 surviving colonies tested, 89 were found to mobilize the *P. aeruginosa* PAO chromosome, the frequency of marker transfer being as high as 10^{-2}/donor parent cell. By examining the ability of a range of these donors to mobilize a range of markers situated at different sites on the PAO chromosome, at least four different origin sites have been identified on the PAO chromosome. These donor strains show essentially the same kinetics of marker transfer as do Hfr strains of *E. coli*. Each has a single origin site, displays oriented transfer of chromosome markers from that origin site and the intact integrated plasmid is not recovered from the chromosomal recombinants. This behaviour of pMO190 when integrated is in marked contrast in the situation with R68.45 which has a wide range of origin sites on the *P. aeruginosa* PAO chromosome (13, 14).

3. *Some Have Acquired Additional Auxotrophic Requirements.* This suggests that integration of the plasmid has occurred at the site of a gene for biosynthesis of the new requirement. In each case where such auxotrophy is found, the origin of chromosome transfer is either at or close to the chromosomal location of the new auxotrophic marker.

4. *Cotransduction of Plasmid and Chromosomal Markers Has Been Demonstrated.* PAO709 is a derivative of PAO2 (*ser-3*) in which pMO190 has apparently integrated into the chromosome to cause a requirement for lysine. By prototroph reduction tests using the transducing phage F116L, the lysine marker of PAO709 (*lys-16*) was shown to be at, or closely linked to *lys-12* situated at 20' (9). To demonstrate linkage of *lys-16* and the plasmid borne carbenicillin resistance determinant (Cb^r), phage F116L was propagated on PAO709 and used to transduce the auxotrophic recipient PAO362 (*argH32*). Carbenicillin resistant transductants were selected on nutrient agar plus 500 μg/ml carbenicillin.

The following classes and frequencies of transductants were found:

	%
$Arg^- \; Lys^- \; Cb^r \; Tc^r \; Km^r$	86
$Arg^+ \; Lys^- \; Cb^r \; Tc^s \; Km^s$	10
$Arg^- \; Lys^- \; Cb^r \; Tc^s \; Km^s$	4

Thus there is 100% cotransduction between the selected marker (Cb^r) and the *lys-16* marker found in PA0709, as well as 10% cotransduction between *argH* and the Cb^r determinant, and between *argH* and *lys-16*. *argH* and *lys-12* are 45% cotransducible (15) and the reduced frequency of cotransduction of *argH32* and *lys-16* probably is a result of the integration of the plasmid genome into the PAO chromosome. The high (88%) cotransduction of all three plasmid resistance determinants shows that a substantial amount, if not all, of the pMO190 genome has been integrated into the PAO chromosome.

Similar results were obtained with an independently isolated survivor of growth at 43°C, PA0708 in which pMO190 has apparently integrated into a site very close to *leu-8*$^+$ (48'), resulting in a requirement for leucine for the integrated strain. With both PA0708 and PA0709, the origin of chromosome transfer in conjugation is located close to the *leu-8* and *lys-12* regions of the chromosome respectively.

5. *There Is No Free Plasmid DNA.* The DNA of pMO190 cannot be demonstrated as supercoiled CCC DNA in lysates of those strains which have chromosome mobilization ability and which have also completely lost transfer function of plasmid borne antibiotic resistances. With PAO2 (pMO190), free plasmid DNA can be readily demonstrated from bacterial cultures grown at 28°C.

6. *Plasmid DNA Has Been Shown to Be Physically Present in the Bacterial Chromosome.* By use of Southern hybridization techniques (16), IncP-1 plasmid DNA has been shown to be integrated in PAO2 (pMO190) strains which survive growth at 43°C on nutrient agar containing carbenicillin. The chromosomal DNA of such cultures was isolated, digested with either *HindIII* or *BamHI* and the fragments separated by gel electrophoresis and the DNA fragments transferred to nitrocellulose filters. Probe DNA prepared from R68 or pMO190 was labelled with ^{32}P by nick translation and, after treatment with either *HindIII* or *BamHI*, was examined for its ability to hybridize to the bacterial chromosomal DNA. The results indicated integration of pMO190 DNA into the PAO chromosome, there being no unique site of integration (data not shown).

From all this evidence, we conclude that pMO190 can integrate into the *P. aeruginosa* PAO chromosome. We cannot determine whether integration is mediated either by TnA or by IS21 and experiments to distinguish between these two possibilities are in progress. The isolation of donors with Hfr-like properties will be of value in the comprehensive mapping of the *P. aeruginosa* PAO chromosome. Furthermore, the ability of IncP-1 plasmids to integrate into the chromosome provides a new approach to intergeneric hybridization and this is described below.

B. The Use of pMO190 for Intergeneric Hybridization

The integration of pMO190 into the *P. aeruginosa* chromosome which can be detected by growth at 43° provides a technique for the integration of a segment of heterologous bacterial chromosome into the *P. aeruginosa* chromosome. One approach by which this could be achieved is by the isolation of an Enhanced Chromosome Mobilzing (ECM) (7) variant of pMO190, with the expectation that it would have the same ability to form prime plasmids as has been found with R68.45 (17). Despite extensive efforts, no ECM variants have been isolated from pMO190. Attempts to isolate ECM variants from the temperature sensitive mutant of RP4, pTH10, isolated by Harayama *et al.* (12) have also been without success. Had either of these attempts been successful, R prime plasmids from the temperature sensitive ECM plasmid would have been isolated and used as vectors to insert bacterial chromosome of species other than *P. aeruginosa* into the PAO chromosome.

As an alternative, we have constructed *in vitro* prime plasmids using pMO190. Plasmid DNA was isolated from pMO190 and bacterial DNA isolated from *E. coli* K12. Each DNA preparation was fragmented with the restriction endonuclease *HindIII*, the two preparations mixed, and after ligation, the mixture was transformed into *P. aeruginosa* with selection for carbenicillin resistance (18). Insertion of bacterial chromosome fragments into the single *HindIII* site of pMO190 should result in inactivation of the kanamycin resistance determinant so that those transformants most likely to carry inserted fragments of *E. coli* chromosome were those which were kanamycin sensitive. Amongst a range of prime plasmids so constructed, we have identified one plasmid which carries a segment of *E. coli* chromosome which includes the entire tryptophan operon. This plasmid, denoted pMO1120, has been shown to have the following properties:

(1) Codes for carbenicillin resistance and tetracycline resistance, is Tra^{+} and has the temperature sensitive characteristics of pMO190.

(2) The ability to complement mutants of all tryptophan biosynthetic genes in *E. coli* as indicated by growth on minimal medium of tryptophan auxotrophs of *E. coli* carrying pMO1120. Thus, pMO1120 has the wide host range characteristics of pMO190, and is Tra^{+} for Trp^{+}.

(3) The ability to complement all tryptophan biosynthetic genes in *P. aeruginosa* PAO by a similar test involving mutants of each *trp* gene of that organism. In such

complemented strains of *P. aeruginosa*, it was shown by transduction that the *trp* mutant allele was still present in the chromosome.

When *P. aeruginosa* PAO316 (pMO1120) - (*trp-6,lys-12,met-28, his-4*), was grown at 43°C on nutrient agar + 500 μg/ml carbenicillin, colony survival was 10^{-5} compared to that of the same culture plated on the same media at 28°C. Amongst the survivors at 43°C, some had integrated the prime plasmid pMO1120 including the *E. coli* chromosome fragment into the PAO chromosome. The evidence for this is provided by the following:

(1) Such strains were still Trp^+, but had lost the ability to transfer Trp^+, Cb^r or Tc^r. Each of these phenotypes can be transferred at frequencies of 10^{-1}-10^{-2}/donor cell from PAO316 (pMO1120) which has only been grown at 28°C.

(2) The Trp^+ survivors still carried the chromosomal PAO *trp-6* allele as shown by its cotransduction with *argC* at the expected frequency.

(3) Some survivors at 43°C had acquired additional auxotrophic requirements, suggesting integration of pMO1120 into other biosynthetic genes with subsequent loss of function.

(4) Some survivors at 43°C have chromosome mobilizing ability. Strains such as PAO316 (pMO1120) in which pMO1120 is not integrated do not show any ability to mobilize PAO chromosome.

We have thus been able to construct *P. aeruginosa* strains carrying a segment of *E. coli* chromosome integrated into the *P. aeruginosa* chromosome by means of the temperature sensitive plasmid vector pMO190. This integrated *E. coli* chromosomal segment can express the *E. coli* genes for tryptophan biosynthesis in *P. aeruginosa* PAO.

C. Integration of R91-5 into the Chromosome of P. putida

It has been a part of our general approach to develop systems of genetic analysis which are applicable to the widest range of *Pseudomonas* species and if possible to other bacterial genera. Unfortunately, pMO190 and pMO1120 cannot be used to generate chromosome donor strains in most species of *Pseudomonas* because only a few species of this genus can grow at 43°C, and at lower temperatures, variants with integrated plasmids cannot easily be selected.

However, with *P. putida* it has been possible to obtain integration of derivatives of the IncP-10 plasmid R91-5 which have acquired donor ability. R91-5 has a host range limited to *P. aeruginosa*. It codes for carbenicillin resistance and promotes chromosome transfer from two origins in *P. aeruginosa* strain PAT (19,20). Matings of *P. aeruginosa* PAO (R91-5) x *P. putida* PPN do not show any transfer of the plasmid. However, in view of the results obtained with transfer of R91-5::Tn*501* matings (see below) it is very likely that R91-5 does transfer from *P. aeruginosa* to *P. putida* but the plasmid is unable to replicate. Derivatives of R91-5 loaded with Tn*501*, coding for mercury resistance, have been isolated (21) and it has been found that these plasmid derivatives carried by a *P. aeruginosa* PAO donor can be transferred in conjugation to *P. putida* PPN at a frequency of 10^{-6} per donor cell, selection being made for transfer of carbenicillin resistance.

The *P. putida* exconjugants are resistant to both mercury and carbenicillin. Evidence that R91-5::Tn*501* has integrated into the PPN chromosome is provided by the following data:

(1) More than 90% of the carbenicillin and mercury resistant exconjugants have chromosome mobilizing ability. They can transfer chromosomal markers at a frequency as high as 10^{-1}/donor parent cell. The kinetics of marker trnasfer is entirely analogous to that found with the Hfr forms of *E. coli*. There is a single origin of marker transfer, high frequency of proximal marker transfer, reduced frequency of marker transfer with increasing distance from the origin and very low frequency transfer of plasmid genes. From nearly a thousand independently isolated donor strains, seven different origins have been identified. From a series of matings using these various donors, it has been possible to construct a circular map of the *P. putida* PPN chromosome. Figure 1 displays the PPN chromosome map and the site of the various origins of transfer found with R91-5::Tn*501* integrates.

(2) In one case, integration of R91-5::Tn*501* has produced an auxtrophic requirement for serine, suggesting that plasmid integration has affected the function of a serine biosynthetic gene.

(3) It has been possible to transduce an R91-5::Tn*501* Hfr origin site from one PPN strain to another. The transducing phage pf16h2 (22) was used, with the serine requiring Hfr strain described above as the donor and another PPN auxotroph as the recipient. Selection was made for carbenicillin resistance on media containing serine to allow growth of Ser^- transductants. Of 200 transductants tested, all had acquired serine auxotrophy and carbenicillin resistance, indicating close linkage of these two markers. Nineteen of the 200 had inherited mercury resistance, indicating transfer of Tn*501* and 18 of these had chromosome mobilizing ability (Cma). The direction of transfer and location of the site of origin of each of the 18 Cma^+ transductants was the same as that of the

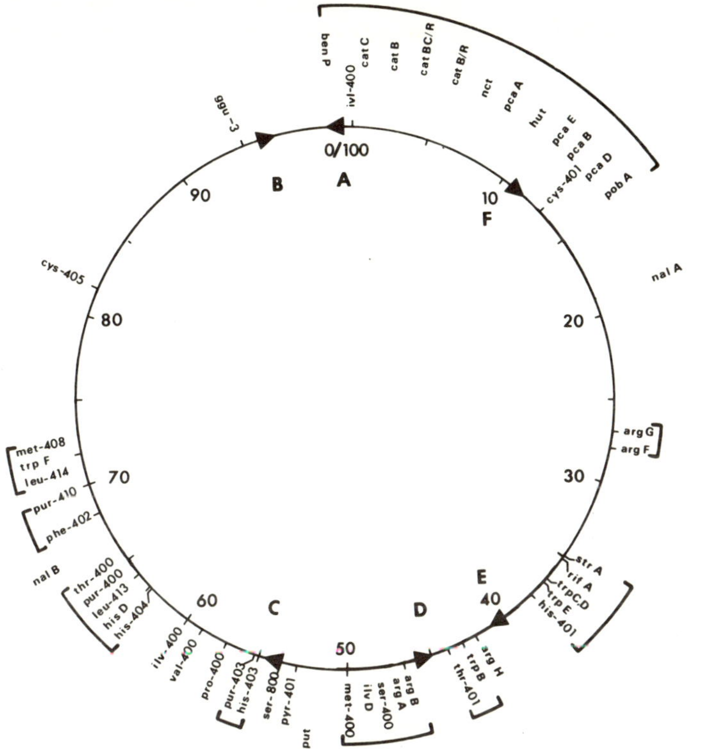

Fig. 1. Chromosome map of P. putida *PPN. The following marker designations have been used:* arg, *arginine requirement;* ben, *benzoate utilization;* cat, *catechol utilization;* cys, *cysteine requirement;* ggu, *glucose uptake;* his, *histidine requirement;* ilv, *isoleucine plus valine requirement;* ivl, *isoleucine plus valine leucine requirement;* leu, *leucine requirement;* met, *methionine requirement;* nal, *nalidixic acid resistance;* nct, *nicotinate utilization;* pca, *protocatechuate utilization;* phe, *phenylalanine requirement;* pob, *p-hydroxy benzoate utilization;* pro, *proline requirement;* pur, *adenine requirement;* put, *proline utilization;* pyr, *uracil requirement;* rif, *rifampicin resistance;* ser, *serine requirement;* str, *streptomycin resistance;* thr, *threonine requirement;* trp, *tryptophan requirement;* val, *valine requirement.*

donor used in the transduction (origin C, see Fig. 1).

The results of this transduction clearly demonstrate the role of Tn*501* in promoting bacterial chromosome transfer, suggesting that it is this transposon, rather than Tn*1* which is essential for the formation of the Hfr like donor strains by R91-5::Tn*501* in *P. putida*.

(4) No supercoiled CC R91-5::Tn*501* plasmid DNA has been found in lysates of 6 *P. putida* donor strains with different origins of transfer.

(5) Hybridization of R91-5::Tn*501* DNA to chromosomal DNA obtained from two donor strains by the Southern technique provided direct evidence of integration of the plasmid into the *P. putida* chromosome (Willetts, personal communication).

III. DISCUSSION

Using two entirely different approaches, we have demonstrated the integration of different plasmids into *P. aeruginosa* and *P. putida*. The common feature is that in each case the bacterial strain with the integrated plasmid has acquired chromosome mobilizing ability. There is specificity of integration but in each case, a range of origin sites has been identified which has meant that the use of such donors has benefited chromosome mapping studies. As yet we do not know the means by which integration has occurred. With pMO190 it could be due to either Tn*1* or IS21 and experiments to distinguish between these two possibilities are underway. Likewise with R91-5::Tn*501*, either Tn*1* or Tn*501* could be the means by which integration is achieved. The fact that Tn*501* must be transduced for an origin of chromosome transfer to be transduced suggests that Tn*501* is necessary for chromosome mobilization but this does not necessarily imply that integration is achieved by means of Tn*501*.

The isolation of pMO190, the identification of techniques which select for its integration into the chromosome and the derivation of pMO1120 from pMO190 has enabled a different approach for the construction of intergeneric bacterial hybrids. It should be recognized that the only example studied so far involves the *E. coli trp* operon and this may be a special case because of the high efficiency of the *trp* promoter. Other hybrids are under study to test the generality of this method. However, we have demonstrated an alternative means of incorporating foreign DNA into the bacterial genome to that of the more widely used method involving plasmid vectors.

The use of R91-5::Tn*501* integrates for chromosomal mapping in *P. putida* has enabled the construction with the map of *P. aeruginosa*. The method is not dependent upon Tn*501* and the transposons Tn*5* and Tn*7* are equally effective in selection for integrates of loaded R91-5 plasmids into the *P. putida* chromosome. By means of R prime plasmid complementation test (23) it is possible to identify many of the mutant lesions of the *P. putida* auxotrophic markers which have been mapped. A comparison of the *P. putida* and *P. aeruginosa* maps shows that clusters of the same markers are found in each species but that the relative relationships on the circular map of each species differs. While it is possible to devise a theoretical series of inversions and translocations of chromosomal fragments by which the gene arrangement of one species can be related to the gene arrangement of the other species, nothing is known of the actual mechanism which has resulted in the existing two arrangements of chromosomal genes. Hopefully R91-5::Tn*501* derivatives can be used to form chromosome donors in other species of *Pseudomonas* which could enable a comparison of the gene distribution to be made for these species. Such information is unavailable for any other bacterial genus and data obtained by this approach should make possible a new approach for studies on the taxonomy and evolution of the genus *Pseudomonas*.

ACKNOWLEDGMENTS

We wish to thank Anabel Sutherland and Linda Bell for enthusiastic technical assistance and Beechams (Australia) for their generous gift of carbenicillin.

REFERENCES

1. Holloway, B.W. (1979) Plasmid *2*,1-19
2. Willetts, N.S., Crowther, C. & Holloway, B.W. (1981) Plasmid *6*,30-52
3. Riess, G., Holloway, B.W. & Pühler, A. (1980) Genet. Res. *36*,99-109
4. Leemans, J., Villaroel, R., Silva, B., Van Montagu, M. & Schell, J. (1980) Gene *10*,319-328
5. Krishnapillai, V., Royle, P.L. & Lehrer, J. (1981) Genetics *97*,495-511
6. Dean, H.F., Royle, P. & Morgan, A.F. (1979) J. Bacteriol. *138*,249-250
7. Holloway, B.W., Krishnapillai, V. & Morgan, A.F. (1979) Microbiol. Rev. *43*,73-102
8. Haas, D., Watson, J., Krug, R. & Leisinger, T. (1981) Molec. Gen. Genet. *182*,240-244
9. Royle, P.L., Matsumoto, H. & Holloway, B.W. (1981) J. Bacteriol. *145*,145-155
10. Danilevich, V.N., Stepanshin, Y.G., Volozhantsev, N.V. & Golub, E.I. (1978) Molec. Gen. Genet. *166*,313-320
11. Robinson, M.K., Bennett, P.M., Falkow, S. & Dodd, H.M. (1980) Plasmid *5*,343-347

12. Harayama, S., Tsuda, M. & Iino, T. (1980) Molec. Gen. Genet. *180*,47-56
13. Haas, D. & Holloway, B.W. (1976) Molec. Gen. Genet. *144*, 243-251
14. Haas, D. & Holloway, B.W. (1978) Molec. Gen. Genet. *158*, 229-237
15. Haas, D., Holloway, B.W., Schamböck, A. & Leisinger, T. (1977) Molec. Gen. Genet. *154*,7-22
16. Southern, E. (1975) J. Mol. Biol. *98*,503-517
17. Holloway, B.W. (1978) J. Bacteriol. *133*,1078-1082
18. Sinclair, M.I. & Morgan, A.F. (1978) Aust. J. Biol. Sci. *31*,679-688
19. Chandler, P.M. & Krishnapillai, V. (1974) Genet. Res. *23*, 251-257
20. Watson, J.M. & Holloway, B.W. (1978) J. Bacteriol. *133*, 1113-1125
21. Krishnapillai, V. (1979) Plasmid *2*,237-246
22. Chakrabarty, A.M., Gunsalus, C.F. & Gunsalus, I.C. (1968) Proc. Nat. Acad. Sci. (U.S.A.) *60*,168-175
23. Morgan, A.F. (1982) J. Bacteriol. (in press)

BIOCHEMISTRY: *β*-LACTAMASE

DEOXYAMINOGLYCOSIDES ACTIVE AGAINST RESISTANT STRAINS

Hamao Umezawa

Institute of Microbial Chemistry
Shinagawa-ku, Tokyo, Japan

In 1967, I discovered 3'-O-phosphotransferase and 6'-N-acetyltransferase which transferred the terminal phosphate of ATP to the 3'-hydroxyl group of kanamycin, neomycin, paromomycin etc. (1) or transferred the acetyl group of acetyl-CoA to the 6'-amino group (2) and these enzymes were suggested to be involved in the mechanism of resistance. In order to proof this enzymic mechanism of resistance conclusively, in collaboration with S. Umezawa et al., we synthesized 3'-deoxykanamycin A (3) and 3',4'-dideoxykanamycin B (4). These derivatives inhibited the growth of resistant strains and thus the enzymic mechanism of resistance was conclusively proved. Moreover, not only the deoxygenation but also the modification of the 1-amino group which binds to the enzymes has provided us with derivatives which inhibit the growth of both sensitive and resistant strains.

At present, as described in the author's review (5, 6), besides 3'-O-phosphotransferase [APH(3')] and 6'-N-acetyltransferase [AAC(6')], 4'-O-adenylyltransferase [AAD(4')], 2"-O-phosphotransferase [APH(2")], 2"-O-adenylyltransferase [AAD(2")], 3-N-acetyltransferase [AAC(3)], 2'-N-acetyltransferase [AAC(2')] are known to be involved in mechanism of resistance to kanamycin and other 2-deoxystreptamine-containing aminoglycoside antibiotics. All these enzymes have two binding sites: the one directed to the adenosine moiety of ATP or acetyl-CoA, the other to aminoglycosides. They may have relationships in their generation or evolution. At the present knowledge, the possibility cannot be denied that resistant strains which produce other O-phospho or adenylyltransferases may appear in future. I was, therefore, interested in the derivative which can inhibit the bacterial growth and has the least number of hydroxyl group, and with my colleagues we synthesized various deoxy derivatives of kanamycins A and B. On the other hand, we continued the screening for aminoglycoside antibiotics and found istamycins A and B. These are the members of fortimicin group antibiotics, and from the structural viewpoint this group of antibiotics are deoxyaminoglycosides. The antibacterial activity of deoxy derivatives of kanamycin and fortimicin group antibiotics indicate that the amino groups have predominant roles in the antibacterial action of these antibiotics and their binding to bacterial ribosomes.

In this paper, I will review our studies on deoxy derivatives and istamycin, and discuss the structure-activity relationships. I will also discuss the structure-activity relationships of deoxyaminoglycosides with a peptie antibiotic, negamycin.

DEOXY DERIVATIVES OF KANAMYCINS A AND B

Compared with 3'-deoxykanamycin A, 3',4'-dideoxykanamycin B has a stronger antibacterial activity. It has also a low ototoxicity. Therefore, this was clinically studied and since 1975 this derivative which is called dibekacin has widely been used for the treatment of bacterial infections. It was also shown by these studies that the 3'- and 4'-hydroxyl groups are not involved in the binding of kanamycin to bacterial ribosomes.

The X-ray crystal analysis of kanamycin A monosulfate showed the whole molecular structure (7) and also the presence of a hydrogen bonding between 5- and 2'-hydroxyl groups. If this hydrogen bonding is strong, even in a solution, the 4-(6-amino-6-deoxy-D-glucopyranosyl)-2-deoxystreptamine moiety can have a rigid structure as shown by the X-ray crystal analysis, that is, no bond rotation around the glycosidic linkage. Although this was not right as later described, for some years I thought that this rigid structure of this part of the molecule might be necessary for the strong antibacterial action of kanamycins. Therefore, we synthesized first deoxy derivatives which still contained the 5- and 2'-hydroxyl groups. Starting from 3',4'-dideoxykanamycin B (dibekacin), 4"- and 6"-deoxy derivatives were synthesized. The amino groups were protected by <u>tert</u>-butoxycarbonylation, the 4"- and 6"-hydroxyl groups were first protected by cyclohexylidene group and the 2"-hydroxyl group was acylated with benzoyl chloride. Thereafter, the cyclohexylidene group was removed and for the preparation of the 4",6"-dideoxydibekacin, the 4"- and 6"-hydroxyl groups were mesylated. The deoxygenation of the 4"- and 6"-hydroxyl groups was achieved by halogenation followed by hydrogenolysis. The removal of the protective groups gave 4",6"-dideoxydibekacin, that is, 3',4',4",6"-tetradeoxykanamycin B (8). For the preparation of 6"-deoxydibekacin the 6"-hydroxyl group of the product after the removal of the cyclohexylidene group was tosylated. By the similar procedure as described above 6"-deoxydibekacin, that is, 3',4',6"-trideoxykanamycin B was obtained (8).

The minimal inhibitory concentrations of these deoxy

derivatives against 19 typical strains including resistant ones are shown in Table 1. Dibekacin and 6"-deoxydibekacin showed almost the same activity against all strains, indicating that the 6"-hydroxyl group is not involved in the antibacterial action. 4",6"-Dideoxydibekacin showed also almost the same activity against all strains except Ps. aeruginosa. The removal of the 4"-hydroxyl group markedly reduced the activity to inhibit the growth of Ps. aeruginosa. At first, the 4"-hydroxy

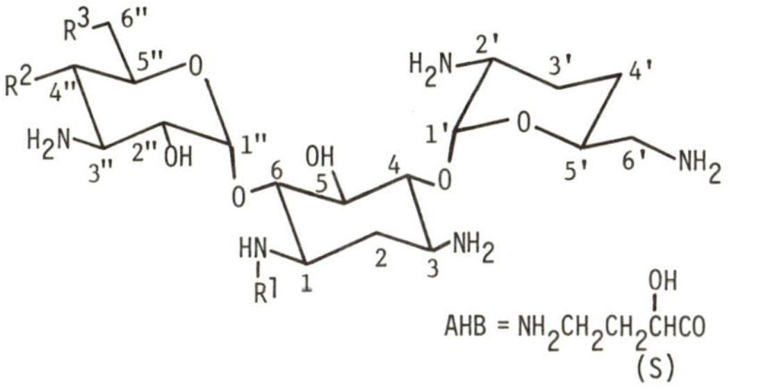

1): R^1 = H, R^2, R^3 = OH
2): R^1 = H, R^2 = OH, R^3 = H
3): R^1, R^2, R^3 = H
4): R^1 = AHB, R^2 = OH, R^3 = H
5): R^1 = AHB, R^2, R^3 = H

Table 1. *4",6"-Deoxy Derivatives of Kanamycin B*

Test organisms	*MIC mcg/ml* 1	2	3	4	5
S. aureus	0.78	0.78	0.39	1.56	0.78
Ap01[a)]	1.56	3.13	3.13	1.56	1.56
S. epidermidis 109[a)]	1.56	1.56	3.13	1.56	1.56
E. coli K12	1.56	1.56	1.56	1.56	1.56
R5[b)]	>100	>100	100	50	100
ML1629[c)]	3.13	1.56	3.13	1.56	1.56
R55[d)]	100	>100	100	1.56	1.56
JR225[f)]	>100	>100	>100	1.56	1.56
Kl. pneumoniae	0.78	1.56	0.78	1.56	0.70
Serratia sp. 4[d)]	100	100	100	6.25	6.25
Pr. rettgeri GN466	1.56	3.13	1.56	12.5	3.13
Providencia Pv16[g)]	>100	>100	>100	6.25	6.25
Ps. aeruginosa A3	3.13	1.56	3.13	3.13	1.56
H9[e)]	6.25	12.5	100	6.25	6.25
H11	12.5	12.5	100	50	50
TI-13[c)]	6.25	12.5	100	6.25	25
B13[c,e)]	6.25	50	100	25	12.5
99[f)]	50	12.5	100	12.5	12.5
K-Ps102[i)]	6.25	6.25	25	12.5	12.5

Resistance mechanisms: *a) AAD(4'), b) AAC(6'), c) APH(3')-I,*
d) AAD(2"), e) APH(3')-II, f) AAC(3),
g) AAC(2'), h) APH(3')-III, i) permeability

group which is located adjacent to the 3"-amino group was thought to make the molecule permeable into Ps. aeruginosa. But as later described, the weak activity of 4",6"-dideoxydibekacin against Ps. aeruginosa was shown to be due to its weak effect on protein synthesis on Ps. aeruginosa ribosomes.

We introduced (S)-4-amino-2-hydroxybutyryl (AHB) group into the 1-amino group of these deoxy derivatives of dibekacin. The 1-amino group of the 2-deoxystreptamine moiety has been suggested to be involved in the binding to the type I of 3'-O-phosphotransferase and some types of 3-N-acetyltransferase as first shown by the effect of amikacin. The introduction of the AHB group into this amino group has been known to produce derivatives inhibiting resistant strains. This group on the 1-amino group produces a steric hindrance against 2"-O-phospho or adenylyltransferase and such derivatives inhibit the growth of resistant strains producing these enzymes. Therefore, the 6"-deoxydibekacin derivatives containing the AHB group on the 1-amino group expanded the antibacterial spectrum against resistant strains. The introduction of the AHB group to the 1-amino group of 4",6"-dideoxydibekacin increased the activity against Ps. aeruginosa and 1-N-AHB-4",6"-dideoxydibekacin showed only a slightly weaker activity against Ps. aeruginosa than 1-N-AHB-6"-deoxydibekacin.

As already described, X-ray crystal analysis of kanamycin A monosulfate (7) indicated a hydrogen bonding between the 5- and 2'-hydroxyl groups. But this hydrogen bonding does not concern the antibacterial activity. As reported in several papers (9-11), epimerization of the C-5 of sisomicin increased the potency against Pseudomonas, Providencia and Proteus. 5-Deoxygentamicin C complex (12), 5-deoxykanamycin A (13), 5-deoxykanamycin B (14), 5,3',4'-trideoxykanamycin B (15, 16), and 5,6"-dideoxykanamycin B (17) have been reported to have an antibacterial good activity.

As reproted by Fukatsu (16), during the study of industrual synthetic process of dibekacin, 5,3',4'-trideoxykanamycin B was obtained. This trideoxy derivative showed a similar antibacterial activity as dibekacin, but it had a higher toxicity than dibekacin. Starting from this trideoxykanamycin B, that is, 5-deoxydibekacin, we synthesized 5,3',4',6"-tetradeoxykanamycin B and 5,3',4',4",6"-pentadeoxykanamycin B and their 1-N-AHB[(S)-4-amino-2-hydroxybutyryl] derivatives as shown in Table 2 (18). Then, the removal of the 5-hydroxyl group reduced slightly the antibacterial activity as shown by the comparison of the minimal inhibition concentrations in Table 2 with those in Table 3. As in the previous case (Table 1), the removal of both 4"- and 6"-hydroxyl groups markedly lowered the antibacterial activity against Ps. aeruginosa. The introduction of AHB group into the 1-amino group of these deoxy derivatives markedly increased the activity against all strains including Ps. aeruginosa. The

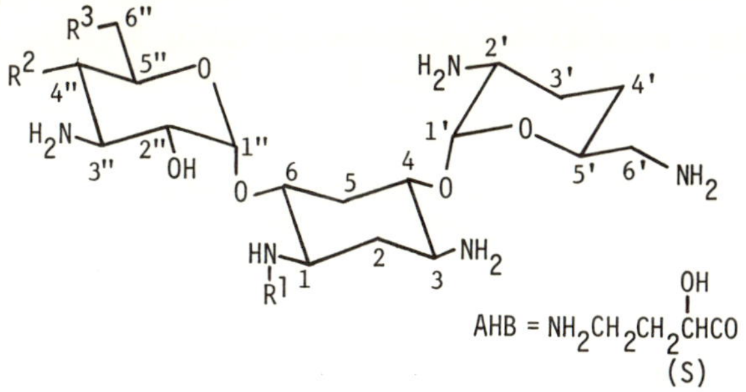

1): R^1 = H, R^2, R^3 = OH
2): R^1 = H, R^2 = OH, R^3 = H
3): R^1, R^2, R^3 = H
4): R^1 = AHB, R^2, R^3 = OH
5): R^1 = AHB, R^2 = OH, R^3 = H
6): R^1 = AHB, R^2, R^3 = H

Table 2. *5-Deoxy Derivatives of Kanamycin B (I)*

Test organisms	*MIC mcg/ml* 1	2	3	4	5	6
S. aureus	3.13	3.13	3.13	<0.20	<0.20	<0.20
Ap01[a)]	6.25	6.25	12.5	0.78	0.78	<0.20
S. epidermidis 109[a)]	6.25	6.25	25	1.56	0.78	0.20
E. coli K12	12.5	25	100	1.56	0.78	1.56
R5[b)]	>100	>100	>100	100	50	100
ML1629[c)]	12.5	6.25	6.25	1.56	0.78	1.56
R55[d)]	>100	>100	>100	3.13	1.56	1.56
JR225	50	25	12.5	1.56	0.78	0.78
Kl. pneumoniae	6.25	3.13	6.25	0.78	0.78	0.78
Serratia sp. 4[d)]	100	50	50	25	25	6.25
Pr. rettgeri GN466	25	25	25	1.56	1.56	3.13
Providencia Pv16[g)]	>100	>100	>100	6.25	25	12.5
Ps. aeruginosa A3	12.5	3.13	6.25	0.39	0.78	0.78
H9[e)]	50	100	50	3.13	3.13	25
H11	12.5	25	50	12.5	12.5	3.13
TI-13[c)]	12.5	12.5	25	1.56	3.13	1.56
B13[c,e)]	50	50	100	6.25	12.5	12.5
99[f)]	25	>100	25	3.13	6.25	6.25
K-Ps102[i)]	12.5	12.5	50	1.56	3.13	1.56

a) AAD(4'), b) AAC(6'), c) APH(3')-I, d) AAD(2"), e) APH(3')-II, f) AAC(3), g) AAC(2'), h) APH(3')-III, i) permeability

antibacterial spectrum of 1-N-AHB-dibekacin has been tested by Dr. Mitsuhashi and his associates of Medical School of Gumma University against more than 300 strains clinically isolated and the results suggested the potential usefulness of this dibekacin derivative. Therefore, its 4"-deoxy or 4",6"-dideoxy derivatives (5 and 6 in Table 2) seem to exhibit a wide spectrum against organisms clinically isolated.

We also synthesized 6'-N-methyl and 6"-chloro derivatives of 5,3',4' trideoxy and 5,3',4',6"-tetradeoxykanamycin B and their 1-N-AHB derivatives as shown in Table 3 (18). Although the data are not shown in Table 3, 6'-N-methylation produced the antibacterial activity against some strains producing 6'-N-acetyltransferase. 6"-Chlorination increased the antibacterial activity of 5,3',4',6"-tetradeoxykanamycin B against most strains except Ps. aeruginosa. 6"-Chlorination increased the activity of 6'-N-methyl-5,3',4'-trideoxykanamycin B. The introduction of AHB group into the 1-amino group enhanced the activity.

As indicated by the antibacterial activities of 5-deoxy

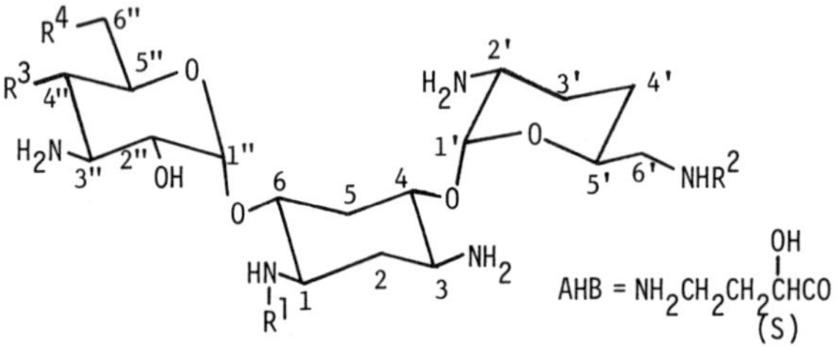

1): R^1, R^2 = H, R^3, R^4 = OH
2): R^1, R^2 = H, R^3 = OH, R^4 = Cl
3): R^1 = H, R^2 = CH_3, R^3, R^4 = OH
4): R^1 = H, R^2 = CH_3, R^3 = OH, R^4 = Cl
5): R^1 = AHB, R^2 = H, R^3, R^4 = OH
6): R^1 = AHB, R^2 = H, R^3 = OH, R^4 = Cl
7): R^1 = AHB, R^2 = CH_3, R^3 = OH, R^4 = Cl

Table 3. *5-Deoxy Derivatives of Kanamycin B (II)*

Test organisms	*MIC mcg/ml* 1	2	3	4	5	6	7
S. aureus	3.13	0.78	1.56	0.78	<0.20	<0.20	<0.20
Ap01[a)]	6.25	0.78	25	0.78	0.78	<0.20	0.39
S. epidermidis 109[d)]	6.25	0.20	6.25	0.39	1.56	<0.20	<0.20
E. coli K12	12.5	3.13	12.5	3.13	1.56	1.56	1.56
R5[b)]	100	100	50	25	100	50	6.25
ML1629[c)]	12.5	3.13	25	3.13	1.56	3.13	3.13
R55[d)]	100	100	100	50	3.13	3.13	1.56
JR225[f)]	50	50	100	50	1.56	0.78	1.56
Kl. pneumoniae	6.25	3.13	12.5	3.13	0.78	1.56	1.56
Serratia sp. 4[d)]	100	25	50	25	25	25	12.5
Pr. rettgeri GN466	25	3.13	25	3.13	1.56	3.13	1.56
Providencia Pv16[g)]	>100	>100	>100	>100	6.25	12.5	12.5
Ps. aeruginosa A3	12.5	3.13	3.13	1.56	0.39	6.25	0.78
H9[e)]	50	>100	25	12.5	12.5	6.25	12.5
H11	12.5	>100	25	12.5	12.5	12.5	12.5
TI-13[c)]	12.5	50	12.5	6.25	1.56	6.25	1.56
B13[c,e)]	50	50	100	25	6.25	12.5	50
99[f)]	25	12.5	50	12.5	3.13	3,13	6.25
K-Ps102[i)]	12.5	12.5	25	6.25	1.56	6.25	6.25

a) AAD(4'), b) AAC(6'), c) APH(3')-I, d) AAD(2"), e) APH(3')-II, f) AAC(3), g) AAC(2'), h) APH(3')-III, i) permeability derivatives shown in Table 2 and 3, the 5-, 3'-, 4'-, 4"- and 6"-hydroxyl groups do not concern the antibacterial activity of 1-N-AHB derivatives. Daniels <u>et al</u>. (1979) (19) reported that 1-N-[(S)-3-amino-2-hydroxypropionyl]-2'-deaminogentamicin C1a and 1-N-[(S)-3-amino-2-hydroxylpropionyl]-2'-deaminosisomicin have a good antibacterial activity and spectrum. Except against resistant strains producing 3-N-acetyltransferase and 6'-N-acetyltransferase, these 2'-deoxy derivatives exhibit a strong inhibition against all strains tested. This indicates that the 2'-amino or 2'-hydroxyl group is not involved in the antibacterial action. We synthesized 2'-deoxykanamycin A and its 1-N-AHB derivative (2'-deoxyamikacin). These 2'-deoxy derivatives showed almost the same antibacterial activity as the parent compounds, that is, the removal of the 2'-hydroxyl group of kanamycin A and amikacin does not influence the antibacterial activity.

On the basis of the experimental results described above,

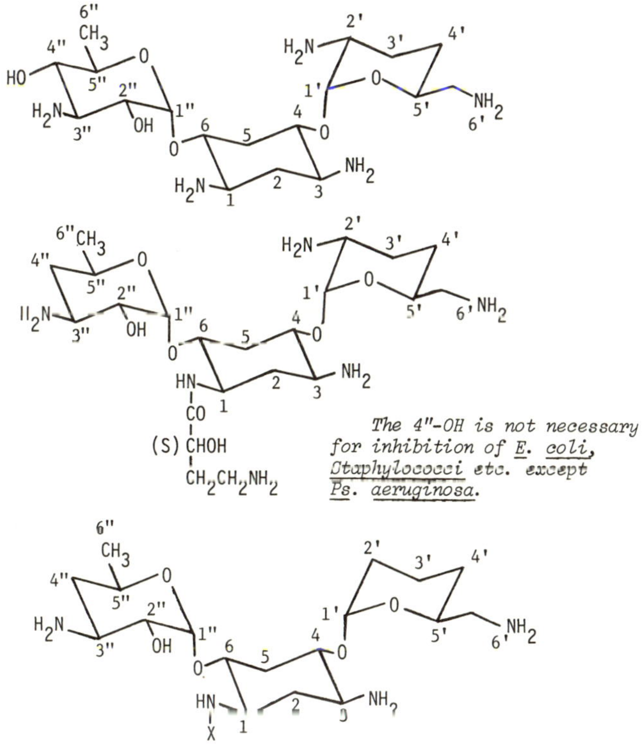

Possible structure with the least number of hydroxyl group. The necessity of the 2"-OH has not been fully studied.

Fig. 1 Active Structure with Low Number of Hydroxyl Group

the structure which has the least number of hydroxyl group can as shown in Fig. 1. It has not been fully studied whether the 2"-hydroxyl group is the absolute requirement for the antibacterial activity or not. The removal of the 4"-hydroxyl group reduced the antibacterial activity against Ps. aeruginosa. 4",6"-Dideoxydibekacin showed a significantly weaker inhibition against protein synthesis on Ps. aeruginosa ribosomes than dibekacin. In case of Pseudomonas, the 4"-hydroxyl group of dibekacin may be involved in the binding to ribosomes. But in case of 1-N-AHB-dibekacin or 1-N-AHB-5-deoxydibekacin, the removal of the 4"-hydroxyl group did not decrease the activity. Based on the experimental results, it can be concluded that the amino groups have predominant roles in the antibacterial action, that is, in the binding to bacterial ribosomes.

STUDIES ON ISTAMYCIN

The antibiotics which can be recognized as deoxyaminoglycosides have been found by the screening. From the structural viewpoint, fortimicin found by Nara et al. (1977) (20),

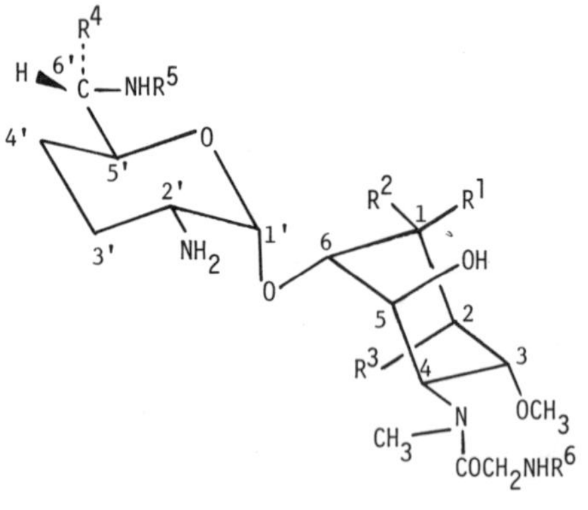

	R^1	R^2	R^3	R^4	R^5	R^6
Fortimicin A	NH_2	H	OH	CH_3	H	H
Fortimicin C	NH_2	H	OH	CH_3	H	$CONH_2$
Fortimicin D	NH_2	H	OH	H	H	H
Sporaricin A	H	NH_2	H	CH_3	H	H
Istamycin A (Sannamycin A)	NH_2	H	H	H	CH_3	H
Istamycin B	H	NH_2	H	H	CH_3	H
Dactimicin	NH_2	H	OH	CH_3	H	CH=NH

Fig. 2. Fortimicin Group Antibiotics

sporaricin by Deushi *et al.* (1979) (21), istamycin by Okami *et al.* (1979) (22) and dactimicin by Inoue *et al.* (1979) (23) (Fig. 2) are deoxyaminoglycosides. Moreover, as later described, starting from neamine, istamycin A has been chemically synthesized. It is interesting that all these fortimicin group antibiotics were discovered during the last 5 years by 4 Japanese research groups from soil sample collected in Japan.

The istamycin-producing strain was isolated by spreading a soil sample of a sea shore mud (Sagami Bay at Tenjin Island) on an agar medium containing kanamycin at 20 μg/ml (24). This strain produced several aminoglycosides. We extracted and purified two of them and named them istamycins A and B (22). This strain is resistant to istamycin, fortimicin A, neamine, kanamycins A and B, dibekacin, butirosin and ribostamycin at 400 μg/ml and sensitive to gentamicin, lividomycin, neomycin B, paromomycins and streptomycins (the growth was inhibited at 50 μg/ml) (24). It is also resistant to amikacin. This resistance spectrum is very interesting in comparison with that of resistant bacteria. We confirmed that the resistance of this istamycin-producing strain is due to the resistant character of the ribosomes but not to the formation of enzymes inactivating istamycin A, kanamycin etc. (25). 3-N-Acetyltransferase which transfers the acetate of acetyl-CoA to the 3-amino group of the 2-deoxystreptamine moiety of aminoglycoside antibiotics transfers the acetate also to the 1-amino group of fortimicin group antibiotics. The istamycin-producing strain does not produce this enzyme.

We synthesized istamycin A from neamine by the process shown in Fig. 3 (26). 3',4'-Dideoxyneamine was synthesized from neamine and starting from this dideoxyneamine, istamycin A was synthesized through 14 steps. Thus, it can also be said that istamycin A is a deoxy derivative of neomycin A (neamine). During the course of this synthesis, 4-N,6'-N,3-O-tridemethylistamycin A and 6'-N,3-O-dideomethylistamycin A were synthesized (27). Then, as shown in Table 4, the tridemethyl derivative had no antibacterial activity, but the didemethyl derivative showed almost the same activity as istamycin A. It suggests that the 4-N-methyl group is absolutely necessary for the activity to inhibit the growth of bacteria. *E. coli* R5 which produced 6'-N-acetyltransferase was resistant to the 6'-N,3-O-didemethyl derivative (compound 2 in Table 4) but sensitive to istamycin A. It indicates that 6'-N-acetyltransferase which transfers the acetate of acetyl-CoA to the 6'-amino group of kanamycin, neomycin etc. can transfer the acetate also to the 6'-amino group of 6'-N-demethylistamycin. On the other hand, the 6'-N,3-O-didemethyl derivative showed a little stronger inhibition against many strains of *Ps. aeruginosa*

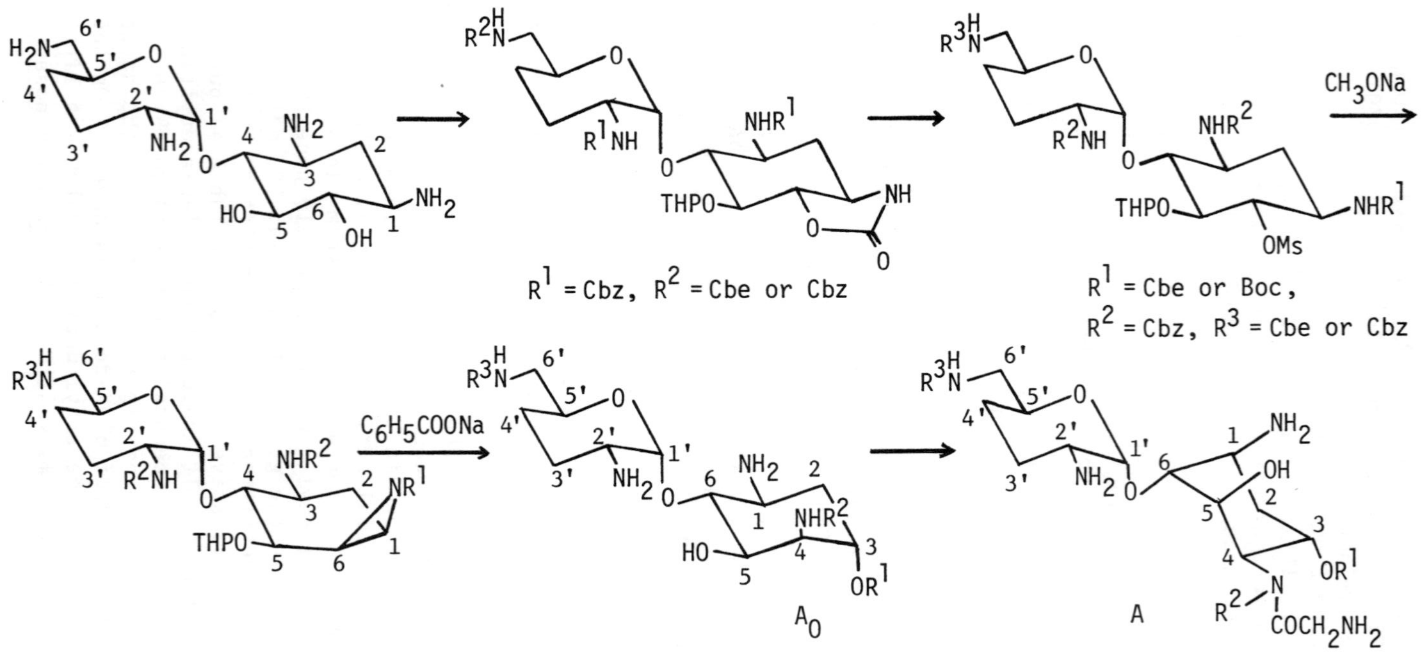

Istamycin A_0 or A: R^1, R^2, R^3 = CH_3

6'-N,3-O-Didemethyl: R^1, R^3 = H, R^2 = CH_3

4-N,6'-N,3-O-Tridemethyl: R^1, R^2, R^3 = H

Fig. 3. Synthesis of Istamycin A and its Demethyl Derivatives from 3',4'-Dideoxyneamine

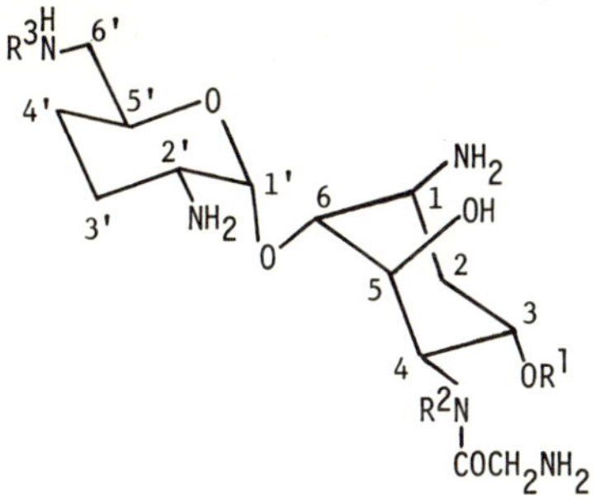

1): 4-N,6'-N,3-O-Tridemethylistamycin A: R^1, R^2, R^3 = H
2): 6'-N,3-O-Didemethylistamycin A: R^1, R^3 = H, R^2 = CH_3
3): Istamycin A: R^1, R^2, R^3 = CH_3

Table 4. *Antibacterial Activiries of Demethyl Derivatives of Istamycin A*

Test organisms	MIC mcg/ml 1	2	3
Staph. aureus FDA209P	>100	1.56	0.78
B. subtilis PCI219	6.25	<0.20	<0.20
E. coli K12	>100	6.25	1.56
K12 R5[a]	>100	100	3.13
K12 ML1629[b]	>100	3.13	3.13
K12 R55[c]	>100	12.5	3.13
K12 C600 R135[d]	>100	>100	>100
JR225[e]	>100	3.13	1.56
Kl. pneumoniae	>100	3.13	1.56
Sh. dysenteriae	>100	12.5	3.13
Proteus rettgeri GN466	>100	12.5	6.25
Serratia marcescens	>100	12.5	6.25
Providencia sp. Pv10[f]	>100	12.5	6.25
Ps. aeruginosa A3	>100	3.13	6.25
No. 12	>100	25	100
H9[g]	>100	12.5	25
H11	>100	50	100
TI-13[b]	>100	12.5	25
GN315[a]	>100	25	25

a) AAC(6'), b) APH(3')-I, c) AAD(2"), d) AAC(3)-I, e) AAC(3)-IV, f) AAC(2'), g) APH(3')-II

than istamycin A. It suggested that the 3-O-demethylation of istamycin A or B would produce the derivatives which might show a good activity against Ps. aeruginosa.

Istamycin B had a little stronger antibacterial activity than istamycin A and, moreover its equatorial 1-amino group seemed to be more resistant to 3-N-acetyltransferase than the axial 1-amino group of istamycin A, because E. coli C600R135 was much more sensitive to istamycin B than A. This resistant strains carries a R factor and produced 3-N-acetyltransferase which transfers the acetyl group to the 3-amino group of the 2-deoxystreptamine-containing antibiotics. Therefore, we synthesized 3-O-demethyl derivatives of istamycin B (28). Then, as shown in Table 5, 3-O-demethylistamycin B (compound 2 in Table 5) showed a stronger inhibition against many strains

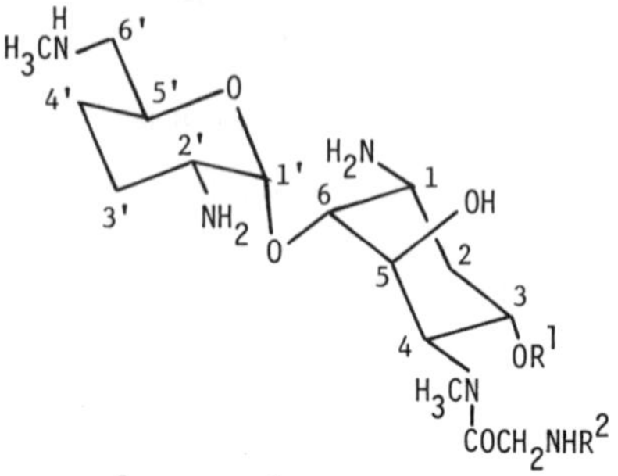

1): Istamycin B: $R^1 = CH_3$, $R^2 = H$

2): 3-O-Demethylistamycin B: R^1, $R^2 = H$

3): 2"-N-Formimidoylistamycin B: $R^1 = CH_3$, $R^2 =$ CH=NH

4): 3-O-Demethyl-2"-N-formimidoylistamycin B: $R^1 = H$, $R^2 =$ CH=NH

Table 5. *Antibacterial Activity of 3-O-Demethyl Derivatives of Istamycin B*

Test organisms	*MIC mcg/ml* 1 (Hemicarbonate hemihydrate)	2 (Sesqui-carbonate)	3 (Disulfate dihydrate)	4 (Disulfate) trihydrate)
Staph. aureus FDA209P	0.39	0.39	0.39	0.20
B. subtilis PCI219	<0.20	<0.20	<0.20	<0.20
E. coli K12	1.56	3.13	1.56	1.56
K12 R5[a]	3.13	6.25	3.13	3.13
K12 ML1629[b]	1.56	3.13	1.56	1.56
K12 LA290 R55[c]	3.13	6.25	1.56	1.56
K12 C600 R135[d]	12.5	6.25	50	12.5
JR225[e]	0.78	1.56	1.56	0.78
Kl. pneumoniae	1.56	3.13	1.56	1.56
Sh. dysenteriae	3.13	6.25	3.13	3.13
Proteus rettgeri GN466	6.25	6.25	6.25	3.13
Serratia marcescens	6.25	12.5	6.25	3.13
Providencia sp. Pv16[f]	6.25	25	12.5	25
Ps. aeruginosa A3	6.25	3.13	12.5	1.56
No. 12	100	12.5	100	12.5
H9[g]	50	25	100	12.5
H11	50	25	>100	25
TI-13[b]	25	12.5	100	6.25
GN315[a]	50	12.5	50	6.25

a) AAC(6'), b) APH(3')-I, c) AAD(2"), d) AAC(3)-I, e) AAC(3)-IV, f) AAC(2'), g) APH(3')-II of *Ps. aeruginosa* than istamycin B. We also synthesized 3-O-demethyl-2"-N-formimidoylistamycin B (28). This showed a stronger activity than the corresponding derivative which had 3-O-methyl group. We sent 3-O-demethylistamycin B to Dr. Mitsuhashi of Medical School, Gumma University. Then, this compound showed a stronger inhibition against 100 strains of *E. coli* and *Ps. aeruginosa*, in comparison with fortimicin group antibiotics containing the 3-O-methyl group. Especially it shows a strong action against *Ps. aeruginosa*. Most strains were inhibited by 1.6 - 12.5 μg/ml. Fortimicin group antibiotics have only a weak activity against *Ps. aeruginosa*. In a published patent, it is described that 3-O-demethylsporaricin has a stronger activity against *Ps. aeruginosa* than sporaricin (29).

Istamycin has one hydroxyl group. This hydroxyl group may be the requirement of the antibacterial activity. Suami *et al*. (1980) (30) synthesized deoxy derivatives of fortimicin A and reported that the 2,5-dideoxy derivative had almost no antibacterial activity, but 2-deoxyfortimicin A had a little stronger activity than fortimicin A. It suggests the necessity of the 5-hydroxyl group for the antibacterial activity. The structure of 2-deoxyfortimicin A is similar to istamycin A except the 5'-(1-aminoethyl) group in the former and the 5'-(methylamino)methyl group in the latter.

The protein synthesis on *Ps. aeruginosa* ribosomes is significantly more strongly inhibited by 3-O-demethylistamycin B than istamycin B. The antibacterial activity of fortimicin group antibiotics and the stronger inhibition of protein synthesis *in vitro* by 3-O-demethylistamycin B suggests that the amino groups and 5-hydroxyl group have the predominant role in the antibacterial activity of fortimicin group antibiotics.

RELATION OF ANTIBACTERIAL DEOXYAMINOGLYCOSIDES TO PEPTIDE ANTIBIOTICS

As discussed above, between the hexose moieties and 2-deoxystreptamine moiety in kanamycin, and gentamicin, there is no hydrogen bonding which is necessary for the exhibition of the antibacterial activity and the amino groups have a predominant role in the antibacterial action. It suggests that there may be closed relationships in the action mechanism between aminoglycosides and basic water-soluble peptide antibiotics. In fact, it is known that in *Mycobacterium tuberculosis* there is a cross resistance between aminoglycoside antibiotics and viomycin group antibiotics.

A peptide antibiotic, negamycin (Fig. 4) which we found in

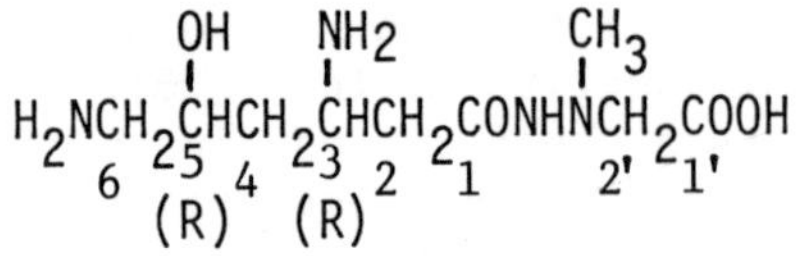

Fig. 4. Negamycin

culture filtrates of Streptomyces purpeofuscus inhibits the growth of gram negative and positive bacteria (31-33). As we reported, it inhibits the protein synthesis of E. coli ribosomes and causes a miscoding as well as aminoglycoside antibiotics (34). It has one hydroxyl group, and its methylation reduces the activity only by about 50%. The deoxygenation also produces only a small influence on the activity (35, 36). 3-Epi-deoxynegamycin which has the different configuration at the carbon atom binding to the 3-amino group has been synthesized and thereafter found in culture filtrates of a strain of Streptomyces goshikiensis has only a weak antibacterial activity in comparison with negamycin (37). It indicates again an important role of the amino group in the antibacterial action. Although the cross resistance has not been observed between negamycin and kanamycin or streptomycin, the experimental results described above indicate a closed relationships in the actions and the mechanisms between this peptide antibiotic and aminoglycoside antibiotics. By the use of CPK space filling models, it was observed that the 6-amino group (the terminal amino group), the 3-amino group and the hydrozide group of negamycin could be superimposed on the 2'-amino group, 3-amino group and 1-amino group of kanamycin (34).

CONCLUSION

Studies of deoxy derivatives and new deoxyaminoglycoside antibiotics have indicated that the amino groups have predominant roles in exhibiting antibacterial action and if resistant strains which form new O-phospho or adenylyltransferases will appear, then useful chemotherapeutic agents will be found in deoxy derivatives.

In order to develop derivatives which can inhibit the growth of resistant strains producing N-acetyltransferases, it is necessary to investigate the modification of aminoglycoside molecules further in detail. In this case, molecular structures of peptide antibiotics inhibiting bacterial protein synthesis may give us useful information.

REFERENCES

1. Umezawa, H., Okanishi, M., Kondo, S., Hamana, K., Utahara, R., Maeda, K. & Mitsuhashi, S. (1967) Science *157*, 1559-1561
2. Umezawa, H., Okanishi, M., Utahara, R., Maeda, K. & Kondo, S. (1967) J. Antibiotics *A-20*, 136-141
3. Umezawa, S., Tsuchiya, T., Muto, R., Nishimura, Y. & Umezawa, H. (1971) J. Antibiotics *24*, 274-275
4. Umezawa, H., Umezawa, S., Tsuchiya, T. & Okazaki, Y. (1971) J. Antibiotics *24*, 485-487
5. Umezawa, H. (1974) in Advances in Carbohydrate Chemistry and Biochemistry (Tipson, R.S. & Horton, D., eds.) Vol. 30, pp.183-225, Academic Press, New York
6. Umezawa, H. (1979) Jap. J. Antibiotics *32* (Suppl.), S1-S14
7. Koyama, G., Iitaka, Y., Maeda, K. & Umezawa, H. (1968) Tetrahedron Lett. 1875-1879
8. Miyasaka, T., Ikeda, D., Kondo, S. & Umezawa, H. (1980) J. Antibiotics *33*, 527-532
9. Waitz, J.A., Miller, G.H., Moss, E. & Chiu, P.J.S. (1978) Antimicrob. Agents & Chemother. *13*, 41-48
10. Kabins, S.A. & Nathan, C. (1978) Antimicrob. Agents & Chemother. *14*, 391-397
11. Fu, K.P. & Neu, H.C. (1978) Antimicrob. Agents & Chemother. *14*, 194-200
12. Rosi, D., Gross, W.A. & Daum, S.J. (1977) J. Antibiotics *30*, 88-97
13. Kavadias, G., Dextraze, R., Massé, R. & Belleau, B. (1978) Can. J. Chem. *56*, 2086-2092
14. Suami, T., Nishiyama, S., Ishikawa, Y. & Umemura, E. (1978) Bull. Chem. Soc. Jpn. *51*, 2354-2357
15. Hayashi, T., Iwaoka, T., Takeda, N. & Ohki, E. (1978) Chem. Pharm. Bull. *26*, 1786-1797
16. Fukatsu, S. (1979) Jap. J. Antibiotics *32* (Suppl.), S178-S186
17. Suami, T. & Nakamura, K. (1980) Bull. Chem. Soc. Jpn. *53*, 3655-3657
18. Umezawa, H., Miyasaka, T., Iwasawa, H., Ikeda, D. & Kondo, S. (1981) J. Antibiotics, in press
19. Daniels, P.J.L., Cooper, A.B., McCombie, S.W., Nagabhushan, T.L., Rane, D.F. & Wright, J. (1979) Jap. J. Antibiotics *32* (Suppl.), S195-S204
20. Nara, T., Yamamoto, M., Kawamoto, I., Takayama, K., Okachi, R., Takasawa, S., Sato, T. & Sato, S. (1977) J. Antibiotics, *30*, 533-540
21. Deushi, T., Iwasaki, A., Kamiya, K., Kunieda, T., Mizoguchi, T., Nakayama, T., Itoh, H., Mori, T. & Oda, T. (1979) J.

Antibiotics *32*, 173-179

22. Okami, Y., Hotta, K., Yoshida, M., Ikeda, D., Kondo, S. & Umezawa, H. (1979) J. Antibiotics *32*, 964-966
23. Inoue, S., Ohba, K., Shomura, T., Kojima, M., Tsuruoka, T., Yoshida, J., Kato, N., Ito, M., Amano, S., Omoto, S., Ezaki, N., Ito, T., Niida, T. & Watanabe, K. (1979) J. Antibiotics *32*, 1354-1356
24. Hotta, K., Saito, N. & Okami, Y. (1980) J. Antibiotics *33*, 1502-1509
25. Yamamoto, H., Hotta, K., Okami, Y. & Umezawa, H. (1981) Biochem. Biophys. Res. Commun. *100*, 1396-1401
26. Ikeda, D., Miyasaka, T., Yoshida, M., Horiuchi, Y., Kondo, S. & Umezawa, H. (1979) J. Antibiotics *32*, 1365-1366
27. Ikeda, D., Horiuchi, Y., Kondo, S. & Umezawa, H. (1980) J. Antibiotics *33*, 1281-1288
28. Horiuchi, Y., Ikeda, D., Kondo, S. & Umezawa, H. (1980) J. Antibiotics *33*, 1577-1580
29. Watanabe, T., Iwasaki, A. & Mori, T. (1980) Japan Kokai 55-55198, April 22
30. Suami, T., Tadano, K. & Matsuzawa, K. (1980) J. Antibiotics *33*, 1289-1299
31. Hamada, M., Takeuchi, T., Kondo, S., Ikeda, Y., Naganawa, H., Maeda, K., Okami, Y. & Umezawa, H. (1970) J. Antibiotics *23*, 170-171
32. Kondo, S., Shibahara, S., Takahashi, S., Maeda, K., Umezawa, H. & Ohno, M. (1971) J. Am. Chem. Soc. *93*, 6305-6306
33. Shibahara, S., Kondo, S., Maeda, K., Umezawa, H. & Ohno, M. (1972) J. Am. Chem. Soc. *94*, 4353-4354
34. Uehara, Y., Kondo, S., Umezawa, H., Suzukake, K. & Hori, M. (1972) J. Antibiotics *25*, 685-688
35. Kondo, S., Iinuma, K., Yoshida, K., Yokose, K., Ikeda, Y., Shimazaki, M. & Umezawa, H. (1976) J. Antibiotics *29*, 208-211
36. Uehara, Y., Hori, M., Kondo, S., Hamada, M. & Umezawa, H. (1976) J. Antibiotics *29*, 937-943
37. Kondo, S., Yoshida, K., Ikeda, T., Iinuma, K., Honma, Y., Hamada, M. & Umezawa, H. (1977) J. Antibiotics *30*, 1137-1139

CLINICAL SIGNIFICANCE OF β-LACTAMASE IN THE TREATMENT OF URINARY TRACT INFECTION

M.Kanematsu, N.Kato, Y.Shimizu
Y.Kawada and T.Nishiura

Department of Urology
Gifu University School of Medicine
Gifu, Japan

I. CLINICAL STUDY

A. Overall Clinical Efficacy

Clinical efficacies of 15 antimicrobial agents, ampicillin, carbenicillin, carindacillin, piperacillin, cefazolin, cefoxitin, cefmetazole, cefamandole, ceftizoxime, cefotaxime, cefmenoxime, moxalactam, gentamicin, amikacin and pipemidic acid on 1370 patients with complicated urinary tract infections(UTI) were studied. All patients had pyuria of over 5 WBCs per hpf, bacteriuria of over 10^4 bacteria per ml of urine and underlying urinary diseases, but none had indwelling catheters. Overall clinical efficacy of the treatment was evaluated by the criteria proposed by the UTI Committee in Japan as excellent, moderate or poor based on the combination of changes in pyuria and bacteriuria (1-3).

Excellent and moderate responses were obtained in 61.8% of the 1053 patients with monomicrobial infections and in 47.6% of the 317 patients with polymicrobial infections. This difference was statistically significant. As shown in Table 1, the differences of clinical efficacies between patients with monomicrobial infections and those with polymicrobial infections were significant only in patients treated with penicillins and cephem antibiotics of the first generation (cefazolin) but were not significant in patients treated with cephem antibiotics of the second and third generation, aminoglycosides or pipemidic acid.

B. Bacteriological Response

Bacteriological responses to the treatment were determined separately in sensitive, intermediate and resistant groups. Bacteriological response was defined as eradicated if bacteria identical to the original strain were not isolated after treatment, and as persisted if bacteria identical to the original strain were isolated after treatment regardless of bacterial count. Susceptibility of bacteria to the given drugs was regarded as sensitive when MICs were 12.5 mcg/ml or less, as intermediate when MICs ranged from 25 to 100 mcg/ml and as resistant when MICs were over 100 mcg/ml.

Bacterial eradication rates of penicillin or cefazolin sensitive strains were significantly lower in patients with polymicrobial infections than those with monomicrobial infections. But no significant differences were observed between patients with monomicrobial and polymicrobial infections regarding the bacterial eradication rates of penicillin or cefazolin intermediate or resistant strains. Furthermore, when bacteriological responses to the treatment were compared between patients with monomicrobial and polymicrobial infections, no significant differences were observed in patients treated with cephem antibiotics of second and third generation, aminoglycosides or pipemidic acid in any of the sensitive, intermediate or resistant groups.

Table 1. *Overall Clinical Efficacies*

Drugs	*Type*	*No.of pts.*	*Excellent + Moderate*	*(%)*	*Poor*	X^2 *test*
Penicillins	*mono*[a)]	*274*	*137*	*(50.0)*	*137*	****[c)]
	poly[b)]	*60*	*14*	*(23.3)*	*46*	
Cephems I	*mono*	*206*	*111*	*(53.9)*	*95*	***[d)]
	poly	*92*	*35*	*(38.0)*	*57*	
Cephems II	*mono*	*107*	*65*	*(60.7)*	*42*	*NS*[e)]
	poly	*53*	*30*	*(56.6)*	*23*	
Cephems III	*mono*	*192*	*152*	*(79.2)*	*40*	*NS*
	poly	*66*	*46*	*(69.7)*	*20*	
Aminoglycosides	*mono*	*85*	*61*	*(71.8)*	*24*	*NS*
	poly	*15*	*8*	*(53.3)*	*7*	
Pipemidic acid	*mono*	*189*	*125*	*(66.1)*	*64*	*NS*
	poly	*31*	*18*	*(58.1)*	*13*	

a) mono : monomicrobial infection
b) poly : polymicrobial infection
*c) ** : $p < 0.01$*
*d) * : $p < 0.05$*
e) NS : not significant

II. FUNDAMENTAL STUDY

A. *Biophotometric study*

The response to ampicillin of single and mixed cultures of *Escherichia coli* and *Staphylococcus epidermidis* was studied in an *in vitro* system by use of biophotometer. MIC of ampicillin for test strains was 6.25 mcg/ml for both *E.coli* and *S.epidermidis*. *S.epidermidis* was shown by macroiodometric assay to produce β-lactamase.

As shown in Fig. 1, when ampicillin was added to achieve a concentration of 25 mcg per ml (4 MIC for both strains), rapid lysis of *E.coli* and delayed lysis of *S.epidermidis* occurred, growth of both organisms was suppressed for the whole of the 24 hour period of observation. When the mixture of the two organisms was exposed to a similar concentration of ampicillin, rapid lysis was again observed, but growth resumed after several hours. A subculture of the mixture after overnight incubation yielded a pure culture of *E.coli*. The effect of ampicillin on *E.coli* was systematically reduced in the presence of *S.epidermidis* at all concentrations tested (from 6.25 mcg/ml to 25 mcg/ml).

B. *β-Lactamase Activity*

Ampicillin was incubated with filtrates of benzylpenicillin-

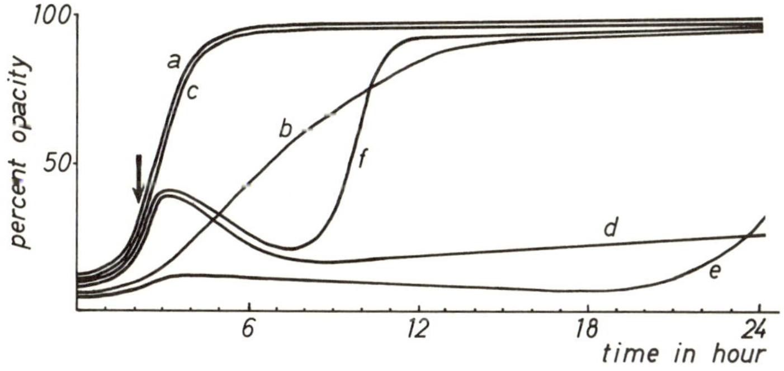

Fig. 1. Biophotometric records of E.coli (a and d) S.epidermidis (b and e), and mixed cultures of both strains (c and f). a, b, c : no antibiotic ; d, e, f : sufficient ampicillin to achieve a concentration of 25 mcg/ml (4 MIC for both strains) added at arrow

induced and non-induced cultures of *S.epidermidis*. The amount of ampicillin remaining in each case after overnight incubation at 37°C, as determined by cup method, was 0 % in induced case and 73.1 ± 7.3 % in non-induced cases.

III. CONCLUSION

β-Lactamase production by another organism in the mixture was considered one possible reason why fully susceptible bacteria which would normally be eradicated when present as the single infecting organism, frequently persisted when present in mixed infection.

REFERENCES

1. Kawada, Y. & Nishiura, T.(1979) Jap. J. Urol. *70*,534-545
2. Ohkoshi, M. & Kawamura, N. (The UTI Committee, Japan) (1980) Chemotherapy (Tokyo) *28*,321-341
3. Ohkoshi, M. (The UTI Committee, Japan) (1980) Chemotherapy (Tokyo) *28*,1351-1358

BIOCHEMICAL MECHANISMS OF β-LACTAM RESISTANCE IN *Streptomyces*

H. Ogawara

Second Department of Biochemistry
Meiji College of Pharmacy
Setagaya, Tokyo, Japan

Streptomyces species are known to be producers of various kinds of antibiotics including β-lactams. Being prokaryotes, *Streptomyces* must protect themselves from the attack of β-lactams, their own metabolites. They attain this by using three mechanisms: β-lactamase, a change of targets (penicillin-binding proteins (PBP)) and a permeability barrier. Here, the properties of these mechanisms in *Streptomyces* are reviewed, comparing with those in pathogenic bacteria (1).

I. β-LACTAMASE

As reported in a previous Symposium (2), most *Streptomyces* strains produce β-lactamases constitutively and extracellularly, irrespective of their resistance to β-lactam compounds. From an evolutionary point of view of β-lactamases in pathogenic bacteria, it is interesting to know the similarities and differences in their properties from those in pathogenic bacteria (3).

A β-lactamase of *S. cellulosae* was purified by CM-52 cellulose chromatography and affinity chromatography on Blue Sepharose to the extent that the final preparation gave a single band on SDS-polyacrylamide gel electrophoretograms stained with Coomassie Brilliant Blue. The enzyme, comprised of a single polypeptide chain of 27,000 dalton, showed pI of 9.5 and an optimum pH of 7.0. In substrate specificity it is similar to some enzymes from gram-negative bacteria, where these enzymes catalyze the hydrolysis of carbenicillin and cloxacillin as well as benzylpenicillin and consist of two subunits of over 20,000 dalton. However, the *Streptomyces* enzyme cannot be inactivated by NaCl and consists of a single polypeptide chain of a smaller molecular weight. Another and completely novel property of this β-lactamase is that this enzyme binds a dinucleotide, $NADP^+$ (4), and has a dinucleotide fold (5). Furthermore, the β lactamase activity is slightly but definitely lessened by the presence of $NADP^+$. These results indicate that for their survival *Streptomyces* species hydrolyze their own metabolites, β-lactams, and escape from their attack upon the targets by employing enzymes derived from those which use $NADP^+$ as a co-factor (6). This indication is supported by the fact that the enzyme binds benzylpenicillin (K_m = 5 x 10^{-4} M) and $NADP^+$ (K_{ass} = 7 x 10^{-4} M) only weakly.

A purified β-lactamase of *S. cacaoi* is also similar to some enzymes in gram-negative bacteria in that both enzymes hydrolyze the "penicillinase-resistant" penicillins, methicillin and cloxacillin as well as benzylpenicillin (7). However, they are different from each other in many aspects. The relative hydrolysis rates of methicillin and cloxacillin to benzylpenicillin are quite different between the *Streptomyces* enzyme and that mediated by the R-factor in gram-negative bacteria. Furthermore, the enzymes from gram-negative bacteria are resistant to inhibition by pCMB and cloxacillin but susceptible to inhibition by NaCl, whereas the *Streptomyces* enzyme is not sensitive to NaCl but is relatively resistant to inhibition by pCMB. The turnover number in the *Streptomyces* enzyme (2.84 x 10^4 mol of benzylpenicillin hydrolyzed /mol of enzyme/min) is comparable to that in *S. aureus* enzymes and TEM-type enzymes, but much higher than that in cloxacillin-hydrolyzing enzymes from gram-negative bacteria. However, both enzymes can hydrolyze cephalosporins and cephamycins only very slowly. Consequently, the arrangement of the functional groups in the active site differs from enzyme to enzyme, although the essential features are similar. The molecular weight of the two is also different: the bacterial enzymes have molecular weights of 24,000 or 40,000, whereas that of the *Streptomyces* enzyme is 34,000.

Another point which should be pointed out here is that this β-lactamase catalyzes the hydrolysis of benzylpenicillin and cloxacillin in different mechanisms. K_m values for benzylpenicillin increased with decreasing pH. In contrast, the pattern of K_m versus pH vurve for cloxacillin showed a parabolic shape with a minimum at pH 7.5. In this, the enzyme is different from *B. cereus* β-lactamase I, where K_m is independent of pH. Different behavior towards the two substrates was also observed in V_{max} versus pH curve. Furthermore, the activation energy for the catalytic reaction of hydrolysis of benzylpenicillin and cloxacillin was quite different. Thus, it is strongly suggested that β-lactamase can change the construction mode of the active site groups to fit particular substrates, or that it has more than two catalytic sites for penicillins which can be used discriminately for individual penicillins such as benzylpenicillin and cloxacillin (8).

In summary, although β-lactamases in *Streptomyces* can be classified into roughly five groups on the basis of their substrate specificities and physico-chemical properties (1,6), it is better to say that they are species- and strain-specific, as in the cases in gram-negative bacteria, are derived from many origins such as target proteins and dehydrogenases, and are closely related to β-lactamases in pathogenic bacteria.

In at least some *Streptomyces* strains, the trait of β-lactamase biosynthesis or its control is lost spontaneously or after treatment with mutagenic agents at high frequency (9, 10). In *S. lavendulae* such mutation was followed by pleiotropic effects on secondary metabolism (aerial mycelium and spore formation, antibiotic production, pigment formation and so on) and by auxotrophy to argininosuccinate or arginine (11). Moreover, in mutants with low β-lactamase activity, production is controlled by nitrogen catabolite repression, and consumption of glycerol as a carbon source is also dependent on the nitrogen source. Reversion of mutants to arginine non-requiring prototrophs restores the ability to form aerial mycelia and spores. A most possible explanation for these phenomena is that a common regulatory gene for secondary metabolism forms a transposable elements, although it is impossible to explain all the observed phenomena by a single theory at the present time. Similar phenomena were observed with *Streptomyces kasugaensis* (12).

II. PBP

Examining PBP by fluorography, at least five PBP could be detected in the membrane fractions of β-lactam non-producers such as *S. cacaoi*, *S. felleus*, *S. lavendulae* and mutants of *Streptomyces* E750-3. In contrast, two to five PBP at most could be detected in the membrane fractions of β-lactam producers such as *S. olivaceus*, *S. clavuligerus* and *Streptomyces* strain 7371 (13, 14). In general, all the PBP in *Streptomyces* have very low affinity for benzylpenicillin. During the growth cycle the PBP patterns in a β-lactam non-producer did not change significantly, whereas a band of molecular weight of 120,000 was observed only in the early logarithmic phase in *Streptomyces clavuligerus*; PBP-1 (Mr = 83,000) and PBP-2 (Mr = 79,000), a presumed lethal target of many β-lactams (15), appeared only slightly in this phase. Furthermore, heat stability for binding benzylpenicillin to PBP and the releasing curve of [^{14}C] from [^{14}C]benzylpenicillin·PBP complexes are quite different in the two groups of *Streptomyces* species. These differences are quite significant, considering that the microorganisms belong to the same genus; PBP patterns in *E. coli*, *S. typhimurium*, *P. aeruginosa*, *Proteus* and *B. subtilis* are very much alike, although they belong to different genera, and the exact correlation in physiological roles of these bands remains to be clarified. The PBP pattern in *Streptomyces cacaoi*, a β-lactam non-producer, is quite similar to that in *B. subtilis*, suggesting that low affinity PBP and their changes during the cell cycle are the main mechanisms of their protection against β-lactam antibiotics. The accessibility barrier to targets of β-lactams contributes to some extent to a higher level of β-lactam resistance in *Streptomyces cacaoi*.

REFERENCES

1. Ogawara, H. (1981) *Microbiol. Rev.* in press (December issue)
2. Ogawara, H., Horikawa, S., Shimada-Miyoshi, S. & Yasuzawa, K. (1978) in Microbial Drug Resistance (Mitsuhashi, S., ed.) Vol. 2, pp.331-333, Japan Scientific Societies Press, Tokyo
3. Hamilton-Miller, J. M. T. & Smith, J. T. (1979) Beta-lactamases, Academic Press, New York
4. Ogawara, H. & Horikawa, S. (1979) *J. Antibiot. 32*, 1328-1335
5. Thompson, S. T., Cass, K. H. & Stellwagen, E. (1975) *Proc. Natl. Acad. Sci. USA 72*, 669-672
6. Ogawara, H. (1981) in Actinomyces Biology (Kutzner, H. J., Pulverer, G. & Schaal, K. P., ed.) Gustav Fischer Verlag, Berlin, in press
7. Ogawara, H., Mantoku, A. & Shimada, S. (1981) *J. Biol. Chem. 256*, 2649-2655
8. Ogawara, H. & Umezawa, H. (1975) in Microbial Drug Resistance (Mitsuhashi, S. & Hashimoto, H.ed.) Vol. 1, pp.375-389
9. Ogawara, H. & Nozaki, S. (1977) *J. Antibiot. 30*, 337-339
10. Matsubara-Nakano, M., Kataoka, Y. & Ogawara, H. (1980) *Antimicrob. Agents Chemother. 17*, 124-128
11. Nakano, M. M. & Ogawara, H. (1980) *J. Antibiot. 33*, 420-425
12. Nakano, M. M., Ozawa, K. & Ogawara, H. (1980) *J. Bacteriol. 143*, 1501-1503
13. Ogawara, H. & Horikawa, S. (1980) *Antimicrob. Agents Chemother. 17*, 1-7
14. Nakazawa, H., Horikawa, S. & Ogawara, H. (1981) *J. Antibiot. 34*, 1070-1072
15. Ogawara, H. & Horikawa, S. (1980) *J. Antibiot. 33*, 620-624

MECHANISM OF ANTIBACTERIAL ACTION OF CEFMENOXIME AGAINST *PROTEUS VULGARIS* WHICH PRODUCES β-LACTAMASE THAT HYDROLYZES THE DRUG

K. Okonogi, M. Kida, and M. Kuno

Central Research Division
Takeda Chemical Industries Ltd.
Yodogawa-ku, Osaka, Japan

S. Mitsuhashi

Department of Microbiology, School of Medicine
Gunma University
Maebashi, Gunma, Japan

Cefmenoxime, 7β-[2-(2-aminothiazol-4-yl)-(Z)-2-methoxyiminoacetamido]-3-[(1-methyl-1H-tetrazol-5-yl)thiomethyl] ceph-3-em-4-carboxylic acid, is a broad-spectrum cephalosporin (1). It is resistant to hydrolysis by various β-lactamases, except that of *Proteus vulgaris* (2). In spite of its susceptibility to the β-lactamase, cefmenoxime is very active against *P. vulgaris*. Therefore, we investigated the basis of the antibacterial activity of cefmenoxime against *P. vulgaris*.

ANTIBACTERIAL ACTIVITY AGAINST *P. VULGARIS*

Seventy percent of the clinical isolates of *P. vulgaris* were inhibited by 0.1 μg of cefmenoxime per ml, and 95% were inhibited by less than 1 μg/ml. *P. vulgaris* GN4818, which was inhibited by 0.05 μg of cefmenoxime per ml, was used in the following experiments.

STABILITY TO β-LACTAMASE

As shown in Table 1, the rate of hydrolysis of cefmenoxime by the β-lactamase of *P. vulgaris* GN4818 was 58% of that of cephaloridine. The Km value of cefmenoxime for the enzyme was 196 μM (105 μg/ml), and was almost the same as those of

Table 1. Hydrolysis of cefmenoxime and several cephalosporins by β-lactamase of P. vulgaris GN4818

Substrate	Vmax (relative)	Km (μM)	Vmax/Km (relative)
Cephaloridine	100	89	100
Cefmenoxime	58	196	26
Cefotiam	292	272	95
Cefazolin	432	133	289
Cephalothin	139	32	391
Cefuroxime	270	185	130

several other cephalosporins.

INDUCTION OF β-LACTAMASE

The β-lactamase of *P. vulgaris* is an inducible enzyme (3). Therefore, we investigated the ability of cefmenoxime to induce the β-lactamase of *P. vulgaris*. When cefmenoxime was

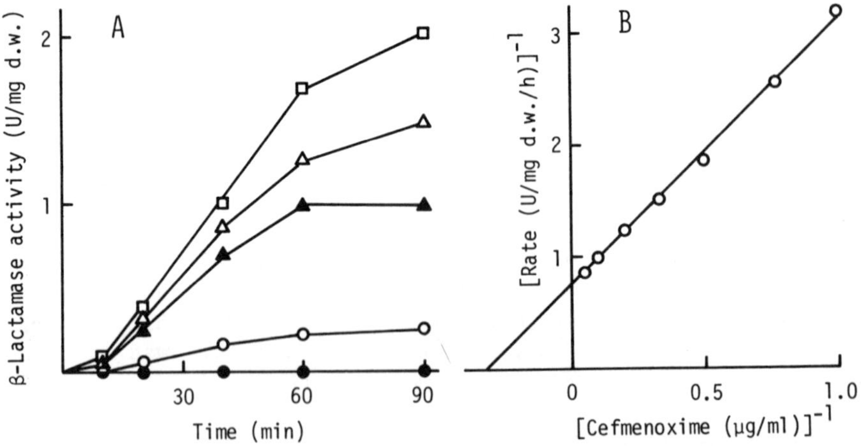

Fig. 1. Induction of P. vulgaris β-lactamase by cefmenoxime. (A) P. vulgaris GN4818 was grown to the logarithmic phase, and β-lactamase was induced for the time indicated by the addition of 0.1 (●), 1 (○), 10 (▲), 100 (△), and 1,000 (□) μg of cefmenoxime per ml. (B) The reciprocal of the rate of the enzyme synthesis was plotted against the reciprocal of the concentration of cefmenoxime.

Table 2. Affinities of cefmenoxime for penicillin-binding proteins of P. vulgaris

Concentration (μg/ml) required to inhibit binding of ^{14}C-benzylpenicillin by 50% to PBP:						
1A	1B	2	3	4	5	6
0.37	6.0	0.52	0.0088	3.1	>30	>30

added to the logarithmically growing cells, the organism began to synthesize β-lactamase after 10 minutes lag time, and continued to synthesize the enzyme for at least 30 minutes (Fig. 1A). The rate of enzyme synthesis during this 30 minutes depended on the concentration of cefmenoxime. A plot of the reciprocal of the rate of enzyme synthesis against the reciprocal of the concentration of the inducer was found to be linear (Fig. 1B). From Figure 1B, the maximum rate of enzyme synthesis (Rmax) and the concentration of cefmenoxime required for enzyme synthesis at half of Rmax were determined as 1.4 U/mg of dry weight/h and 3.2 μg/ml, respectively. The rate of enzyme synthesis induced by 0.05 μg of cefmenoxime per ml, which was a minimum inhibitory concentration (MIC) for this strain, was calculated to be 0.021 U/mg of dry weight/h. This rate was 1.5% of the Rmax.

AFFINITIES FOR PENICILLIN-BINDING PROTEINS

Penicillin-binding proteins (PBPs) located on the inner membrane of *P. vulgaris* were separated on sodium dodecyl sufate/polyacrylamide gel electrophoresis. The seven major proteins were named PBP-1A, 1B, 2, 3, 4, 5, and 6 in order of increasing electrophoretic mobility. The patterns on electrophoresis of the PBPs of *P. vulgaris* was similar to those of *Proteus mirabilis* and *Proteus morganii* (4). Cefmenoxime showed remarkably high affinity for PBP-3, and good affinities for PBPs-1A and 2 (Table 2). Cefmenoxime induced filamentous cells at 0.05 μg/ml.

SUSCEPTIBILITIES OF MUTANT STRAINS

The β-lactamase-defective and constitutive strains were obtained from *P. vulgaris* GN4818. The MIC of cefmenoxime for the β-lactamase-defective strain, CS4035 was 0.05 μg/ml and was the same as that for the wild strain, whereas the MIC for the β-lactamase-constitutive strain, CS4017 was 12.5 μg/ml.

DISCUSSION

Cefmenoxime was slightly more resistant to hydrolysis by the β-lactamase of *P. vulgaris* than several other cephalosporins. This degree of stability to the β-lactamase, however, cannot explain the strong antibacterial activity of cefmenoxime against *P. vulgaris*, even if the Km value is taken into account. Cefmenoxime scarcely induced β-lactamase formation by *P. vulgaris* at 0.1 μg/ml or less, although it was an active inducer at higher concentrations. On the other hand, cefmenoxime inhibited [^{14}C]benzylpenicillin binding to the PBP-3 of *P. vulgaris* by 50% at 0.0088 μg/ml. Cefmenoxime induced filamentous cells at 0.05 μg/ml, at which concentration it did not bind to PBPs other than PBP-3. This suggests that PBP-3 of *P. vulgaris* is responsible for the cell division and is a primary target of cefmenoxime. Induction of filamentous cells by 0.05 μg of cefmenoxime per ml also suggests that cefmenoxime can bind to the PBP-3 in intact cells as well as in isolated membranes at this concentration.

From the above observations, we concluded that cefmenoxime had a higher affinity for the antibacterial target site than for the β-lactamase repressor and thus inhibited the growth of *P. vulgaris* before it acted as a β-lactamase inducer. The susceptibilities of the β-lactamase defective and constitutive strains to cefmenoxime support this conclusion.

REFERENCES

1. Tsuchiya, K., Kondo, M., Kida, M., Nakao, M., Iwahi, T., Nishi, T., Noji, Y., Takeuchi, M. & Nozaki, Y. (1981) *Antimicrob. Agents Chemother.* *19*, 56-65
2. Okonogi, K., Kuno, M., Kida, M. & Mitsuhashi, S. (1981) *Antimicrob. Agents Chemother.* *20*, 171-175
3. Sawai, T., Mitsuhashi, S. & Yamagishi, S. (1968) *Jap. J. Microbiol.* *12*, 423-434
4. Ohya, S., Yamazaki, M., Sugawara, S. & Matsuhashi, M. (1979) *J. Bacteriol.* *137*, 474-479

A NEW TYPE PENICILLINASE PRODUCED BY *BACTEROIDES FRAGILIS* AND THE TRANSFERABILITY OF THE PENICILLINASE PRODUCTION

K. Sato, Y. Matsuura, M. Inoue, and S. Mitsuhashi

Laboratory of Drug Resistance in Bacteria,
School of Medicine, Gunma University,
Maebashi, Japan

INTRODUCTION

The β-lactamase produced by gram-positive and gram-negative bacteria have been regarded as significant roles in the resistant organisms against β-lactam antibiotics (7,11). The enzymes in gram-positive bacteria are mostly known to be of inducible- and extracellar-type, and much more active to penicillins than cephalosporins (6,8). The chromosome-mediated β-lactamases in gram-negative bacteria are species-specific in their physicochemical and immunological properties and are classified into several groups by substrate profiles, inhibitory studies, physicochemical and immunological properties (7,11).

The strains of *Bacteroides fragilis* which have been recognized as important pathogens were known to be moderately or highly resistant against penicillins and cephalosporins (3,7). These resistances have been ascribed to the production of β-lactamase by several investigators (1,2,4,9,12). We reported the presence of chromosome-mediated β-lactamase in *B. fragilis* and classified the enzyme as a new group, i.e., cefuroximase (CXase) in enzymological and immunological properties (7). Garrod (5) found penicillinase-like activity in *B. fragilis*, and Pinkus et al. (5) also reported penicillin-hydrolyzing enzyme. However, the enzymes mainly hydrolyzed more actively cephalosporins than penicillins (1,2,4,9,12), and were sensitive to inhibitors such as cefoxitin, cloxacillin, and p-chloromercuribenzoate (2,3). Therefore resistance to β-lactam antibiotics in *B. fragilis* strains is attributed to the production of cefuroximase, penicillinase or both enzymes like gram-negative aerobes.

This paper deals with the studies on some properties of penicillinase from a strain of highly penicillin-resistant *B. fragilis* GN11499 which was found from 80 clinical isolates of *B. fragilis*. We also describe the conjugal transferability of the penicillinase production by filter mating method.

SUSCEPTIBILITY TEST AND β-LACTAMASE ACTIVITY

Susceptibility of 80 *B. fragilis* strains against various β-lactam antibiotics was examined by agar dilution method. It was found that the minimal inhibitory concentrations (MICs) of 52 strains were 50 μg/ml, but 28 strains were resistant to 100 μg/ml of cephaloridine and ampicillin. Especially, *B. fragilis* GN11499 showed high resistance to penicillins and its MICs for penicillin G, ampicillin, carbenicillin, and cloxacillin were 800, 3200, 3200, 3200 or more, respectively. *B. fragilis* GN11499 was moderately resistant to cephaloridine and cephalothin. The crude enzyme from GN11499 was purified by column chromatography and SDS-polyacrylamide slab gel electrophoresis gave a single protein band.

A. *Pelationship between bacterial growth and formation of the penicillinase*

The production of penicillinase at various phases of growth of GN11499 was studied in brain heart infusion broth (Difco) supplemented with 0.5% yeast extract, 0.05% cysteine hydrochloride and 0.01% hemin (BHI-CH broth) at 37°C by shaking culture method. Maximal activities in the intact cells were detected at the late logarithmic phase after about 6 h of incubation. The penicillinase activities decreased rapidly thereafter with the time of incubation. Release of the enzyme from bacterial cells into surrounding media was parallel with the growth of bacteria and the activity was found between 7 and 24 h after incubation. This result was similar to the report of Olsson et al. (9), indicating that *B. fragilis* strain produced cephalosporinase and released the enzyme at the rate of 50 percent or more in the stationary phase of shaked-culture. These facts indicated that *B. fragilis* strains produced extracellulary β-lactamase as like penicillinase in gram-positive aerobes.

B. *Specific activity and substrate profile of the penicillinase*

The substrate profiles of penicillinase were studied comparing with the enzyme of *B. fragilis* GN11477 which was known to produce cephalosporinase (12). Substrate profiles were expressed by the relative rate of hydrolysis taking the rate of penicillin G hydrolysis as 100 (Table 1).
The penicillinase from GN11499 hydrolyaed ampicillin and

cloxacillin at much higher rate than penicillin G, carbenicillin, cephaloridine, and cephalothin. Substrate profile indicated a unique type of penicillinase. This is very similar to oxacillin-hydrolyzing penicillinase mediated by R factor reported by Yamagishi et al. (13). The enzyme from GN11477, on the contrary, hydrolyzed cephaloridine and cephalothin at a high rate, but hydrolyzing activity of penicillins was rather low.

C. Inhibitory study

Inhibitions of the penicillinase by various compounds are shown in Table 2. Clavulanic acid and CP45899 which were known as penicillinase inhibitors showed high inhibitory effects. Similarly MK0787 and cefoxitin were proved to be the strong inhibitors. The sulfhydryl inhibitor, p-chloromercuribenzoate also inhibited the enzyme activity.

Table 1. *Specific activities and substrate profiles in the crude enzymes from B. fragilis GN11499 and GN 1477*

Substrates	*GN11499*		*GN11477*	
	β-Lactamase activity[a]	*Substrate profile*[b]	*β-Lactamase activity*[a]	*Substrate profile*[b]
Penicillin G	*0.007*	*100*	*0.015*	*100*
Ampicillin	*0.250*	*357*	*0.004*	*27*
Carbenicillin	*0.030*	*43*	*0.001*	*7*
Cloxacillin	*0.190*	*271*	*0.001*	*7*
Cepfaloridine	*0.062*	*71*	*0.500*	*3333*
Cephalothin	*0.041*	*57*	*0.413*	*2765*
Cefoxitin	*0.001*	*1*	*0.001*	*7*

[a] *Specific activities of crude enzymes, units/mg of protein.*

[b] *Substrate profiles are expressed by the relative rate of hydrolysis of each substrate taking the absolute rate of penicillin G as 100.*

TRANSFERABILITY OF THE PENICILLINASE PRODUCTION

The penicillinase production and tetracycline resistance

in *B. fragilis* GN11499 were found to be conjugally transferable by using a filter-mating method (13). As shown in Table 3, the ability of penicillinase production and tetracycline resistance were transferred from GN11499 to two drug-sensitive strains of *B. fragilis* GN11495 and *B. vulgatus* GN11496 at frequencies between 10^{-6} to 10^{-7}. All of the ampicillin-resistant transconjugants produced penicillinase and showed tetracycline resistance. Transfer of the resistance was not observed when donor and recipient cells were mixed in BHI-CH medium.

Table 2. *Effects of various inhibitors on the penicillinase from B. fragilis GN11499*

Inhibitors	*Percentage of inhibition*		
	0.10	*1.0*	*10.0 (μM)*
Clavulanic acid	*96*[a]	*100*	*100*
CP45899	*75*	*100*	*100*
MK0787	*100*	*100*	*100*
Cefoxitin	*90*	*100*	*100*
p-chloromercuri-benzoate	*23*	*84*	*100*

[a] *The enzyme was preincubated in 0.05 M phosphate buffer (pH 7.0) with each inhibitor at indicated concentration for 5 min at 30°C and the remaining activity was assayed using 200 μM of ampicillin as a substrate.*

Table 3. *Transfer of penicillinase production and tetracycline resistance by filter mating*

Donor	*Recipient*	*Selective*[a] *drug*	*Frequency*	*Transconjugants*	
				Drug resistance	*β-lactamase*[b]
GN11499	*GN11495*	*APC or TC*	*6 x* 10^{-6}	$(APC.TC)^r$	*0.1*
GN11499	*GN11495*	*APC*	*2 x* 10^{-6}	$(APC.TC)^r$	*0.12*
GN11499	*GN11496*	*TC*	*1 x* 10^{-7}	$(APC.TC)^r$	*0.08*
GN11499	*GN11496*	*APC or TC*	*5 x* 10^{-6}	$(APC.TC)^r$	*0.07*
-	*GN11495*	*APC*	10^{-8}	$(APC.TC)^s$	-
-	*GN11495*	*TC*	10^{-8}	$(APC.TC)^s$	-
-	*GN11496*	*APC or TC*	10^{-8}	$(APC.TC)^s$	-

[a] *GN11495 and GN11496 were resistant to rifampicin(RFP) and clindamycin(CLD) and used as the recipients of R plasmids.*

The selective media contained both RFP(200 μg/ml) and CLD (100 μg/ml), and each of the selective drugs ampicillin (APC, 50 μg/ml) and tetracycline(TC, 12.5 μg/ml).

b) *β-Lactamase activity: units/mg of protein; (-), no β-lactamase production.*

CONCLUSION

The penicillinase produced by *B. fragilis* GN11499 has a unique substrate profile. The enzyme hydrolyzed ampicillin and cloxacillin at much higher rate than penicillin G, carbenicillin and cephalosporins. The enzyme activity was inhibited by various inhibitors including cefoxitin. Cefoxitin was known to be an inhibitor against typhical cephalosporinases. These properties of penicillinase mediated by plasmid was different from those reported by Richmond(11) and by Mitsuhashi and Inoue (7). Resistance to ampicillin and tetracycline of GN11499 was conjugally transferred to two recipients of *B. fragilis* and *B. vulgatus*, indicating a wide spread of R plasmids in both aerobes and anaerobes.

REFERENCES

1. Anderson, J.D., and R.B. Sykes. 1973. J. Med. Microbiol. *6*, 201-206.
2. Britz, M.L., and R.G. Wilkinson. 1978. Antimicrob. Agents Chemother. *13*, 373-382.
3. Darland, G., and J. Birrnbaum. 1977. Antimicrob. Agents Chemother. *11*, 725-734.
4. Del Bene, V.E., and W.E. Farrar, Jr. 1973. Antimicrob. Agents Chemother. *3*, 369-372.
5. Garrod, L.P. 1955. Br. Med. J. *2*, 1529-1531.
6. Kuwabara, S. 1970. Biochem. J. *118*, 457-465.
7. Mitsuhashi, S. and M. Inoue. In S. Mitsuhashi(ed), Beta-lactam antibiotics. 1981. Japan Sci. Soc. Press, Tokyo/ Springer-Verlag, Berlin Heidelberg, New York. pp. 15-37.
8. Novic, R.P., and D. Bouanchaud. 1971. Ann. N.Y. Acard. Sci. *182*, 279-294.
9. Olsson, B., C.E. Nord, and T. Wadstom. 1976. Antimicrob. Agents Chemother. *9*, 727-735.
10. Pinkus, G., G. Veto, and A.I. Braude. 1968. J. Bacterol. *96*, 1437-1438.
11. Richmond, M.H., and R.B. Sykes. 1973. Adv. Microbiol. Physiol. *9*, 31-38.

12. Sato, K., M. Inoue, and S. Mitsuhashi. Antimicrob. Agents Chemother. *17*, 736-737.
13. Yamagishi, S., K. Ohara, T. Sawai, and S. Mitsuhashi. 1969. J. Biochem. *66*, 11-20.

β-LACTAMASE FROM PSEUDOMONAS MALTOPHILIA

Y. SAINO, M. INOUE, S. MITSUHASHI

Department of Microbiology
School of Medicine, Gunma University
Maebashi, Gunma, Japan

F. KOBAYASHI

Tokyo Research Laboratories
Kowa Co., Ltd.
Higashimurayama, Tokyo, Japan

Pseudomonas maltophilia has been reported to be generally resistant to many aminoglycosides and β-lactams including newly introduced N-formimidoyl thienamycin (MK0787) (1-2). Epidemiological studies have disclosed that resistance to β-lactams is mainly due to the production of β-lactamase by resistant strains (3). In spite of many papers on β-lactamase production, there has been no report on the β-lactamase of *P. maltophilia*. We have found that seven strains of *P. maltophilia* produced two types of β-lactamase. Present paper deals with the purification and biochemical properties of the penicillin β-lactamase produced by *P. maltophilia GN12873*.

EXPERIMENTAL

A. β-Lactamase activity in P. maltophilia

All of seven strains tested produced two kinds of β-lactamase, which were separable by CM-Sephadex column chromatography. One is an inducible penicillinase (PCase) (L-1), and the other is an inducible cephalosporinase (CSase) (L-2). The substrate profiles of the L-1 from seven strains were found to be quite similar to each other.

B. Purification of the L-1 from P. maltophilia GN12873

P. maltophilia GN12873 was grown at 30°C in the medium B containing penicillin G (100 μg/ml) as an inducer. After 3 hr

Table 1. *Kinetics of hydrolysis of β-lactam antibiotics by β-lactamase L-1 from P. maltophilia GN12873*

Antibiotic	Km (μM)	Ki (μM)	Vmax (relative)
Penicillin G	210		100
Ampicillin	172		58
Carbenicillin	444		46
Cloxacillin	229		42
Cephaloridine	385		6
Cephalothin	90.9		5
Cephalexin	400		4
Cefoperazone	30.8		3
Cefuroxime	13.2		1
Cefoxitin		3.34	
YM09330		3.21	
Moxalactam		0.59	
MK0787	242		24
Carpetimycin A		0.94	
Carpetimycin B		0.06	
Clavulanic acid		>100	
CP-45899		24.5	

of incubation, the cells were harvested by centrifugation and washed twice with 50 mM phosphate buffer (pH 7.0). The washed cells were resuspended in the same buffer and disrupted by ultrasonic disintegrator. From the cell-free extract, the L-1 was purified about 140-fold by means of QAE-Sephadex and Sephadex G-200 column chromatography. The purified enzyme preparation gave a single protein band on the polyacrylamide gel electrophoresis (PAGE) and SDS-PAGE.

C. Physicochemical properties

The molecular weight of the purified L-1 was estimated to be 118,000 by gel filtration through Sephadex G-200 and to be 26,000 by SDS-PAGE. Neutral sugar was not detected in the enzyme. The isoelectric point was 6.9.

D. Enzymological and immunochemical properties

The optimal pH and optimal temperature for the activity was 8.0 and 35°C, respectively. The kinetic parameters of the purified L-1 for various β-lactams are shown in Table 1. The enzyme showed much higher activity against penicillin G, ampicillin, carbenicillin, cloxacillin, and MK0787 than cephaloridine, cephalothin, and cephalexin. The activity of

L-1 was competitively inhibited by carpetimycins A and B (4) and cephamycins, and not by clavulanic acid. The enzyme was also inhibited by EDTA but recovered almost completely by addition of zinc ion. Anti-L-1 serum prepared by immunizing mice with the purified L-1 of *P. maltophilia GN12873* could neutralize the activity of L-1 from seven strains. However, the activity of L-2 and those of other β-lactamases previously reported (5) were not inhibited by this antiserum.

DISCUSSION

It was found that *P. maltophilia* produced two types of β-lactamase (L-1 and L-2). The L-1 seemed to be a metallo-enzyme with four identical subunits. The substrate profile of the L-1 suggested that the enzyme was of PCase type, whereas its activity was not inhibited by clavulanic acid which is known to be a PCase inhibitor (5). Especially, the L-1 could hydrolyze MK0787 at a significant rate. Thus, the resistance of *P. maltophilia* against MK0787 may be due to the production of this enzyme. It is interesting to note that a renal dipeptidase, which inactivates MK0787 and related carbapenem antibiotics, is a zinc metalloenzyme (6-7).

REFERENCES

1. Fass, R. J. & Barnishan, J. (1979) Antimicrob. Agents Chemother. *16*, 434-438
2. Kesado, T., Hashizume, T. & Asahi, Y. (1980) Antimicrob. Agents Chemother. *17*, 912-917
3. Mitsuhashi, S., Yamagishi, S. Sawai, T. & Kawabe, H. (1977) in R factor drug resistance plasmid (Mitsuhashi, S., ed.) pp.195-251, University of Tokyo Press, Tokyo
4. Nakayama, M., Iwasaki, A., Kimura, S., Mizoguchi, T., Tanabe, S., Murakami, A., Watanabe, I., Okuchi, M., Ito, H., Saino, Y., Kobayashi, F. & Mori, T. (1980) J. Antibiot. *33*, 1388-1390.
5. Mitsuhashi, S. & Inoue, M. (1981) in Beta-lactam antibiotics (Mitsuhashi, S., ed.) pp.41-56, Japan Scientific Societies Press, Tokyo & Springer-Verlag, Berlin Heidelberg New York.
6. Kropp, H., Sundelof, J. G., Hajdu, R. & Kahan, F. M. (1980) in Program Abstr. 20th Intersci. Conf. Antimicrob. Agents Chemother., abstr. no.272
7. Campbell, B. J. (1970) in Methods in Enzymology (Perlmann, G. E. & Lorand, L., ed.) Vol. *19*, pp.722-729, Academic Press. New York.

PURIFICATION AND PROPERTIES OF β-LACTAMASE FROM CLOSTRIDIUM SYMBIOSUM

M. TAJIMA, Y. TAKENOUCHI, H. DOMON

Biological Research Laboratories, Sankyo Co., Ltd.
Shinagawa-ku, Tokyo, JAPAN

and

S. SUGAWARA

Research Institute, Sankyo Co., Ltd.
Shinagawa-ku, Tokyo, JAPAN

The β-lactamase of *Bacteroides fragilis* was firstly noted in 1968 but received little attention until 1973 (1-2). Recently, the properties such as molecular weight (29000-32000), isoelectric point (pI 4.6-5.4), and substrate profile of the enzyme were clarified (3-5). While there are few report of β-lactamases of genus *Clostridium*. The β-lactamases produced by *C. ramosum* and *C. clostridiiforme* are markedly different from *Bacteroides* enzymes, both in substrate profile and the lack of inhibition by *p*CMB and cloxacillin (6). The authors have purified a β-lactamase from *C. symbiosum* T-1. The purified β-lactamase hydrolysed all of cephalosporins tested but not hydrolysed cephamycins and semi-synthetic penicillins at all.

EXPERIMENTAL

A. Purification of β-lactamase from C. symbiosum T-1

The organism studied was obtained from pus and it was identified according to the Bergey's Manual.

Cultures of *C. symbiosum* in the Trypto-soy broth (30 g) supplemented with yeast extract (5 g), cysteineHCl (0.6 g) and glucose (7.5 g) were centrifuged, washed once with 0.1 M phosphate buffer, pH 7.2 and resuspended in the same buffer. The cells were disrupted by an ultrasonic disintegrator and the cell-free extract was passed through a CM-Sephadex column equilibrated with 0.01 M phosphate buffer, pH 6.0. The first peak contained a large amount of inactive protein. The β-lactamase appeared in the second peak after elution with a linear gradient of 0.01 to 0.2 M buffer, pH 6.0. The active fractions (eluted with 0.08 to 0.09 M) were pooled and used for isoelectric focusing as previously reported (7). Fractions around the peak of enzyme activity (pH 6.34 to 6.60) were pooled and applied to a Sephadex G-200 column equilibrated with 0.05 M Tris-HCl-0.1 M NaCl buffer, pH 8.0, and eluted with the same buffer. The active fractions were dialysed against distilled water and lyophilized.

The purity of the lyophilized preparation was checked by polyacrylamide gel electrophoresis (PAGE) and SDS-PAGE. In PAGE the enzyme activity and the stained band coincided, and SDS-PAGE gave a single protein band.

B. *Physicochemical properties of the β-lactamase*

The molecular weight of the purified enzyme, determined by gel filtration through Sephadex G-200, was 32000, and, from its electrophoretic mobility in the SDS-PAGE, the subunit molecular weight was also estimated to be 32000. Neutral sugar was not detectable in the enzyme. Cysteine and tryptophan were not found among its amino acids. The isoelectric point was 6.50.

Table 1. *Kinetics of hydrolysis of β-lactam antibiotics by β-lactamase from Clostridium symbiosum T-1*

Antibiotic	*Vmax (Relative)*	*Km (μM)*	*Ki (μM)*
Cephaloridine	*100*	*65*	
Cefazolin	*115*	*133*	
Cephalothin	*87*	*30*	
Cephalexin	*5*	*26*	
Cephradine	*5*	*19*	
Cefsulodin	*29*	*71*	
Cefotiam	*22*	*15*	
Cefamandole	*23*	*10*	
Cefoperazone	*37*	*44*	
Cefuroxime	*49*	*54*	
Cefotaxime	*21*	*19*	
Cefmenoxime	*19*	*14*	
Ceftizoxime	*3*		*2.0*
Cefmetazole	*0*		*0.03*
Cefoxitin	*0*		*0.21*
Moxalactam	*0*		*0.14*
Penicillin G	*0.8*		
Ampicillin	*0*		*0.24*
Carbenicillin	*0*		*1.1*
Cloxacillin	*0*		*0.09*
Methicillin	*Not tested*		*5.4*

C. *Enzymic properties of the β-lactamase*

The pH-activity curve for the enzyme was rather broad and had a maximum at about 6.5, and the optimal temperature was 50 °C. The kinetic parameters of the enzyme activity towards cephalosporins, cephamycins and penicillins are given in Table 1. The enzyme hydrolysed all of the cephalosporins tested including cefuroxime, cefotaxime and cefmenoxime but not hydrolysed cephamycins and semi-synthetic penicillins at all. Its activity was inhibited by both cephamycins and semi-synthetic penicillins.

DISCUSSION

The purified β-lactamase from *C. symbiosum* T-1 hydrolysed cephalosporins including cefuroxime and cefotaxime and its activity was inhibited by both cloxacillin and cephamycins. These properties are similar to the substrate profile of *B. fragilis* enzymes, however, isoelectric point was different from each other. The *C. ramosum* enzyme hydrolysed both penicillins and cephalosporins at approximately equal rates and *C. clostridiiforme* enzyme hydrolysed penicillins but not cephalosporins and it was not inhibited by cloxacillin and cefoxitin (6). Thus, the β-lactamase from *C. symbiosum* was classified as the cefuroxime-hydrolysing β-lactamase (8).

REFERENCES

1. Pikus, G., Veto, G. & Braude, A.I. (1968) J. Bacteriol. *96*, 1437-1438
2. Del Bene, V.E. & Farrar, Jr., W.E. (1973) Antimicrob. Agents Chemother. *3*, 369-372
3. Britz, M.L. & Wilkinson, R.G. (1978) Antimicrob. Agents Chemother. *13*. 373-382
4. Sato, K., Inoue, M. & Mitsuhashi, S. (1980) Antimicrob. Agents Chemother. *17*, 736-737
5. Olsson-Liljequist, B., Dornbusch, K. & Nord, C.E. (1980) Antimicrob. Agents Chemother. *18*, 220-225
6. Weinrich, A.E. & Del Bene, V.E. (1976) Antimicrob. Agents Chemother. *10*, 106-111
7. Tajima, M., Takenouchi, Y., Sugawara, S., Inoue, M. & Mitsuhashi, S. (1980) J. Gen. Microbiol. *121*, 449-456
8. Hirai, K., Iyobe, S., Inoue, M. & Mitsuhashi, S. (1980) Antimicrob. Agents Chemother. *17*, 355-358

BIOCHEMISTRY : PENICILLIN BINDING PROTEINS AND OUTER MEMBRANES

PENICILLIN BINDING PROTEINS AS TARGETS OF THE LETHAL ACTION OF β-LACTAM ANTIBIOTICS[1]

Jack L. Strominger

Biochemistry and Molecular Biology
Sherman Fairchild Biochemistry Building
7 Divinity Avenue Cambridge, Massachusetts 02138, U.S.A.

Bacterial cell walls are made in three stages. The first stage in the cytoplasm ends in the synthesis of UDP-acetylmuramyl-pentapeptide (the nucleotide which accumulates in penicillin inhibited bacteria). The second stage occurs in the cell membrane. In the membrane the nucleotide precursors are used to form, on a lipid intermediate, a disaccharide-pentapeptide which is then transferred to a growing cell wall to add a new unit to the wall. The lipid was identified as a C_{55}-isoprenoid alcohol (undecaprenol). The third step is the cross linking of nascent peptidoglycan strands in a transpeptidation reaction in which the terminal D-alanine of one strand is eliminated in formation of a cross link. This is the penicillin sensitive reaction in bacterial cell wall synthesis which leads to the lethality of penicillins. A second penicillin sensitive enzyme, D-alanine carboxypeptidase, was also identified.

A mechanism for the inhibition of the transpeptidation reaction by penicillin was proposed in 1965, *viz.* that penicillins are substrate analogs. In the normal reaction the substrate, *i.e.* the nascent peptidoglycan strand, reacts with the enzyme to form an acyl enzyme intermediate. That acyl enzyme intermediate can then react with the second strand to form the cross link and regenerate the inhibited enzyme. If the intermediate is attacked by water, it becomes a carboxypeptidase. If penicillins are substrate analogs of the nascent peptidoglycan strand, they would then acylate this enzyme forming a penicilloyl enzyme intermediate.

I would like to present evidence for the correctness of this hypothesis. It has three important predictions: 1) the inhibited enzyme is a penicilloyl enzyme; 2) acyl enzymes are intermediates in the reaction sequence with substrate and 3) if penicillins are in fact substrate analogs, then the penicilloyl residue and the acyl derived from substrate must be substituted on the same amino acid in the inhibited enzyme.

The binding of penicillin to bacterial membranes was described early in the history of studies of the mechanism of action of penicillin. These early workers described saturation of the membranes of sensitive bacteria

[1] *This paper is an edited transcript of the talk delivered at the Tokyo Symposium. Supported by a research grant from the NIH (AI-09152) and NSF (PCM-78-24129).*

by radioactive penicillin and the fact that membranes of resistant bacteria are saturated with much more difficulty. Initially we thought that it would only be necessary to isolate the protein to which penicillin was bound in order to isolate transpeptidase, but the situation turned out to be much more complex. A clue to the complexity had in fact been published by Gardner in Fleming's laboratory in the 1940s. He observed that the morphological effect depended on the concentration of penicillin G. That is, at low penicillin concentrations filaments are formed; at higher concentrations bulges are formed which lead to lysis. So Gardner appreciated that the mechanism of action of penicillin was quite complex.

More modern biochemical techniques have revealed that all bacterial species examined contain multiple penicillin binding proteins. Bacilli, *S. aureus.* and *E. coli* contain from 4 to 8 protein components which bind radioactive penicillin. The discovery of multiple penicillin binding proteins raised many questions: what are their functions, how many killing sites are there, what enzymatic reactions are catalyzed, what is the mechanism by which the enzymes are inhibited and are these enzymes related to penicillinases? If the penicilloyl enzymes were attacked by water, then penicilloic acid would be released. The question therefore arises as to whether the penicilloyl enzymes are related to penicillinases; penicillinases may be evolutionary developments of the penicillin binding proteins.

Affinity chromatography was employed to purify these proteins. A Sepharose column was substituted by 6-amino-penicillanic acid attached through a side arm and solubilized bacterial membranes were applied to this column. Penicillin binding proteins attach covalently to the column and can then be eluted by hydroxylamine. This procedure yields a mixture of the multiple PBPs of the organism employed. A variety of methods have been introduced to separate the penicillin binding proteins from each other. For example, the selectivity of β-lactams for different penicillin binding proteins can be used. The length of the spacer arm is very important and various solubility properties can be used to separate the penicillin binding proteins. In our laboratory, we have separated all of the 5 PBPs of *S. aureus,* and several of those of *E. coli* (PBPs 2,5,6) and *B. subtilis* (PBP la and 5). For example, PBP2 of *E. coli* is present in an amount of only 20 molecules per cell. It represents only 1% of the total penicillin binding proteins which are only 1% of the membrane protein, but in one step using an appropriate procedure a 10,000 fold purification of PBP2 to homogeneity was achieved.

Some interesting questions can be asked and answered with the purified proteins. For one thing it was shown that the penicillin bound is in fact the penicilloyl residue, attached as an ester to a serine hydroxyl in the enzyme. That is one important prediction of the Tipper-Strominger 1965 hypothesis. A very interesting phenomenon is the release of the penicilloyl residue, autocatalyzed by these penicillin binding proteins. Two forms of release occur. One is hydrolysis to give penicilloic acid. A very unusual release reaction was also discovered which involves the scission of the 5,6 bond in penicilloyl to give phenylacetylglycyl enzyme (which releases phenylacetylglycine) and a compound which we believe is dimethylthiazoline carboxylate. The latter then decomposes to

N-formyl penicillamine.

The 5 PBPs of *S. aureus* release penicilloyl in different ways. PBPs 3a and 3b don't release it at all; they form a very stable penicilloyl-protein intermediate. PBPs 1 and 2 release at a slow rate and the products formed in this slow release are phenylacetyl glycine and N-formyl penicillamine. PBP4 releases penicilloyl at a relatively rapid rate and the product formed is penicilloic acid, *i.e.* PBP4 is a weak penicillinase. We therefore see the possibility that penicillinases could have evolved from the penicillin binding proteins.

Next acyl enzyme intermediates in the reaction which are derived from substrates were detected. The problem in demonstrating acyl enzyme intermediates is that K_3, the rate constant for hydrolysis, is much more rapid than K_2 the rate constant for acylation. Therefore, the acyl enzyme intermediate never accumulated in any reasonable amount. In order to accumulate this compound, advantage was taken of the information that had been gathered in studies of proteolytic enzymes such as chymotrypsin, namely that synthetic ester substrate analogs are much better acylating agents then the normal amides. An ester analog of a synthetic substrate was therefore synthesized. A good synthetic substrate is L-lysyl-D-alanyl-D-alanine. We synthesized L-lysyl-D-alanyl-D-lactate (D-lactate replacing D-alanine). The prediction proved to be correct. The efficiency of the reaction in every case was much greater with the lactate ester substrate than with the alanine amide substrate due to an acceleration of K_2. We could in fact accumulate acyl enzyme intermediates using the ester substrate in three different penicillin binding proteins from three different bacteria. The effectiveness of penicillin as an antibiotic is due to the high reactivity of the β-lactam ring. That is K, the binding constant (K_1/K_{-1}) is practically the same for the β-lactam antibiotic and substrates. However, K_2, the acylation constant, is several orders of magnitude greater for penicillin than for substrates and K_3, the acylation constant, is very much slower for β-lactam antibiotics than for substrates. Penicillins owe their effectiveness to two properties: 1) they are extraordinarily effective acylating agents for the active sites of the PBPs and 2) K_3, the deacylation constant for the proteins is very slow for the penicillin.

In order to prove that the acyl group and the penicilloyl group were located on the same residue the substrate-labeled acyl enzyme and the penicillin-labeled penicilloyl enzyme were degraded and peptides containing the active sites obtained. Amino acid sequences showed that the peptides to which the substrate was attached and to which penicillin was attached were identical in two PBPs studied. Both are attached to serine 36, close to the N-terminal end. Thus, the third facet of the 1965 hypothesis, namely that penicilloyl and acyl are attached at the same site on the inhibited enzymes, has been established.

In order to extend these studies to other bacteria and other PBPs, the method of isolating peptides and sequencing is much too laborious. Without going into too much detail, the method of partial proteolysis was therefore employed. The patterns of cleavage as a function of time of the penicillin-labeled and the substrate-labeled PBPs using several different cleavage reagents were identical employing either *B. stearothermophilus*

PBP5 or the *Streptomyces* R61 enzyme. The identity of cleavage pattern with a variety of cleavage agents also proves that the substrate derived acyl group and penicilloyl are attached at the same site. Penicillin binding proteins from five different genera of bacteria have been used in these studies.

Homology between β-lactamases and penicillin binding proteins could be shown at the amino acid sequence level. If penicillinases were compared to each other, about 40% homology of amino acid residues was apparent in the region including serine 36 and in these β-lactamases, the penicilloyl group has been shown to be attached at that same serine residue. When penicillin binding proteins were compared with the penicillinases, the same level of homology was found in the region available for comparison. It remains to be seen if the homology will hold when a comparison can be made over a much larger part of the protein. However at the present time the data indicate that there is homology between penicillinases and penicillin binding proteins and that therefore penicillinases may have originated in evolution from the penicillin binding proteins.

Structural Features of PBPs

Two of the penicillin binding proteins from bacilli have been studied. The membrane binding region, about 2,000 or 3,000 daltons in size, is at the C-terminal end. The active site is at the other end, close to the N-terminus. These proteins can be proteolysed to obtain fragments which retain both penicillin binding and catalytic activity. A 35,000 dalton fragment of *B. subtilis* PBP5 has been isolated in high yield. Even smaller fragments, as small as 18,000, can be shown to retain penicillin binding activity but we have not been able to purify them enough to be certain that they still have catalytic activity. In fact at first proteolysis activates catalytic activity and increases thermal stability, but as proteolysis continues thermal stability decreases. A large part of the mass of the protein could be involved in stabilization of the protein and only a small part of the protein may be needed for penicillin binding and catalytic activities.

The hydrophobic peptide at the C-terminus was sequenced and found to contain a number of charged residues. Molecular models showed that all of the hydrophobic residues are on one face of the α helix and the few charged residues are on the other face. These hydrophobic regions could pack in dimers or trimers into the membrane with the charged and hydrophilic residues facing each other in the interior and the hydrophobic regions facing the lipid bilayer.

Enzymatic Activity

What kind of enzymatic reactions do the PBPs catalyze? Following a demonstration of catalytic activity in the high molecular weight binding proteins of *E. coli*, we have been able to do the same with PBPs of bacilli. The incubations have to be carried out *on filter paper.* Under ordinary assay conditions no reaction occurs in the test tube. Even on the filter, a small amount of detergent is needed to demonstrate activity. In *E. coli* these high molecular weight proteins catalyze both transglycysylase and transpeptidase reactions. They are double headed enzymes which catalyze each of the last two steps in cell wall synthesis. Only the last step, transpeptidation, is sensitive to penicillin; the transglycysylase activity is insensitive. On the other hand transglycosylation is inhibited by very low concentrations of macarbomycin, an antibiotic which is known to inhibit the transglycosylation and by several other antibiotics which is known to inhibit the transglycosylation and by several other antibiotics which inhibit the transglycosylation of cell wall synthesis. However, the bacillus enzymes do not transpeptidate under the conditions studied. There are many possible explanations of their failure to transpeptidate under these conditions. Matsuhashi and his colleagues, however, have demonstrated transpeptidation with similar enzymes from *E. coli.*

High concentrations of solvent can be substituted for filter paper. In the presence of 10% ethylene glycol, 10% glycerol, 8% methanol a large reaction proceeded in the test tube. Solvents that do not contain hydroxyl groups were ineffective. High concentrations of sugar could also be substituted for solvent. Possibly filter paper or glycerol or methanol or sugar are acting as an artificial acceptor in the transglycosylation reaction.

That is the present status of studies of the functions of these individual penicillin binding proteins. The mechanism of action of penicillin has been studied in my laboratory for 30 years but two very important problems still remain. One of them is the precise definition of the functions of the individual penicillin binding proteins. Why are there 7 in *E. coli*? What individual reactions do they catalyze that are necessary for the bacterial cell? The other probelm which I find fascinating is the side chain effect–why does the side chain of the β-lactam antibiotics influence the selectivity for binding to different penicillin binding proteins? How does that work? Why does mecillinam at low concentrations acylate *only* PBP2 in *E. coli*? This general area contains very difficult problems and I think they can only be answered by crystallizing several of these proteins and studying the crystal structure of the penicilloyl proteins as well as the acyl proteins. This task is only recently begun and I think the answers will be 10 years or more in finally coming. Thank you very much for the invitation to speak to you in Tokyo.

REFERENCES

The following is a list of relatively recent references from my laboratory. Documentation of the data presented in my talk and references to the many contributions of other laboratories can be found in these original papers.

Kozarich, J.W., Buchanan, C.E., Curtis, S.J., Hammarstrom, S., and Strominger, J.L. *In*: Advances in Experimental Medicine and Biology, Vol. 80, "Membrane Toxicity" (M.W. Miller and A.E. Shamo, eds.) Plenum Publishing Corp., New York, pp. 301–318 (1977).

Nishino, T., Kozarich, J.W., and Strominger, J.L. J. Biol. Chem. 252: 2934–2939 (1977).

Kozarich, J.W., Nishino, T., Willoughby, E., and Strominger, J.L. J. Biol.

Chem. *252*: 7525-7529 (1977).
Strominger, J.L. Microbiology 1977: 177-181 (1977).
Buchanan, C.E. Microbiology 1977: 191-194 (1977).
Kozarich, J.W. Microbiology 1977: 203-208 (1977).
Georgopapadakou, N., Hammarstrom, S., and Strominger, J.L. Proc. Natl. Acad. Sci. USA *74*: 1009-1012 (1977).
Iwaya, M. and Strominger, J.L. Proc. Natl. Acad. Sci. USA *74*: 2980-2984 (1977).
Kozarich, J.W. and Strominger, J.L. J. Biol. Chem. *253*: 1272-1278 (1978).
Buchanan, C.E., Hsia, J., and Strominger, J.L. J. Bacteriol. *131*: 1008-1010 (1977).
Rasmussen, J.R. and Strominger, J.L. Proc. Natl. Acad. Sci. USA *75*: 84-88 (1978).
Curtis, S.J. and Strominger, J.L. J. Biol. Chem. *253*: 2584-2588 (1978).
Kleppe, G. and Strominger, J.L. J. Biol. Chem. *254*: 4856-4862 (1979).
Matsuhashi, M., Tamaki, S., Curtis, S.J., and Strominger, J.L. J. Bacteriol. *137*: 644-647 (1979).
Waxman, D.J. and Strominger, J.L. J. Biol. Chem. *254*: 4863-4875 (1979).
Yocum, R.R., Waxman, D.J., Rasmussen, J.R., and Strominger, J.L. Proc. Natl. Acad. Sci. USA *76*: 2730-2734 (1979).
Yocum, R.R., Waxman, D.J., and Strominger, J.L. Trends in Biochem. Sci., *April*, 97-101 (1980).
Yocum, R.R., Waxman, D.J., and Strominger, J.L. Trends in Biochem. Sci. *5*: 97-101 (1980).
Waxman, D.J., Yocum, R.R., and Strominger, J.L. Phil. Trans. Roy. Soc. Lond. *B 289*: 257-271 (1980).
Waxman, D.J. and Strominger, J.L. J. Biol. Chem. *254*: 12056-12061 (1979).
Waxman, D.J. and Strominger, J.L. J. Biol. Chem. *255*: 3964-3976 (1980).
Yocum, R.R., Rasmussen, J.R., and Strominger, J.L. J. Biol. Chem. *255*: 3977-3986 (1980).
Kleppe, G., Yu, W., and Strominger, J.L. Antimicrobial Agents and Chemotherapy, in press.
Curtis, S.J. and Strominger, J.L. J. Bacteriol. *145*: 398-403 (1981).
Waxman, D.J. and Strominger, J.L. FEBS 13th Meeting, August 1980.
Strominger, J.L. *In*: β-Lactam Antibiotics (Proceedings of the Beta Lactam Symposium, N.Y.C., Sept. 29, 1980-October 2, 1980) Academic Press, New York, pp. 123-125 (1981).
Moews, P.C., Knox, J.R., Waxman, D.J., and Strominger, J.L. Int. J. Pept. Protein Res. *17*: 211-218 (1981).
Waxman, D.J., Yu, W. and Strominger, J.L. J. Biol. Chem. *255*: 11577-11587 (1980).
Waxman, D.J. and Strominger, J.L. J. Biol. Chem. *256*: 2059-2066 (1981).
Waxman, D.J. and Strominger, J.L. J. Biol. Chem. *256*: 2059-2066 (1981).
Amanuma, H. and Strominger, J.L. J. Biol. Chem. 255: 11173-11180 (1980).
Amanuma, H. and Strominger, J.L. In: Beta-Lactam Antibiotics (S. Mitsuhashi, ed.) Japan Sci. Soc. Press, Tokyo/Springer-Verlag, Berlin, pp.179-

184 (1981).
Waxman, D. J., Lindgren, D. M., and Strominger, J. L. J. Bacteriol. *148*: 950–955 (1981).
Waxman, D. J., Amanuma, H., and Strominger, J. L. FEBS Letters *139*: 159–163 (1982).
Yocum, R. R., Amanuma, H., O'Brien, T. A., Waxman, D. J., and Strominger, J. L. J. Bacteriol., in press (1982).
Jackson, G. E. and Strominger, J. L. J. Biol. Chem., in press (1982).

MECHANISM OF PEPTIDOGLYCAN SYNTHESIS BY PENICILLIN-BINDING PROTEINS IN BACTERIA AND EFFECT OF ANTIBIOTICS

M. Matsuhashi[1]
J. Nakagawa
S. Tomioka
F. Ishino
S. Tamaki

Institute of Applied Microbiology
University of Tokyo
Bunkyo-ku, Tokyo, Japan

I. INTRODUCTION

The major targets of β-lactam antibiotics in killing bacteria are penicillin-binding proteins (PBPs).

When bacterial cells are exposed to β-lactam antibiotics, these compounds penetrate the cell surface through the outer membrane in gram-negative bacteria, or through the rigid layer of cell walls in gram-positive bacteria to reach the cytoplasmic membrane and attack the PBPs. There are multiple PBPs differing in physiological functions, and these PBPs are each supposed to be involved in specific steps in morphological development of the cell during the cell cycle: inhibition by a β-lactam antibiotic or defect by a mutation of one of the PBPs may cause specific morphological changes of the cells. However, in many cases this correlation is not clear, because loss of activity of some PBPs seems to be compensated for by other PBPs (1,2), and therefore the function of a PBP cannot always be determined exactly from morphological changes of the cell caused by its inactivation. Moreover, the situation is also complicated by the fact that many antibiotics inhibit two or more PBPs in the cell at once. In *Escherichia coli*, fortunately the correlations seem to be easier to follow than in most other bacteria and substantial progress in this field has been achieved mainly by studies on this organism.

At present, we know that PBPs are enzymes involved in biosynthesis of peptidoglycan, the basal, rigid structure of the bac-

[1]This work was supported in part by a Grant-in-Aid for Scientific Research from the Ministry of Education of Japan.

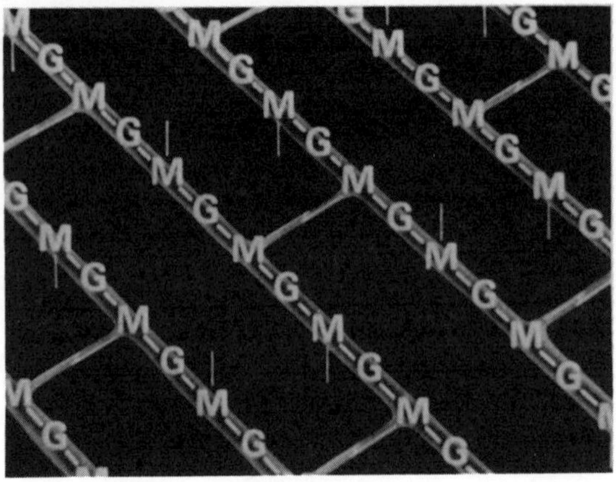

Fig. 1. Structure of peptidoglycan of Escherichia coli. *G: N-acetylglucosamine; M: N-acetylmuramic acid; thick lines: β-1,4-hexosaminoglycan chains; thin lines: tetrapeptide side chains consisting of -L-Ala-D-Glu-meso-A_2pm-D-Ala(COOH) and cross-bridges consisting of octapeptide, -L-Ala-D-Glu-meso-A_2pm-D-Ala -L-Ala-D-Glu-meso-A_2pm-D-Ala(COOH).*

terial cell walls (Fig. 1), namely, side walls and septa. In *E. coli*, most of the higher molecular weight PBPs (Mr 6 x 10^4 to 9 x 10^4) were found to be bifunctional peptidoglycan synthetases (3-8). These PBPs have peptidoglycan transglycosylase activity, which extends the glycan chain, and β-lactam-sensitive transpeptidase activity, which catalyzes the formation of peptide crossbridges (3-8). PBPs in other gram-negative bacteria closely related to *E. coli*, such as *Salmonella*, *Proteus* or *Pseudomonas*, are supposed to function in a similar way. However, in gram-positive bacilli and streptococci the functions of the PBPs seem to be considerably different from those in *E. coli*, because β-lactam antibiotics do not cause the same specific morphological changes as in *E. coli* cells. Finally, in *Staphylococcus aureus* and *Micrococcus luteus*, PBPs seem to function solely as transpeptidases, because most of the transglycosylase in the cell membranes was found to be different from the major PBPs of this organism (9). No enzymatic activities of PBP-1, 2 and 3, the higher molecular weight PBPs in *S. aureus*, have yet been demonstrated (10), but PBP-4, the smallest PBP in *S. aureus*, has DD-carboxypeptidase activity that may function in secondary crosslinking of peptidoglycan (11).

Studies on bacterial cellular components that bind to externally added radiolabeled penicillin started in the late 1940's (12-14) but more extensive work was carried out by Strominger and his coworkers after prediction of the mechanism of action of penicillin through formation of a covalently bound penicilloyl-enzyme complex by Tipper and Strominger in 1965 (15) and the finding of a penicillin-sensitive transpeptidase reaction in *E. coli* by Izaki, Matsuhashi and Strominger in 1966 (16). In 1972 Suginaka, Blumberg and Strominger (17) reported separation of PBPs of several gram-positive bacteria by chromatographies and electrofocusing. Separation of *E. coli* PBPs by sodium dodecylsulfate (SDS) - polyacrylamide gel electrophoresis was reported in 1975 by Spratt and Pardee (18). They (18,19) used genetical and antibiotic-binding techniques to correlate defects or inhibition of some PBPs with morphological changes of *E. coli* cells. Their results showed that PBP-2 (Mr 6.6 x 10^4), the target of mecillinam, an amidinopenicillin, is responsible for formation of the rod-shape of the cell and PBP-3 (Mr 6 x 10^4) for formation of the septum. PBP-1 (Mr ca. 9 x 10^4), separated later into PBP-1A and a triplet of PBP-1Bs (1,2,20), seemed to be involved in cell-elongation (19), the different proteins functioning in compensation for lack of one another (1,2). The lower molecular weight PBPs in *E. coli* seemed to have less important functions than the higher molecular weight PBPs (21-25) (Fig. 2).

Further information on the functions of PBPs was obtained by

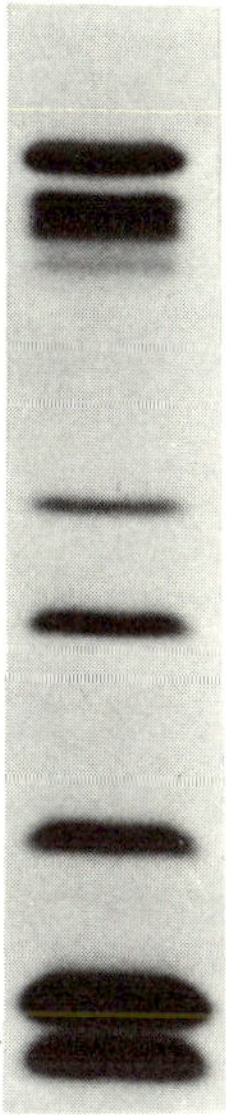

PBP	*Mr* (10^{-3})	*FUNCTION*	*ENZYME ACTIVITIES*
1A	*ca. 90*	*Cell-elongation (1,2,19)*	*Transglycosylase-transpeptidase (4,6,7)*
1Bs α β γ	*ca. 90*	*Cell-elongation (1,2,19)*	*Transglycosylase-transpeptidase (3,6,7,8)*
2	*66*	*Formation of rod shape (19)*	*(Transglycosylase?-) Transpeptidase (26)*
3	*60*	*Formation of septa (19)*	*Transglycosylase-transpeptidase (5)*
4	*49*	*Auxiliary (21,22)*	*DD-peptide hydrolase (DD-carboxypeptidase & DD-endopeptidase) (21,22)*
5	*42*	*Auxiliary (23,25)*	*DD-carboxypeptidase (24,27)*
6	*40*	*Auxiliary?*	*DD-carboxypeptidase (27,28)*

Fig. 2. Functions of E. coli *PBPs (SDS - polyacrylamide gel electrophoretic pattern). The molecular weights of PBP-2 to 6 are cited from B.G. Spratt (29). Transpeptidases and peptide hydrolases are inhibited by β-lactams.*

isolating individual PBPs in pure form and investigating the mechanism of the reactions catalyzed by these proteins. Information on the functions of PBPs in *E. coli* obtained in genetic and enzymatic investigations is summarized in Fig. 2.

The following sections describe the enzyme activities of *E. coli* PBP-1A, 1Bs and 3. Only the enzyme activities of PBP-2 of *E. coli* are not yet fully understood. Recently in collaboration with Spratt, we found peptidoglycan crosslinking enzyme activity, which is very sensitive to mecillinam, in the membranes from *E. coli* cells over-producing PBP-2 (26).

The PBPs were purified from membranes of *E. coli* cells which over-produce the appropriate PBPs (30-32). The proteins were extracted with detergents and were purified on an affinity column after various preliminary purification steps, if necessary (3-7). Purified PBP-1A and 3 each gave a single band on SDS-polyacrylamide gel electrophoresis and were free from contaminating proteins as judged by protein-staining with Coomassie brilliant blue or fluorography of the protein-[^{14}C]penicillin G complexes. Only PBP-1Bs gave a triplet consisting of two major and one minor band with activity for binding penicillin G, on the electrophoresis. For determination of whether the two enzyme activities reside in the same molecule or in different ones, each component of PBP-1Bs thus had to be separated in an enzymatically active state (8).

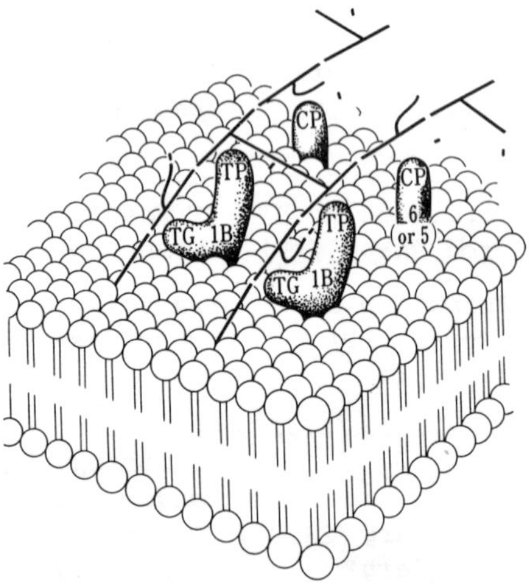

Fig. 3. Proposed mechanism of synthesis of peptidoglycan by PBP-1B in E. coli. *Numbers indicate penicillin-binding proteins. TG, transglycosylase domain; TP, transpeptidase domain; CP, DD-carboxypeptidase protein. The three-dimensional structure of the protein has not been studied.*

II. DUAL ENZYME ACTIVITIES OF PEPTIDOGLYCAN SYNTHESIS IN *E. COLI* PBP-1Bs

Nakagawa, Tamaki and Matsuhashi (3) in 1979 purified PBP-1Bs from crude membranes of *E. coli* and found that the purified preparations of PBP-1Bs possessed two enzyme activities, transglycosylase and penicillin-sensitive transpeptidase, which together carried out synthesis of crosslinked peptidoglycan from lipid-linked precursor (3). A cue for this finding was provided by the fact that the remarkable thermostability and Sarkosyl-resistance of the two enzyme activities in crude membranes correlated with those of the penicillin-binding activity of the same proteins (32). (Other workers (31,33) reported similar results in 1980.) From this finding, we proposed a scheme for the bifunctional enzyme mechanism of these proteins (6,7,34), which is shown diagrammatically in Fig. 3.

The crosslinked peptidoglycan is formed on the two individual active sites of the protein: the one for transglycosylation and the other for transpeptidation. The two active sites may reside in the same molecule of PBP-1B, which is mounted in the cytoplasmic membrane by its hydrophobic moiety. On the active site of transglycosylase, uncrosslinked peptidoglycan is formed from the lipid-linked precursor, GlcNAc-MurNAc(-pentapeptide)-PP-undecaprenol, which is produced in, and supplied from the cytoplasmic membrane. On the active site of transpeptidase, crossbridges consisting of the nonapeptide -L-Ala-D-Glu-meso-A_2pm-D-Ala -L-Ala-D-Glu-meso-A_2pm-D-Ala-D-Ala(COOH) may be formed from the two pentapeptide side chains, -L-Ala-D-Glu-meso-A_2pm-D-Ala-D-Ala(COOH), with release of one terminal D-alanine molecule per crossbridge. Then by a DD-carboxypeptidase reaction, probably of PBP-5 or 6, the excess terminal D-alanine molecules that are not used for crosslinking are removed to form tetrapeptide side chains and octapeptide crossbridges (see legend to Fig. 1). The scheme in Fig. 3 is drawn assuming that peptidoglycan is linked to undecaprenol-diphosphate in the membrane by its reducing terminals. Formation of glycan with a four-fold helical structure, as proposed by Burge, Fowler and Reaveley (35) for the structure of peptidoglycan, may be favorable for keeping the extent of crosslinkage to about 25%; that is, two peptide side chains participate to crosslinkage per four peptide side chains (Fig. 1).

If the active site of the transpeptidase is blocked by a β-lactam antibiotic, only transglycosylation may take place, and the glycan chains formed cannot be crosslinked, causing failure to form a rigid network of peptidoglycan.

Most highly bactericidal β-lactams, such as penicillin G, ampicillin, apalcillin, azlocillin, cephaloridine, cefotaxime, cefoperazone, cefoxitin, cefmetazole or N-formimidoyl thienamycin, are effective on PBP-1Bs. Mecillinam and cephalexin inhibited the transpeptidation catalyzed by PBP-1Bs only at extremely high concentrations and thus their targets are definitely not PBP-1Bs.

The time course of the peptidoglycan-synthetic reactions by PBP-1Bs (Fig. 4) showed that the peptidoglycan formed was 23 to 24% crosslinked for most of the reaction. This indicates that the two reactions, extension of glycan chains (transglycosylation) and formation of crossbridges (transpeptidation), are cooperative reactions. Assay of the transpeptidase reaction alone has not been achieved, but the transglycosylase reaction can be assayed independently from the transpeptidase reaction. Moenomycin and macarbomycin inhibit the transglycosylase reaction but penicillin G and other β-lactams do not.

The transglycosylase activity of PBP-1Bs did not seem to have any distinct optimal pH under certain reaction conditions and was observed between pH 6 and 9. Under the best conditions,

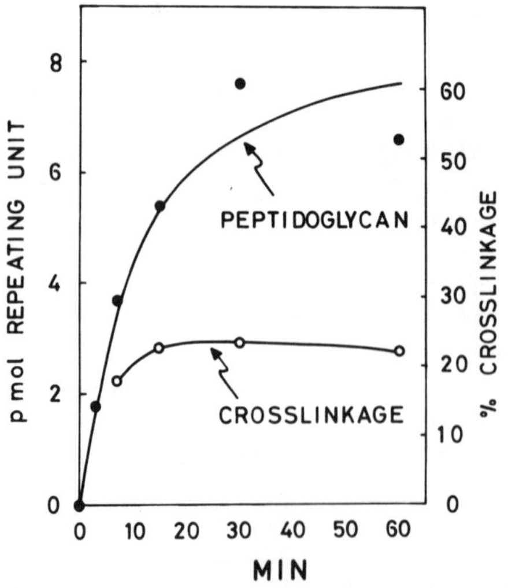

*Fig. 4. Time course of peptidoglycan synthetic reactions by purified PBP-1Bs (*E. coli*). The reaction mixture in a final volume of 35 µl consisted of 46 p mols GlcNAc-MurNAc(-L-Ala-D-Glu-meso-[^{14}C]A$_2$pm-D-Ala-D-Ala)-PP-undecaprenol in 5 µl methanol, 0.06 M Tris-HCl, pH 8.5, 0.03 M MgCl$_2$, 1.4 mM 2-mercaptoethanol, 3.7 µg purified PBP-1Bs (*E. coli*) and 0.04% (wt/vol) Sarkosyl. Temperature, 37°C. The extent of crosslinkage (%) was calculated from the ratio (by radioactivity), 100 x [bis-(disaccharide-peptide)/2] / [disaccharide-peptide + bis(disaccharide-peptide)].*

the purified enzyme formed about 0.7 mol glycan (as repeating units) per mol protein per min and bound 0.2 mol penicillin G per mol protein. These values are too low to explain the cell wall peptidoglycan synthetic activity *in vivo*. Higher activities *in vitro* should be obtainable on improving the assay conditions.

However, since PBP-1Bs are stable to changes in many external conditions such as temperature, anionic and neutral detergents, sulfhydryl agents, and pH, they could be responsible for most of the cell-wall peptidoglycan synthetic activities in *E. coli* in a wide range of growth conditions. Moreover, the enzyme activities of PBP-1Bs are identical with those observed in 1966 by Izaki, Matsuhashi and Strominger (16) in crude membranes of *E. coli*, because under the conditions used previously, other PBPs, 1A, 2 and 3, would have been virtually inactive (1). The product of the cooperative reactions of transglycosylase and transpeptidase of PBP-1Bs was insoluble in water and when subjected to paper chromatography, it appeared as a condensed spot at the origin, whereas the product formed in the reaction in the presence of an effective β-lactam was water-soluble and appeared as a spread spot. Similar phenomena were observed previously with crude membranes of *E. coli* (16).

III. DUAL ENZYME ACTIVITIES OF PEPTIDOGLYCAN SYNTHESIS OF EACH PBP-1B COMPONENT OF *E. COLI*

Separation of the three components of *E. coli* PBP-1Bs (named α, β and γ) from each other in enzymatically active states was finally achieved by Nakagawa and Matsuhashi (8) using the recently developed technique of treatment with guanidine-HCl to renature proteins that have been separated by SDS-polyacrylamide gel electrophoresis (36). Each of PBP-1Bs showed activities of transglycosylase, transpeptidase, and penicillin-binding. The recovery of transglycosylase activity was about 1/3 of the initial amount for each component. The peptidoglycan formed by each component was also crosslinked (1/2.5 to 1/2 of that with the original PBP-1Bs), indicating recovery of transpeptidase activity, too. The components separated also bound [^{14}C]penicillin G (about 1/6 to 1/5 of that of the original PBP-1Bs).

These results indicate that PBP-1Bs are a group of bifunctional peptidoglycan synthetic enzymes with activities of transglycosylase and transpeptidase. As tryptic digestion of each of the three components caused formation of a doublet of smaller penicillin-binding proteins of similar molecular weights (about 5 x 10^4), the three protein fractions seem to have similar molecular structures, but details of their structures and the reasons for the molecular differences are unknown.

IV. ANOTHER BIFUNCTIONAL CELL WALL-PEPTIDOGLYCAN SYNTHETIC ENZYME IN *E. COLI*, PBP-1A

Similar dual enzyme activities for the synthesis of cross-linked peptidoglycan were found in *E. coli* PBP-1A by Ishino, Mitsui, Tamaki and Matsuhashi (4). PBP-1A is supposed to constitute a detour enzyme system for that of PBP-1Bs, and one system can compensate for lack of the other (1,2). However, PBP-1A is more easily extracted than PBP-1Bs with detergents and the reactions catalyzed by PBP-1A seem to be slightly different from those catalyzed by PBP-1Bs (4,37). The transglycosylase activity of PBP-1A was heat- and Sarkosyl-sensitive, under certain conditions was greatly stimulated by divalent cations (about 100-fold stimulation by Co^{++} ion), had a distinct pH optimum at about pH 6 and was active in the presence

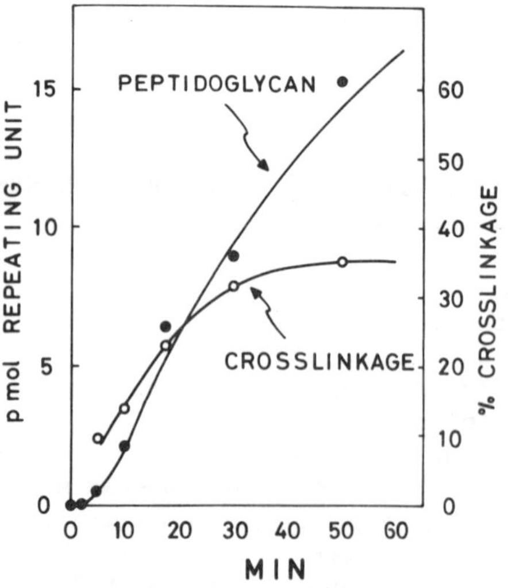

*Fig. 5. Time course of peptidoglycan synthetic reactions by purified PBP-1A (*E. coli*). The reaction mixture in a final volume of 35 µl consisted of 26 p mols [^{14}C]GlcNAc-MurNAc(-L-Ala-D-Glu-meso-A$_2$pm-D-Ala-D-Ala)-PP-undecaprenol in 5 µl methanol, 10 mM Tris-maleate buffer, pH 6.0, 1 mM MgCl$_2$, 2.4 µg purified PBP-1A (*E. coli*) and 0.05% (wt/vol) Triton X-100. Temperature, 37°C. The extent of crosslinkage (%) was calculated from the ratio (by radioactivity), 100 x [bis(disaccharide-peptide)/2 + tris(disaccharide-peptide) x 2/3] / [disaccharide-peptide + bis(disaccharide-peptide) + tris(disaccharide-peptide)]. The amount of putative tetramer was negligible.*

of high concentrations of Triton X-100, when reaction was performed on filter paper. The difference between the transpeptidase reactions was even greater. The extent of crosslinking of the peptidoglycan formed enzymatically by PBP-1A increased with time of incubation to as much as 35% (Fig. 5). Therefore, the two reactions, transglycosylation and transpeptidation of PBP-1A, are not so tightly cooperative as in PBP-1Bs. Lysozyme digestion of the product by PBP-1A resulted in isolation of compounds with lower mobilities on paper chromatography (solvent: isobutyric acid -.1 M ammonia, 1 : 0.6) than bis(disaccharide-tetrapeptide), which consists of two disaccharide units crosslinked by octapeptide (38). These compounds are supposed to be the trimer (37) and tetramer of the repeating unit. Probably PBP-1A is responsible for the hypercrosslinking of peptidoglycan in *E. coli* cells. The highest transglycosylase activity was measured as 10 mols of repeating unit per mol protein per min at 37°C.

The transglycosylase reaction of PBP-1A was inhibited by moenomycin and macarbomycin (4,6,7) and the transpeptidase reaction was inhibited by most β-lactams.

V. SEPTUM-PEPTIDOGLYCAN SYNTHETIC DUAL ENZYME ACTIVITIES IN *E. COLI* PBP-3

Dual enzyme activities for the formation of crosslinked peptidoglycan were found in purified *E. coli* PBP-3 by Ishino and Matsuhashi (5). The reactions were similar to those of *E. coli* PBP-1A and 1Bs. However, under the reaction conditions used, the product of the reaction, peptidoglycan, was crosslinked only 6 to 10%. Its transglycosylase reaction was inhibited by moenomycin and macarbomycin, like those of PBP-1A and 1Bs, but its transpeptidase reaction was inhibited specifically by β-lactams such as penicillin G, apalcillin, azlocillin or cephalexin that at low concentrations bound to it and inhibited septum formation of *E. coli in vivo*, causing formation of filamentous cells (5). N-Formimidoyl thienamycin and nocardicin A, which do not bind to PBP-3 and do not cause formation of filamentous cells at low concentrations, also did not seem to inhibit its transpeptidase activity at these low concentrations.

VI. DISCUSSION

Table 1 summarizes the sensitivities of transpeptidase reactions of higher molecular weight *E. coli* PBPs. Most β-lactams inhibited the transpeptidase reaction of PBP-1A at very low concentrations (in most cases less than 0.2 μg antibiotic per ml was required for 50% inhibition) but none of the β-lactams tested inhibited the transpeptidase reaction of PBP-1Bs at concentrations of much less than 1 μg per ml. The transpeptidase activity of PBP-2 was measured in a crude membrane system (26), because no assay method has yet been established for purified PBP-2. From these results it seems that the primary targets of penicillin G (and probably also apalcillin, azlocillin and cephalexin in Table 1) are PBP-3 and 1A, while its secondary targets are PBP-1Bs and 2. Thus the antibiotic(s) cause formation of unseptated, filamentous cells at low concentrations and cell lysis at higher concentrations. PBP-2 is the primary killing target of mecillinam and probably also of N-formimidoyl thienamycin, while PBP-1A and 1Bs are also important targets for the latter compound; therefore both antibiotics cause formation of round cells at lower concentrations. The results of the enzymatic inhibition experiments correlate qualitatively with those of binding experiments (ref. 29 and unpublished data), although much lower concentrations of antibiotics were sometimes required for inhibiting transpeptidase reactions than for binding to the respective PBPs. Probably antibiotics inhibit the enzymes by forming reversible complexes before or without forming irreversible, covalent linkages (39). As a whole, we can conclude that the killing targets in *E. coli* of most β-lactams listed in Table 1 are higher molecular weight PBPs.

Table 1. *Concentrations (μg per ml) of β-lactams required for 50% inhibition of transpeptidase reactions of higher molecular weight PBPs of* E. coli

Antibiotic	*PBP-1A*	*PBP-1Bs*	*PBP-2*[a]	*PBP-3*
Penicillin G	*< 0.1*	*1-2*	*1*	*0.1-0.3*
Apalcillin	*0.2*	*3*	*nd*[b]	*≦ 0.1*
Azlocillin	*0.1*	*1.5*	*nd*	*0.1-0.2*
Mecillinam	*> 50*	*> 1,000*	*0.2*	*nd*
Cephalexin	*< 2*	*1,000*	*nd*	*≧ 30*
Cefmetazole	*nd*	*0.4*[a]	*> 40*	*nd*
N-formimidoyl thienamycin	*0.05*	*0.8*	*1*[c]	*> 3*
Nocardicin A	*10*	*80-300*	*nd*	*1,000*

[a] *Assayed in crude membranes.*
[b] *Not determined.*
[c] *The high pH of the assay condition was unsuitable for this antibiotic.*

*Fig. 6. Proposed scheme for synthesis of the network of peptidoglycan (*E. coli*). The scheme was drawn assuming that the glycan chains are oriented vertically to the long axis of the rod-shaped cells (43) and that the major peptidoglycan-nicking enzyme is DD-endopeptidase (44). Theoretically, however, endomuramidase and N-acetylmuramyl-L-alanine amidase could also be the nicking enzymes.*

However, several examples are known of β-lactams that do not seem to exert their killing effects simply by inhibiting the transpeptidase activities of PBPs. We observed the following two unusual cases. N-Formimidoyl thienamycin inhibited not only the transpeptidase reaction, but also the transglycosylase reaction of *E. coli* PBP-1A; the concentrations required for 50% inhibition of the two reactions were about the same (40). A new cephamycin possessing a 7-(D-2-amino-2-carboxy-ethylthio)-acetamide side chain was reported to lyse early-stationary cells[2] of *E. coli* and several other gram-negative bacteria at concentrations as low as its MICs (41), whereas most other β-lactams lyse early stationary cells only at much higher concentrations.

Peptidoglycan is an entity that covers the whole cell surface of bacteria. Its extension is supposed to be preceded by the action of a lytic enzyme(s) that forms a nick in the network of old peptidoglycan (Fig. 6). Then new pieces of peptidoglycan are synthesized by the action of higher molecular weight PBPs, and these pieces are joined to the old peptidoglycan network by the action of certain transpeptidases or transglycosylases. A

[2]Similar phenomena have been observed to certain extents with nocardicin A (42) and cefmetazole (unpublished data).

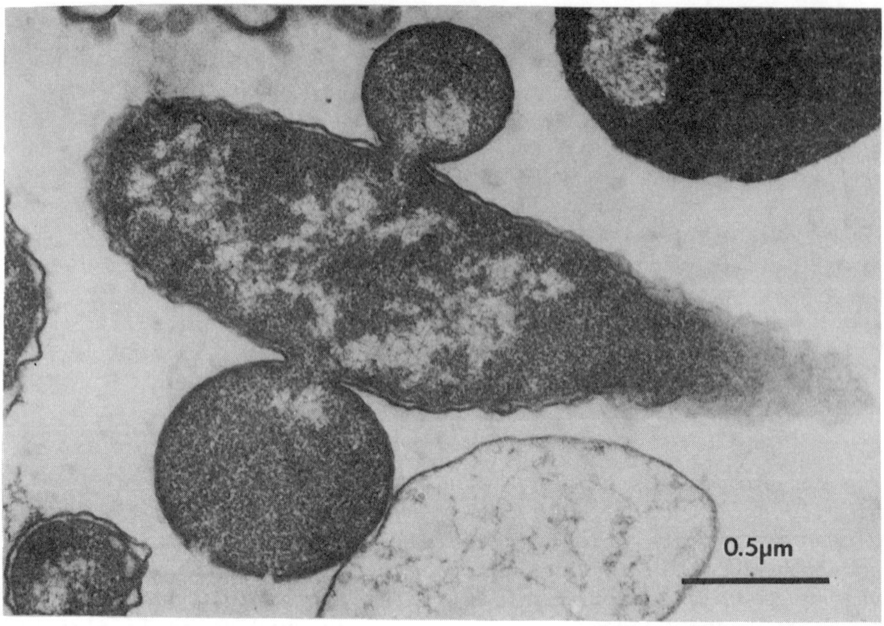

Fig. 7. Electron microscopic appearance of an ultrathin section of an E. coli *cell showing twin bulges. The photograph was provided by Ms. A. Hirata.*

triggering effect on an autolytic enzyme(s) by β-lactams has been proposed by Tomasz and his coworkers (45) but no evidence for this has yet been obtained in experiments using purified enzymes. The peculiar cell-lytic mechanism of the cephamycin described above is unknown. One possibility is that it lyses early-stationary cells not simply by attacking their PBPs but also by simultaneously triggering or amplifying the nicking enzyme activity. Another possibility is that the antibiotic is concentrated on the peptidoglycan or its precursor in the cells owing to the structural similarity of its side chain, and thus exerts its inhibitory effect on PBPs more efficiently. In any case, the cells attacked by low concentrations of this antibiotic form bulges at several places on the cell surface. An example of twin bulges about 180° apart on the growing zone of an *E. coli* cell formed by this way is shown in Fig. 7. Study of this novel mechanism of cell lysis will be useful for understanding the over-all mechanisms of action of β-lactam antibiotics, and thus will be important for developing new β-lactam antibiotics.

REFERENCES

1. Tamaki, S., Nakajima, S. & Matsuhashi, M.(1977) Proc. Natl. Acad. Sci. USA *74*,5472-5476
2. Suzuki, H., Nishimura, Y. & Hirota, Y.(1978) Proc. Natl. Acad. Sci. USA *75*,664-668
3. Nakagawa, J., Tamaki, S. & Matsuhashi, M.(1979) Agric. Biol. Chem. *43*,1379-1380
4. Ishino, F., Mitsui, K., Tamaki, S. & Matsuhashi, M.(1980) Biochem. Biophys. Res. Commun. *97*,287-293
5. Ishino, F. & Matsuhashi, M.(1981) Biochem. Biophys. Res. Commun. *101*,905-911
6. Matsuhashi, M., Ishino, F., Nakagawa, J., Mitsui, K., Nakajima-Iijima, S., Tamaki, S. & Hashizume, T.(1981) in β-Lactam Antibiotics --- Mode of Action, New Developments, and Future Prospects (Salton,M.R.J. & Shockman,G.D.,eds.) pp. 169-184, Academic Press, New York
7. Matsuhashi, M., Nakagawa, J., Ishino, F., Nakajima-Iijima, S., Tomioka, S., Doi, M. & Tamaki, S.(1981) in Beta-Lactam Antibiotics (Mitsuhashi,S.,ed.) pp.203-223, Japan Scientific Societies Press, Tokyo, & Springer-Verlag, Berlin, Heidelberg, New York
8. Nakagawa, J. & Matsuhashi, M.(1982) Biochem. Biophys. Res. Commun. in press
9. Park, W. & Matsuhashi, M., manuscript in preparation
10. Waxman, D.J. & Strominger, J.L.(1979) J. Biol. Chem. *254*, 12056-12061
11. Wyke, A.W., Ward, J.B., Hayes, M.V. & Curtis, N.A.C.(1981) Eur. J. Biochem. *119*,389-393
12. Cooper, P.D.(1956) Bacteriol. Rev. *20*,28-48
13. Maass, E.A. & Johnson, M.J.(1949) J. Bacteriol. *57*,415-422
14. Pasinsky, A. & Kastorskaya, T.(1947) Biochimia *12*,465-476
15. Tipper, D.J. & Strominger, J.L.(1965) Proc. Natl. Acad. Sci. USA *54*,1133-1141
16. Izaki, K., Matsuhashi, M. & Strominger, J.L.(1966) Proc. Natl. Acad. Sci. USA *55*,656-663
17. Suginaka, H., Blumberg, P.M. & Strominger, J.L.(1972) J. Biol. Chem. *247*,5279-5288
18. Spratt, B.G. & Pardee, A.B.(1975) Nature *254*,516-517
19. Spratt, B.G.(1975) Proc. Natl. Acad. Sci. USA *72*,2999-3003
20. Spratt, B.G., Jobanputra, V. & Schwarz, U.(1977) FEBS Lett. *79*,374-378
21. Matsuhashi, M., Takagaki, Y., Maruyama, I.N., Tamaki, S., Nishimura, Y., Suzuki, H., Ogino, U. & Hirota, Y.(1977) Proc. Natl. Acad. Sci. USA *74*,2976-2979
22. Iwaya, M. & Strominger, J.L.(1977) Proc. Natl. Acad. Sci. USA *74*,2980-2984
23. Matsuhashi, M., Maruyama, I.N., Takagaki, Y., Tamaki, S.,

Nishimura, Y. & Hirota, Y.(1978) Proc. Natl. Acad. Sci. USA *75*,2631-2635
24. Matsuhashi, M., Tamaki, S., Curtis, S.J. & Strominger, J.L.(1979) J. Bacteriol. *137*,644-647
25. Tamaki, S., Nakagawa, J., Maruyama, I.N. & Matsuhashi, M.(1978) Agric. Biol. Chem. *42*,2147-2150
26. Ishino, F., Tamaki, S., Spratt, B.G. & Matsuhashi, M., manuscript in preparation
27. Spratt, B.G. & Strominger, J.L.(1976) J. Bacteriol. *127*, 660-663
28. Amanuma, H. & Strominger, J.L.(1980) J. Biol. Chem. *255*, 11173-11180
29. Spratt, B.G.(1977) Eur. J. Biochem. *72*,341-352
30. Clarke, L. & Carbon, J.(1976) Cell *9*,91-99
31. Tamura, T., Suzuki, H., Nishimura, Y., Mizoguchi, J. & Hirota, Y.(1980) Proc. Natl. Acad. Sci. USA *77*,4499-4503
32. Nakagawa, J., Matsuzawa, H. & Matsuhashi, M.(1979) J. Bacteriol. *138*,1029-1032
33. Suzuki, H., Van Heijenoort, Y., Tamura, T., Mizoguchi, J., Hirota, Y. & Van Heijenoort, J.(1980) FEBS Lett. *110*,245-249
34. Matsuhashi, M., Noguchi, H. & Tamaki, S.(1979) Chemotherapy (Tokyo) *27*,827-840
35. Burge, R.E., Fowler, A.G. & Reaveley, D.A.(1977) J. Mol. Biol. *117*,927-953
36. Hager, D.A. & Burgess, R.R.(1980) Anal. Biochem. *109*,76-86
37. Tomioka, S., Ishino, F., Tamaki, S. & Matsuhashi, M., manuscript in preparation
38. Weidel, W. & Pelzer, H.(1964) Adv. Enzymol. *26*,193-232
39. Ghuysen, J.M., Frère, J.M., Leyh-Bouille, M., Coyette, J., Duez, C., Joris, B., Dusart, J., Nguyen-Distèche, M., Dideberg, O., Charlier, P., Knox, J.R., Kelly, J.A., Moews, P.C. & DeLucia, M.L.(1981) in Beta-Lactam Antibiotics (Mitsuhashi,S.,ed.) pp.185-202, Japan Scientific Societies Press, Tokyo, and Springer-Verlag, Berlin, Heidelberg, New York
40. Hashizume, T., Ishino, F., Nakagawa, J., Tamaki, S. & Matsuhashi, M., manuscript in preparation
41. Yamada, Y., Goi, H., Watanabe, T., Tsuruoka, T., Miyauchi, K., Yoshida, T., Hoshiko, S., Kazuno, Y., Inouye, S., Niida, T. & Matsuhashi, M.(1981) Abstracts of the 21st Interscience Conference on Antimicrob. Ag. Chemother. No.53
42. Nishida, M., Mine, Y., Nonoyama, S. & Kojo, H.(1977) J. Antibiotics *30*,917-925
43. Verwer, R.W.H., Beachey, E.H., Keck, W., Stoub, A.M. & Poldermans, J.E.(1980) J. Bacteriol. *141*,327-332
44. Tomioka, S. & Matsuhashi, M.(1978) Biochem. Biophys. Res. Commun. *84*,978-984
45. Tomasz, A.(1979) Annual Rev. Microbiol. *33*,113-137

OUTER-LAYER PERMEABILITY OF β-LACTAM ANTIBIOTICS IN GRAM-NEGATIVE BACTERIA

T.Sawai, R.Hiruma, M.Sonoda and N.Kawana

Division of Microbial Chemistry, Faculty of Pharmaceutical Sciences, Chiba University, Chiba, Japan

Efficacy of β-lactam antibiotics against gram-negative bacteria depends primarily upon their ability to penetrate the permeability barrier of the outer membrane of the bacterial envelope. Transmembrane channels composed of porin, a major outer membrane protein, are believed to be the route used by β-lactam antibiotics to reach their targets in *Escherichia coli* (1).

We previously devised a method for estimating the outer membrane permeability of β-lactam antibiotics(2). The study with the method demonstrated significantly higher activity of cephalosporins to pass through the outer membrane than penicillins (3). Subsequent investigations revealed another interesting fact that an *E.coli* mutant lacking porin showed high resistance to cephalosporins but not to ampicillin. We are interested in studying the route of β-lactam antibiotic permeation in other pathogenic bacteria, and in knowing whether the informations from *E.coli* cells can be applicable to other pathogenic bacteria.

ISOLATION AND PROPERTIES OF OUTER MEMBRANE MUTANTS FROM *E.coli*, *Proteus mirabilis* AND *Enterobacter cloacae*

Guided by the fact that *E.coli* strain lacking porin shows low susceptibility to cephalosporins, we found that the mutant lacking porin could be easily selected on agar medium containing cefoxitin without addition of any chemical mutagen. We applied this method to *P.mirabilis* and *E.cloacae*, and isolated mutant strains defective in the outer membrane protein(s) functionally corresponding to porin of *E.coli*.

E.coli CS197 is the mutant at *omp*B locus, derived from *E.coli* k12 substrain, CS109; CS197 lacks outer membrane protein Ia and Ib (0-9 and 0-8 proteins, respectively, in Mizushima's

Table 1. *Properties of outer membrane mutants of E.coli, P.mirabilis and E.cloacae*

			Susceptibility to -lactams (MIC, μg/ml)						
	Missing outer	Doubling	Penicillins			Cephalosporins			
Strain	membrane protein(s)	time (min)	PCG	APC	CPC	CEZ	CET	CER	CFX
E.coli									
CS109	none	30.2	25	3.1	3.1	1.6	3.1	3.1	3.1
CS197	0-8, 0-9	33.2	50	6.3	12.5	25	50	25	25
CU204	0-10	42.0	25	3.1	3.1	0.8	3.1	3.1	3.1
P.mirabilis									
N-51	none	34.5	3.1	1.6	0.8	3.1	3.1	25	3.1
N-51C1	40K	46.5	12.5	3.1	1.6	12.5	25	200	25
E.cloacae									
206	none	31.5	100	50	200	1.6	3.1	3.1	3.1
206C1	39-40K	46.8	200	50	400	3.1	25	6.3	25
206C2	37K, 39-40K	54.9	200	100	800	50	100	100	200
206C3	37K	32.4	50	50	200	1.6	3.1	6.3	3.1

Abbreviation of antibiotics: PCG, benzylpenicillin; APC, ampicillin; CPC, carbenicillin; CEZ, cefazolin; CET, cephalothin; CER, cephaloridine; CFX, cefoxitin.

nomenclature). These *E.coli* strains were provided from Dr.Pugsley. CU204 was isolated from CS109 as the phage TuII* resistant mutant lacking the outer membrane protein 0-10 which is the phage receptor. *E.coli* mutant CS197 showed significant lower glucose uptake activity, confirming the finding made by other workers(4,5). However, glucose uptake was not affected by the absence of 0-10 protein.

The *P.mirabilis* N-51C1, which is a lower producer of 40K protein, showed significantly lower activity for glucose uptake. This result suggests that 40K protein is functionally similar to porin of *E.coli*. The 40K protein could not be separated into multiple bands by SDS-Urea-polyacrylamide tube gel electrophoresis, therefore,it seems that *P.mirabilis* has only one kind of protein corresponding to porin.

E.cloacae has two kinds of outer membrane proteins which are responsible for glucose uptake, namely 37K and 39-40K. Although 39-40K protein was separated into two bands by gel electrophoresis, they were always lost together as a single mutational event. The glucose uptake in the *E.cloacae* strain decreased stepwise as 39-40K, 37K and then 37K and 39-40K were lost. The results suggested that 37K played a more important role in glucose uptake than 39-40K, however, *E.cloacae* seems to possess multiple proteins functionally corresponding to porin.

EFFECT OF OUTER MEMBRANE MUTATION ON CEFAZOLIN AND AMPICILLIN PERMEATION

In order to assay outer membrane permeation of β-lactam antibiotics using the method of Sawai et al.(2), all the strains were infected with an R plasmid, RGN823, which mediates type Ib penicillinase(TEM-2 penicillinase) production(6).

The outer membrane of the wild type *E.coli*, CS109, was permeable to cefazolin with high efficacy, and an apploximately linear relationship between periplasmic and external drug concentrations was observed. This result is in agreement with that found in our earlier study(3). When *E.coli* lost porin, cefazolin permeation was significantly decreaed as observed in the case of glucose uptake.

In *P.mirabilis*, the experimental results strongly suggested that cefazolin passed through the outer membrane by the same route as that for glucose, and this permeation function was attributed to the 40K protein. *E.cloacae*, which has two kinds of outer membrane proteins responsible for glucose uptake, presented some whate complicated results. Contrary to the case of glucose uptake, 37K protein made little contribution to cefazolin permeation, and the 39-40K protein played a significant role in the permeation function.

The results of ampicillin permeation in *E.coli* CS109 and its porin lacking mutant indicated that ampicillin passed through the outer membrane via porin. However, considerable ampicillin permeation was also found in the mutant lacking porin. The possibility that ampicillin is capable of passing the outer membrane via a route other than channel of porin cannot be ruled out.

Ampicillin permeability in the *E.cloacae* strains was more complicated. The 39-40K protein appeared to play an important role in ampicillin as in the case of cefazolin, but the barrier effect of the outer membrane against ampicillin was significantly reduced by the lack of 37K.

We colud not performed exact measurements for ampicillin permeation in *P.mirabilis strain* because trace amounts of penicillinase leaked out from the intact cells and interfered with the assay of low-permeable materials such as ampicillin.

The results in this study strongly suggested that cefazolin passess through the outer membrane by the same pathway as that for glucose. Similar results were also obtained for cephaloridine and cephalothin. By analogy with porin of *E.coli*, it can be presumed that 40K protein of *P.mirabilis* and 37K and 39-40K proteins of *E.cloacae* build a pore across the outer membrane, and cephalosporins and glucose pass through this pore. This assumption is supported by the fact that the absence of these proteins resulted in a significant lowering of bacterial susceptibility to cephalosporins.

On the contrary, the ampicillin susceptibility was only slightly reduced by the loss of these proteins. There is a possibility that ampicillin uses another pathway in addition to the pore mentioned above for its outer membrane permeation. Another possible pathway for passage of β-lactams across the outer membrane is passive diffusion through the phospholipid bilayer in the outer membrane. Our preliminary results demonstrated a significant passage of ampicillin through an artificial phospholipid layer(liposome) but not of cephalosporins such as cefazolin and cephaloridine.

REFERENCES

1. Nikaido,H. (1979) in Bacterial outer mmebrane (Inoue,M.,ed.) pp.361-407, John Wiley & Sons Inc. New York
2. Sawai,T., Matsuba,K. and Yamagishi,S. (1977) J.Antibiotics *30*, 1134-1136

3. Sawai,T., Matsuba,K., Tamura,A. and Yamagishi,S. (1979) J.Antibiotics *32*, 59-65
4. Nakae,T. (1976) Biochem.Biophys.Res.Commun. *71*,877-884
5. Nakae.T. (1976) J.Biol.Chem. *251*, 2176-2178
6. Sawai.T., Takahashi,K., Yamagishi,S. and Mitsuhashi,S. (1970) J.Bacteriol. *104*, 620-629

THE ROLE OF OUTER MEMBRANE PERMEABILITY IN THE SENSITIVITY AND RESISTANCE OF GRAM-NEGATIVE ORGANISMS TO ANTIBIOTICS[1]

Hiroshi Nikaido

Department of Microbiology and Immunology
University of California
Berkeley, California, U.S.A.

I. OUTER MEMBRANE PERMEABILITY OF *ESCHERICHIA COLI* AND *SALMONELLA TYPHIMURIUM* AND THE DIFFUSION OF β-LACTAM ANTIBIOTICS

A. *Multiple Pathwasys of Penetration through the Outer Membrane*

All Gram-negative bacteria are characteristically surrounded by the outer membrane, and it is now well established that this membrane acts as a barrier of widely differing efficiency for the penetration of various solutes (1-3). One of the important observations made about the permeability of the outer membrane was that there were more than one pathways of diffusion operating in this membrane. This realization came about because mutants with altered, defective lipopolysaccharide, an amphiphilic lipid component of the outer membrane, were found to be hypersensitive toward hydrophobic antibiotics (4,5), and were indeed shown to have outer membranes unusually permeable to these compounds (6), but still exhibited unaltered sensitivity to hydrophilic antibiotics including common β-lactams (4,5). The subsequent study led to the discovery that the "hydrophobic pathway" involved the penetration through the lipid domains of the outer membrane, whereas the hydrophilic molecules mainly diffused through water-filled channels produced by a special class of proteins, "porins" (7). This multiple, or at least dual nature of the outer membrane penetration must be kept in mind at all times. Efforts to fit the outer membrane permeability data to equations based on some unitary hypotheses, for example, are not only meaningless but also could lead to wrong conclusions.

[1]*This work was supported by a U.S. Public Health Service research grant (AI-09644) and an American Cancer Society research grant (BC-20).*

In view of the presence of these dual pathways, it becomes important to know the relative contribution of each of them for the penetration of a given antibiotic. With the classical β-lactam antibiotics, and especially with most cephalosporin derivatives, we believe that the porin pathway plays a predominant role. This conclusion is based on several lines of evidence. (i) With the wild-type cells of *E. coli* and *S. typhimurium*, the lipid domains of the outer membrane are constructed in such a way that it has an unusually low permeability to hydrophobic compounds (6); we believe that this property is related to the unusual, totally asymmetric structure of this portion, in which the outer half of the bilayer contains almost exclusively lipopolysaccharide molecules with few, if any, detectable phospholipid molecules (8,9). (ii) Furthermore, the penetration rate through this pathway is related to the hydrophobicity of the molecule (6). Thus the contribution of this pathway is further decreased with rather hydrophilic molecules such as the common cephalosporins. (iii) The porin channel can at least accomodate the common β-lactam antibiotics, and it was shown to allow the largely non-specific diffusion of any small, hydrophilic molecules (7). (iv) Finally the rate of diffusion of cephaloridine (10), 6-aminopenicillanic acid (11), cephacetrile, cefazolin, cephamandole, and cephalothin (12) was shown to be greatly diminished, usually by 90% or more, in mutants producing very lowered levels of porins.

These results conclusively show the overwhelming importance of the porin pathway for the "classical" cephalosporins. However, the lipid domain cannot be totally impermeable, and thus its relative contribution may become significant under certain conditions. Some of the more recent β-lactam compounds, for example, are either rather hydrophobic or very large, and are expected to penetrate through the *E. coli* or *S. typhimurium* porin channels extremely slowly. In these cases, it would not be surprising even if the diffusion through the lipid domain became a significant, if not predominant, pathway of diffusion, as predicted earlier (13).

B. *Properties of the* E. coli *Porin Channel Studied in Intact Cells*

1. Solute Size and Diffusion Rate. Our earlier work, in which the distribution of saccharides of various sizes was measured after 10-20 min of equilibration, showed that the *E. coli* porin channel allowed the diffusion of solutes less than about 600 daltons, buth in intact cells (14) and in reconstituted vesicles (7). However, the recent study using intact, growing *E. coli* B showed a striking dependence of diffusion rates of solutes on their sizes, although the solutes used had sizes well within the 600-dalton "exclusion limit" (15). As an example, lactose (342 daltons) was found to penetrate 300 times more slowly than glycerol (92 daltons). This type of effect is due to the fact that solutes fail to enter the channel when they hit the rims of the pore and that greatly increased viscous drag from the walls of the pore is experienced by larger solutes (16). In fact, comparison with theoretical predictions (16) allows us to estimate the pore diameter at about 12 Å or 1.2 nm (15).

2. *Hydrophobicity and Diffusion Rate.* Since β-lactams of about the same size but very different degrees of hydrophobicity are known, they were the ideal set of solutes for studying the effect of hydrophobicity on penetration rates. We used *E. coli* K12 strains producing only one of the two porins normally present (1a or "OmpF", and 1b or "OmpC") or a "new" porin (E or "PhoE") normally repressed in wild type strains, and measured, by using the method developed by Zimmermann and Rosselet (17) as well as by Sawai et al (18), the permeability of the outer membrane to cephacetrile, cefazolin, cephamandole, cephalothin, cephaloram, and 7-benzothienylacetamido analog of cephalothin (12,13). Although the general dependence of diffusion rates on hydrophobicity was already noted by Zimmermann and Rosselet (17), our analysis of the data was somewhat more precise. Firstly, we used only monoanionic cephalosporins in order to avoid the influence of different charged groups. Secondly, since the oil-water partition coefficients at neutral pH are influenced greatly by the presence of charged groups and therefore do not reflect accurately the hydrophobicity of the molecule, the partition coefficients of *unionized forms* (P_u) were determined by extrapolation (19) or by calculation based on the fragmental constant method (20). The results showed clearly that the permeability of cephalosporins had a negative correlation with P_u, and indeed most data points were on a straight line if logarithms of permeability coefficients were plotted against log P_u (12,13). Thus a 10-fold increase in P_u seemed to result in about 6-fold decrease in penetration rate with all of the *E. coli* porin channels tested, within the range of $0.3 < P_u < 45$.

Another interesting finding was that the outer membrane containing porin 1a was always about 5-6 times more permeable to any solute than than containing porin 1b (12,13); it seems that a larger fraction of the porin 1b channels is in the "closed" conformation under physiological conditions.

3. Charges of the Solute and the Diffusion Rate. Our main approach was again the Zimmermann-Rosselet analysis with intact *E. coli* K12 cells. We measured the rates of diffusion of cephalosporins with an additional, positively charged group, such as cephaloridine and cephaloglycin, as well as the diffusion rates of the cephalosporins with an additional, negatively charged group, such as SCE-20. These rates measured with the real compounds were then compared with those predicted for hypothetical, monoanionic analogs with the same structure. This approach showed that with normal K12 porins 1a and 1b, the positive charges and additional negative charges significantly accelerated, and retarded, respectively, the penetration of solutes through the channel.

The slow rate of diffusion of β-lactams with double negative charges was recently confirmed by the study of carbenicillin-resistant mutants of *E. coli* K12 (21). These mutants were found to lack only the more efficient porin, 1a (see above), and can survive well in the laboratory by utilizing the remaining 1b channels. Nevertheless, they are resistant to all the β-lactams with double negative charges tested (carbenicillin, ticarcillin, sulbenicillin, and SCE-20) or with strong hydrophobicity (cephaloram and benzylpenicillin), yet remain fully sensitive to rapidly penetrating agents such as cephacetrile, cefazolin, and cephaloridine. Thus with *E. coli*, the penetration through the outer membrane is probably the rate-limiting step in the action of agents with double negative charges or strong hydrophobicity, and even a fivefold decrease in outer membrane permeability seems to produce significant resistance.

Diffusion through porin E channel, which apparently functions as a channel for phosphate and phosphorylated compounds (22), in contrast to diffusion through 1a and 1b channels, was not retarded by the presence of a second negatively charged group.

C. Properties of the E. coli *Porin Channel Studied with Reconstituted Vesicles*

We have recently devised a method in which the rate of solute flux are determined from the rates of swelling of porin-containing liposomes after their dilution into isotonic solution of test solutes (15,23). This method is quite straightforward for the measurement of diffusion rates of either uncharged or zwitterionic compounds, but its use for anionic compounds introduces several problems. Firstly, the buffer anion tends to flow into the vesicle against its own concentration gradient, if the dilution of vesicles into sodium (or potassium) salts of β-lactams produces huge gradients of these cations. Thus one has to use buffers containing very large anions that are impermeable through the pore. Secondly, creation of large membrane potential through the Donnan effect must be avoided. We have been able to overcome these difficulties recently, and have studied a number of compounds. This method is advantageous in many ways. It is extremely rapid and requires only a few minutes, and is usable for compounds that are resistant to β-lactamases. The preliminary results obtained so far are as follows.

(i) The assay with the "classical" monoanionic cephalosporins gave relative rates very close to what have been determined in intact cells. This confirms the validity of the new method.

(ii) Among the more recent compounds, those that contain a substituted oxime structure as a substituent of the α carbon of the C-7 side chains (i.e. ceftizoxime, cefotaxime, cefuroxime, and ceftazidime) were all found to penetrate at least one order of magnitude more slowly than expected from their hydrophobicity values. The molecular basis of this intriguing finding is not yet clear.

(iii) Zwitterionic compounds all penetrated much faster than monoanionic compounds of comparable P_u values. The fastest penetrating compounds were cephaloridine and MK-0787 (N-formimidoylthienamycin).

(iv) Compounds with two negatively charged groups, such as carbenicillin, SCE-20, and moxalactam, all penetrated quite slowly as predicted from the earlier, intact cell studies. Among them, moxalactam seemed to show the best permeability presumably because of its very hydrophilic structure.

(v) Some of the compounds with bulky side chains, such as cefoperazone, piperacillin, and azlocillin, exhibited very low penetration rates. It is at first surprising that some of these compounds with slow penetration rates are quite effective in killing the cells. However, reasonable explanations are possible, and I have presented my speculations on this topic earlier (13).

II. OUTER MEMBRANE PERMEABILITY OF *NEISSERIA GONORRHOEAE*

Porin was purified from the outer membranes of *N. gonorrhoeae* (24). When reconstituted into liposomes, it produced channels somewhat larger than that of *E. coli*. It is also quite efficient in producing channels, that is, only a very small amount of protein is needed to produce a given degree of permeability. This suggests that most of the channels are "open", an idea that is consistent with the high susceptibility of this organism to various antibiotics, including β-lactams.

III. OUTER MEMBRANE PERMEABILITY OF *PSEUDOMONAS AERUGINOSA*

P. aeruginosa porin has been purified and shown to produce much larger pores than those produced by *E. coli* porins (25). Yet, as is well known, *P. aeruginosa* shows strong "intrinsic resistance" toward various antibiotics, an observation suggesting poor permeability of the outer membrane. We have recently shown that the permeability is indeed very low, on the basis of various observations, including the following (26). (i) Zimmermann-Rosselet assay indicates permeability for cephaloridine and cephacetrile at least 100-fold lower than that of *E. coli* outer membrane. (ii) Although these results could be attributed to the unusual location or exposure of the β-lactamase, the rate of hydrolysis of glucose-6-phosphate and p-nitrophenylphosphate by periplasmic phosphatase in intact cells again indicated a permeability about 100-fold lower than that of *E. coli*. (iii) All nutrients must pass through the outer membrane before they get transported by the active transport systems of the cytoplasmic membrane. Kinetics of the various transport systems in intact cells were again consistent with the notion that the outer membrane permeability was very low.

This low permeability of the outer membrane apparently can be explained by the "closed" nature of most of the porin channels, as reconstitution into liposomes showed that, in order to get a given degree of permeability to small sugars, one had to add 20-50 times more *P. aeruginosa* porin protein than *N. gonorrhoeae* porin.

The large size of the pore, however, means that the rate of diffusion is influenced much less by the properties of the solute, such as size, hydrophobicity, and charges, and this idea was indeed confirmed by performing swelling assays with liposomes reconstituted with purified *P. aeruginosa* porin (12). Thus it appears that the most important feature of the β-lactams effective against *P. aeruginosa* is its resistance against the β-lactamase and its affinity to targets, and as long as these criteria are satisfied other features that are not favorable for enteric organisms with their narrower pores, such as large size, or the presence of the second negatively charged group, can be tolerated.

ACKNOWLEDGMENTS

I am grateful to Emiko Y. Rosenberg, who carried out many of the experiments described here. I also thank various pharmaceutical companies that have donated the samples of β-lactam compounds so generously.

REFERENCES

1. Nikaido, H., and Nakae, T., Adv. Microb. Physiol. 20, 163 (1979).
2. Nikaido, H., Angew. Chem. (Internat. Ed.) 18, 337 (1979).
3. Nikaido, H., in "Bacterial Outer Membranes" (M. Inouye, ed.), p. 361. John Wiley & Sons, New York (1979).
4. Roantree, R. J., Kuo, T. T., MacPhee, D. G., and Stocker, B. A. D., Clin. Res. 17, 157 (1969).
5. Schlecht, S., and Westphal, O., Zentr. Bakteriol. Parasitenk. I Orig. 213, 354 (1970).
6. Nikaido, H., Biochim. Biophys. Acta 433, 118 (1976).
7. Nakae, T., J. Biol. Chem. 251, 2176 (1976).
8. Smit, J., Kamio, Y., and Nikaido, H., J. Bacteriol. 124, 942 (1975).
9. Kamio, Y., and Nikaido, H., Biochemistry 15, 2561 (1976).
10. Nikaido, H., Song, S. A., Shaltiel, L., and Nurminen, M., Biochem. Biophys. Res. Commun. 76, 324 (1977).
11. Bavoil, P., Nikaido, H., and von Meyenburg, K., Mol. Gen. Genet. 158, 23 (1977).
12. Nikaido, H., and Rosenberg, E. Y., to be published.
13. Nikaido, H., in "β-Lactam Antibiotics" (M. R. J. Salton and G. D. Shockman, eds.), p.249. Academic Press, New York (1981).
14. Decad, G., and Nikaido, H., J. Bacteriol. 128, 325 (1976).
15. Nikaido, H., and Rosenberg, E. Y., J. Gen. Physiol. 77, 121 (1981).
16. Renkin, E. M., J. Gen. Physiol. 38, 225 (1954).
17. Zimmermann, W., and Rosselet, A., Antimicrob. Ag. Chemother. 12, 368 (1977).
18. Sawai, T., Matsuba, K., and Yamagishi, S., J. Antibiot. 30, 1134 (1977).
19. Tsuji, A., Kubo, O., Miyamoto, E., and Yamana, T., J. Pharmaceut. Sci. 66, 1675 (1977)
20. Rekker, R. F., "The Hydrophobic Fragmental Constant". Elsevier, Amsterdam (1977).
21. Harder, K. J., Nikaido, H., and Matsuhashi, M., Antimicrob. Ag. Chemother. in press.

22. Tommassen, J., and Lugtenberg, B., J. Bacteriol. 143, 151 (1980).
23. Luckey, M., and Nikaido, H., Proc. Nat. Acad. Sci. USA 77, 167 (1980).
24. Douglas, J. T., Lee, M. D., and Nikaido, H., FEMS Microbiol. Lett. in press.
25. Hancock, R. E. W., Decad, G. M., and Nikaido, H., Biochim. Biophys. Acta, 554, 323 (1979).
26. Yoshimura, F., and Nikaido, H., manuscript in preparation.

BIOCHEMISTRY : RESISTANCE MECHANISM

CONTRIBUTIONS TO BIOLOGY FROM STUDIES ON BACTERIAL RESISTANCE

Bernard D. Davis

Bacterial Physiology Unit
Harvard Medical School
Boston, Massachusetts 02115, U.S.A.

The main reason for studying bacterial resistance to antimicrobial agents has obviously been its importance in limiting the effectiveness of chemotherapy. However, investigations of the mechanisms of resistance have incidentally revealed many novel phenomena of broad biological significance, and mutations to drug resistance have also been of enormous value as selective markers in studies in bacterial genetics. In this lecture I shall survey briefly some of the fundamental discoveries that have arisen from these sources. In addition we might note in passing that the resistance (both natural and acquired) encountered in various pathogens has been the major stimulus to the continued search for new antibiotics. The many resulting selective inhibitors have proven to be powerful reagents in both prokaryotic and eukaryotic cell biology.

The history of these developments illustrates with particular clarity three broad heuristic principles:

First, importance for medicine provides a strong motive for intensive investigation of a phenomenon. With chemotherapy the motive has been especially strong, since we are dealing with millions of lives, and with many different mechanisms of resistance.

Second, bacteria have played a preponderant role in the explosive development of molecular genetics in recent decades; and in this role their most valuable feature has been the possibility of sharp, quantitative selection of very low-frequency mutants or genetic recombinants. Thus in classical genetics one could at best score a few thousand individuals for genetic differences, while in microbial genetics one could screen billions of individuals, in an overnight experiment, for the development of resistance (whether to a chemical inhibitor, a bacteriophage, or host defenses) or for a shift from auxotrophy to prototrophy.

Third, many of the phenomena discovered in microbial genetics have been so unexpected, and so unprecedented, that they have been regarded initially as peculiar, special cases, much like the bizarre shape of the shell of some marine mollusk. For example, after the discovery of the transmission of multiple drug resistance in Japan the subject was reviewed in a prominent American journal, Bacteriological Reviews (1), and yet no one in the West undertook to look for the phenomenon for five years. (Perhaps because of the dramatic impact of the Luria-Delbruck evidence for the selection of single-step mutants, noted below, any deviation from its conclusion then seemed like a regression toward Lamarckism!). No doubt any additional peculiar features of drug resistance being discovered today are also likely to be the first examples of a general class.

Early Key Discoveries Based on Resistance

As Luria remarked, the study of bacterial variation was the last strong-hold of Lamarckism, because variants adapted to the selective agent appeared at a remarkably high rate (in dimensions of time, though not in dimensions of numbers of individuals). To clear up this misconception, and to convert the vague field of bacterial variation into a branch of the precise, coherent field of genetics, it was necessary to demonstrate that heredity in bacteria, as in higher organisms, depended on mutable genes. In this development the most convincing contribution was the ingenious design, by Luria and Delbruck, of a simple statistical method (fluctuation analysis) for testing whether the rare resistant cells emerging in parallel cultures exposed to phage have arisen as random spontaneous mutants or have been directed by the phage to change their heredity adaptively. The results unequivocally demonstrated the role of chance mutations.

These results were soon extended by Demerec to mutations to antibiotic resistance. Nevertheless, for many years it was difficult for many microbiologists to give up the idea that the drugs were somehow "training" the organisms. This anthropomorphic projection of the subjective experience of the investigators is not hard to understand: indeed, it is paralleled today by the reluctance of many scientists to recognize a role of genetic differences, as well as differences in experience, in the observed wide range of human behavior. One might say that human behavioral genetics today is at a stage similar to that of bacterial genetics in 1945–and we do not have tests like fluctuation analysis, and enzyme induction or repression, for precisely estimating the roles of genotypic and phenotypic adaptation.

Resistance mutations also played an essential role in another major development in bacterial genetics: the demonstration, by Avery, MacLeod, and McCarty, that transformation of type in pneumococci is mediated by DNA. In this case the resistance was not to a drug but to host defenses in the mouse (and to agglutination by antibodies in the test tube); and while this kind of difference between strains is not ordinarily designated as resistance, it performs the same function of permitting the experimenter to select the progeny of a rare hereditary variant from the much more numerous progeny of the initial strain. Moreover, Hotchkiss soon achieved similar selections with sulfonamide resistance, and he used the resulting ease of selection to show that the transforming material did not consist of a separate molecule for each gene but contained linked genes.

Avery's great discovery is best known for having launched molecular genetics, by revealing that the material of the gene is DNA. However, it also launched bacterial genetics, for it led to recognition that the 15-year-old, obscure phenomenon of transformation of type is a form of rare gene transfer between bacteria. (These organisms had previously been believed to derive their genomes only by asexual, clonal multiplication.) Without the possibility of gene transfer the discovery of mutable units of inheritance in bacteria could not have blossomed into a science of bacterial genetics. Lederberg recognized the significance of this transfer, and he soon discovered the additional mechanisms of conjugation and transduction, using auxotrophy and prototrophy as the basis for selection. However, it was Hayes's later use of streptomycin resistance, as a selective marker, that revealed the novel mechanisms of conjugation: asymmetric transfer of DNA from donor to recipient cell, rather than zygote formation.

If we broaden the concept of resistance still further, beyond animal host defenses to those of a bacterial host, we see that the highly original development of fine-structure genetics by Benzer depended on the complete resistance (or non-permissiveness) of one strain of *E. coli* to the multiplication of r_{II} mutants of phage T4, contrasted with the permissiveness of another strain. This selective resistance made it possible to grow mutants in the permissive host and then to select for exceedingly rare wild-type recombinants between any two such mutants by mixed infection of the non-permissive host. The results were dramatic: in contrast to the resolving power of classical genetics, which had been limited to chromosomal distances of many thousands of nucleotides (usually between different genes), in these phage crosses the resolving power was abruptly increased, thousands of fold, to its ultimate limit. In this way recombinations could be shown to occur between any two adjacent nucleotides, rather than only at special sites. In addition, mutations were shown not to be strictly random but to vary in frequency from one nucleotide to another. Finally, in a refinement of the important one gene-one enzyme generalization of Beadle and Tatum, the gene could now be rigorously defined as the DNA sequence that codes for a specific polypeptide.

This discovery was later extended by Yanofsky to auxotrophic mutants, employing prototrophy as an equally effective basis for selecting recombinants. The mapping of mutations in the cell could then be correlated with corresponding alterations in DNA or protein, identified *in vitro*. Though the development of fine-structure genetics has not been honored as much as some other discoveries in molecular genetics, it laid a necessary foundation for most subsequent developments in the field.

Phenotypic Mechanisms of Drug Resistance

Drug-resistant mutants have provided not only a tool but a challenge: to understand their origin and their physiological mechanisms. Both these problems have been recognized ever since Paul Ehrlich, early in this century, observed the emergence of "drug-fast" strains of trypanosomes. Nevertheless, they received little attention until the development of antibacterial chemotherapy. Soon thereafter, as we have seen, the origin of resistant strains by random mutation and selection was demonstrated. The tools were not yet available, however, for analyzing the phenotypic mechanisms. At that time Maas and I *(2)* published a theoretical paper outlining seven possible mechanisms, which have since been demonstrated in one or another mutant. Let us consider four that have been particularly important in revealing phenomena of broad biological significance.

An *altered site of binding* of the drug has been observed in resistance to a number of antibiotics, and these alterations have often been useful in mapping the genes and in dissecting the function of the components of various enzymes. Examples include rifampicin, acting on the β subunit of RNA polymerase, and novobiocin and nalidixic acid, each acting on a different subunit of DNA gyrase. Even more valuable have been resistance mutations that alter the ribosome, because such a wide variety of antibiotics act on different sites on this complex body, and because other kinds of mutations in the ribosome have been difficult to isolate (presumably because they are lethal). Accordingly, mutations to resistance to antiribosomal antibiotics opened up several possibilities: mapping the genes for ribosomal components, studying the organization of these genes in operons, and identifying the connections between specific proteins and specific functions of the ribosome. In addition, the discovery of the misreading effect of streptomycin, encountered by chance through the use of streptomycin resistance as a genetic marker, had several fundamental consequences. It revealed that an antibiotic could act not only by blocking but also by distorting the function of its receptor; it initiated serious study of an important variable in molecular genetics, the fidelity of information transfer; and it provided an explanation for the paradoxical phenomenon of streptomycin dependence.

Drug-resistant mutants are also proving valuable in studies of the action of β-lactams. Some mutants that have become resistant to the lytic, bactericidal effect of these drugs are still be inhibited by them, and the study of these strains has revealed that lysis is not a simple, mechanical consequence of the inhibition of peptidoglycan crosslinking but depends on the activity of enzymes that cleave peptidoglycan bonds. Moreover, the bacterial membrane contains a number of different penicillin-binding proteins (PBP), and their mutational alteration provides a promising approach to the understanding of the enzymatic processes that shape the bacterial cell wall.

An increased number of copies of the enzyme inhibited by a drug has not been a prominent mechanism of bacterial mutations to resistance, but it has been shown by Schimke to be a mechanism of resistance to trimethoprim in cultured animal cells. Moreover, the extra copies of the enzyme reflect the production of extra copies of its gene. This development is particularly interesting because the ready selection of cells with such changes in the genome, for a particular, readily selectable marker, strongly suggest that the mammalian genome undergoes frequent variations of this kind in other genes—a phenomenon that seems likely to be important in cell differentiation.

Decreased penetration of a drug has been found to be a very frequent cause of resistance, and it has been a useful tool for studying some aspects of membrane transport. For example, mutants resistant to attack by specific phages were found early to have lost the corresponding receptor, and this finding led to the discovery that the receptors for phages are generally proteins whose primary function is the transport of some specific molecule. Similarly, in the membrane organization and function, which is a particularly lively field today, the selection of resistant mutants is a broad tool for obtaining cells with altered membranes – for example, polymyxin-resistant mutants in gram-negative bacteria are often impaired in toxin secretion.

Increased inactivation of the drug is a most important mechanism, but its most frequent origin is the acquisition of a plasmid-borne gene, rather than increased activity of a gene already present in the bacterial chromosome. The discovery of these plasmids, usually carrying multiple resistance genes, led to what has probably been the most profound of all biological contributions from studies on drug resistance.

Plasmid-borne Resistance

As I noted in the Introduction, enteric bacteria from patients in Japan in the 1950's were discovered to have acquired multiple drug resistance at an unexpectedly high rate, and the mechanism was shown to be conjugative transfer of R factors (a form of infective heredity), rather than successive mutations to resistance to each drug. Eventually, however, study of these factors led to a remarkably wide variety of important basic discoveries, including the following:

1. The first plasmid discovered in bacteria, the F agent, long seemed to be a unique device that had evolved to permit the transfer of long chromosomal sequences from one bacterium to another. However, the resistance plasmids revealed the existence of a great variety of bacterial plasmids that differ in many properties, including host specificity, transmissibility, and pattern of resistance genes. Moreover, these accessory small chromosomes carry not only genes for resistance to antibiotics but also genes that contribute other optional properties to the host, such as utilization of various foodstuffs, production of bacteriocins, production of toxins, and resistance to toxic metal ions. Plasmids thus contribute a great deal to the diversity of the microbial world; and similar small accessory chromosomes have begun to be discovered in cells of higher eukaryotes as well as in many different prokaryotic species and in yeasts.

2. Because plasmids not only can carry genes into a cell but also can transfer them to its chromosome, they clearly play an important role in evolution as agents of gene transfer–definitely in prokaryotes, and very likely also in higher eukaryotes. Moreover, it has become clear that plasmids are closely related to viruses, which suggests that the capacity of the latter to cause disease may be merely a byproduct of a primary evolutionary function in the transmission of genes between hosts.

3. Plasmids have also contributed heavily to the recognition of mobile elements in DNA. The rapid variation in the content of resistance genes in plasmids, during cultivation in the laboratory, revealed the existence of transposons carrying these genes, and these genetic elements were found to be replicated and transferred in much the same way as the independently discovered insertion sequences. This feature of plasmids greatly facilitates their function in gene transfer. In addition, the unexpected lability in DNA that it reveals seems likely to have far-

reaching consequences for cell differentiation in higher organisms.

4. In the powerful methodology of molecular recombination of DNA *in vitro,* followed by cloning in transfected cells, drug resistance markers have proved very useful in the recognition of those clones that harbor a recombinant plasmid.

5. Finally, I would close by noting that new approaches emerging from studies on drug resistance have made major contributions to evolutionary biology. First, in the flora of human populations experiencing wide use of antibiotics resistant bacterial mutants (and resistance plasmids) have increased markedly within a few years, and this epidemiological phenomenon provides a striking, direct example of natural selection, on a time scale that is not generally available for natural selection in higher organisms. Moreover, public acceptance of the implications of evolution is hindered by the imperfections in the fossil record; but molecular genetics now provides quite independent, much more direct evidence. And the most direct evidence is that of the crown gall, induced in various plants by the integration of genes from a plasmid from the infecting bacterium *Agrobacterium tumefaciens.* This transfer of DNA under natural conditions, between organisms very far apart in evolution, points sharply to their common evolutionary source. Finally, though such distant transfers have been detected so far only in tumor formation, where they have a large phenotypic effect, it will be surprising if transfers of this kind are not widespread. It would follow that even though species barriers are necessary for the evolution of complex organisms, they are occasionally breached as a means of adding to the available fund of variation.

In summary, studies of the mechanisms of bacterial resistance, and also the use of mutations to resistance as genetic markers, have made many valuable contributions to basic studies in biology, and there are no doubt more to come.

REFERENCES

1. Watanabe, T. (1966) Infective heredity of multiple drug resistance in bacteria. Bacteriol. Rev. 27, 87.
2. Davis, B. D. and Maas, W. K. (1952) Analysis of the biochemical mechanism of drug resistance in certain bacterial mutants. Proc. Natl. Acad. Sci., USA 38, 775.

NUCLEAR MAGNETIC RESONANCE SPECTROMETRIC ASSAY OF β-LACTAMASE IN BACTERIAL CELLS

M. Kono, K. O'hara, Y. Shiomi, and H. Yoshikoshi

Department of Microbiology
Tokyo College of Pharmacy
Hachioji, Tokyo, Japan

Nuclear magnetic resonance(NMR) spectrometric assay has been shown to be applicable to the various β-lactam antibiotics in determining the rate of hydrolysis using crude extract from β-lactamase producing bacteria(1-3). The present paper indicates that the NMR spectrometric assay method can also be used for determining the β-lactamase activity of intact cells. The hydrolysis of a cephamycin antibiotic, cefmetazole(CMZ), by its resistant bacteria was investigated using this method.

I. β-LACTAMASE ACTIVITY IN LYOPHILIZED CELLS

Escherichia(E.) coli K12 W3110 rifampicin resistant mutant (rif^r) carrying plasmid RP1(2) and *Staphylococcus(S.) aureus* TCP2010 resistant to penicillin G(PCG) were used. Bacteria in nutrient broth(NB) were harvested at the exponential growth phase by centrifugation and washed twice with sterilized distilled water. In *S. aureus*, the culture has been pretreated with 10 μg of PCG/ml. The bacterial cells were suspended in sterilized distilled water, lyophilized, and stored at -20°C until use. Lyophilized cells, β-lactam antibiotic and 0.4 ml of 0.1 M deuterated phosphate buffer(pH 7.0 or 5.8) were put into the NMR tube. The spectra were measured at 37°C. NMR spectra of β-lactam antibiotics(PCG, ABPC, CBPC, TIPC, CET, CER, CEZ, CEX, CED, CFX, and CMZ) incubated with the lyophilized cells of *E. coli* or *S. aureus* were the same as those of the crude extract of *E. coli* which was reported on previously(2). The decrease of the 5-H and 6-H peaks in the spectra of PCs, or that of the 6-H and 7-H peaks in the spectra of cephalosporins, was observed according to the cleavage of β-lactam ring in both. An example is given in Fig. 1 which shows the change in the NMR spectra of PCG with *S. aureus*.

II. β-LACTAMASE ACTIVITY IN LIVING CELLS

Bacterial growth in a deuterated(D-) medium prepared by dissolving the medium constituents in D_2O instead of H_2O was examined. It was found that *E. coli* and *S. aureus* could grow normally in the D-medium of brain heart infusion(BHI), but not in the D-medium of NB. The BHI-D-medium was therefore used exclusively for the bacterial growth in these experiments. The reaction mixture which was used consisted of bacterial culture in a BHI-D-medium, powdered phosphate buffer, 10 mg of β-lactam antibiotic, and fresh BHI-D-medium with a total volume of 0.4 ml. The reaction was performed at 37°C. The kinetics of β-lactamase activity of *E. coli* W3110 rif^r(RP1) cells were examined using ABPC as the substrate, and an obvious quantitative relationship between the hydrolysis of the β-lactam ring and the incubation time was observed up to 90 min.

III. SUBSTRATE PROFILES OF *E. COLI* W3110 RFP^r(RP1) β-LACTAMASE DETERMINED BY NMR SPECTROMETRIC ASSAY AND IODOMETRIC ASSAY(1,2,4)

The β-lactamase was determined using lyophilized cells and living cells. The relative rates of hydrolysis of β-lactam antibiotics were expressed as percentages of the degree of hydrolysis of ABPC, and the results were shown in Fig. 2. It is clear that the results from NMR spectrometric assay using lyophilized cells were in close agreement with those from the iodometric assay, though the relative rate of hydrolysis of CER as measured by the iodometric assay was slightly higher than that in the NMR spectrometric assay. When the relative rates of hydrolysis of β-lactam antibiotics determined by NMR spectrometric assay using lyophilized cells and living cells

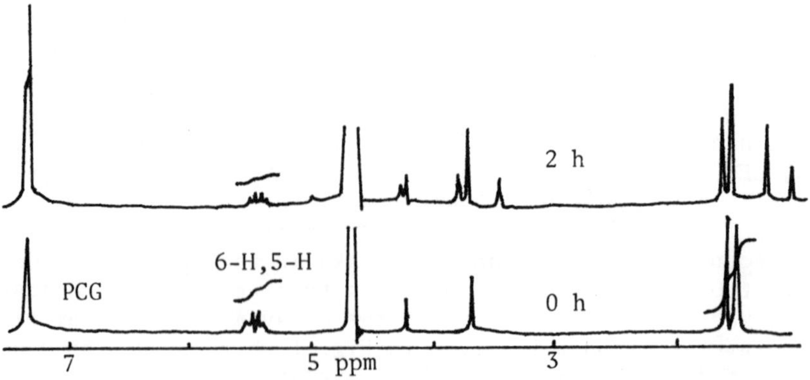

Fig. 1. NMR spectra of PCG in the presence of lyophilized cells of S. aureus TCP2010

were compared, there was no significant difference in their substrate profiles, though the relative rate of hydrolysis of CER was slightly higher for living cells. The hydrolysis rate of CER using living cells was twice as high in the determination by iodometric assay than that by NMR spectrometric assay.

IV. APPLICATION OF NMR SPECTROMETRY FOR THE CLEAVAGE OF CMZ BY ITS RESISTANT *ENTEROBACTER(E.) CLOACAE*

Cefmetazole(CMZ) has been known as a cephamycin antibiotic resistant to β-lactamase. *E. cloacae* strains TCP3179,TCP3831, and TCP3828 were used for the experiment. MICs of these strains to CMZ were 100, 3,200, and more than 12,600 μg/ml, respectively. The cleavage of CMZ by living cells was determined by NMR spectrometry according to the method described in the preceding section, and NMR spectra after 2 h incubation of CMZ with the three strains are presented in Fig. 3. No visible change in the NMR spectrum was observed when CMZ was incubated with TCP3179. On the other hand, a significant change in the NMR spectrum was observed when CMZ was incubated with TCP3831. This change is represented by a

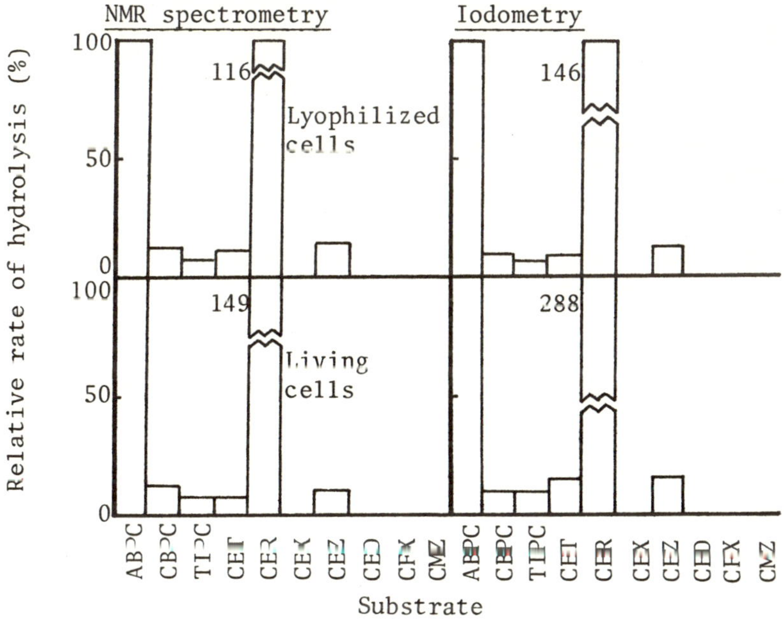

Fig. 2. Comparison of substrate profiles of E. coli W3110 rif^r(RP1) β-lactamase

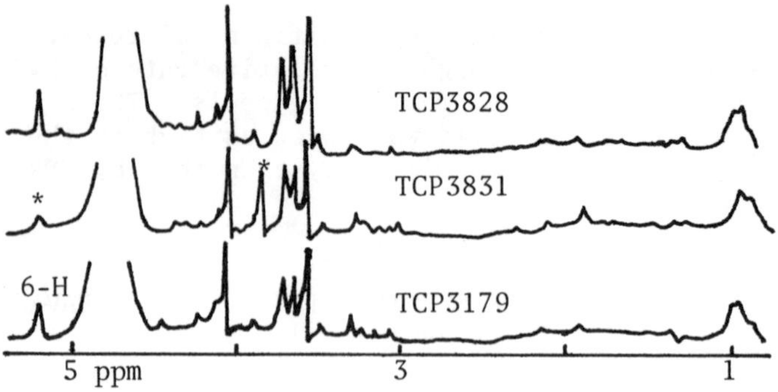

Fig. 3. NMR spectra of CMZ with BHI-D-culture of E. cloacae

decrease in the integration value of 6-H signal at 5.16 ppm and the appearance of a new signal at 3.88 ppm. The new signal was assigned to the N-methyl proton of tetrazol-5-yl thio group released as a concerted reaction from the C-10 position. There was no visible change in the NMR spectrum when CMZ was incubated with TCP3828 for 2 h, but a similar change in the NMR spectrum to that seen in CMZ incubated with TCP3831 was observed after 10 h incubation(data not shown). With PCG in the 0.4 ml of BHI-D-culture the β-lactamase activity of TCP3179 was 6.6 units, of TCP3831 was 35.0 units and of TCP3828 was 111.2 units. The respective results with CMZ were less than 0.2, 3.4 and 0.6 units. From the results presented above, the mechanism of resistance to CMZ in CMZ-resistant *E. cloacae* TCP3831 was found to be due to β-lactamase. The same is known to be the case for CFX resistance in *E. cloacae*(5). However, the very high resistance to CMZ seen in TCP3828 could not be explained by the β-lactamase alone, nor by other structural changes in the CMZ molecule. For this strain, other mechanisms of resistance to CMZ, such as a restriction of the permeability of the drug by changes in the cell membrane or changes in the PC binding protein might be considered.

REFERENCES

1. Nishiura, T., Kawada, Y., Shiomi, Y., O'hara, K. & Kono, M.(1978) Antimicrob. Agents Chemother. *13*,1036-1039
2. Kono, M., O'hara, K. & Shiomi, Y.(1980) Antimicrob. Agents Chemother. *17*,16-19
3. Pratt, R.F., Anderson, E.G. & Odeh, I.(1980) Biochem. Biophys. Res. Commun. *93*,1266-1273
4. Perret, C.J.(1954) Nature *174*,1012-1013
5. Ikeuchi, T. & Osada, Y.(1977) in Microbial Drug Resistance (Mitsuhashi, S., ed.) Vol.2, pp.419-423, Japan Scientific Societies Press, Tokyo & University Park Press, Baltimore

VOLATILIZATION OF MERCURY DETERMINED BY PLASMIDS IN *E. COLI* ISOLATED FROM AN AQUATIC ENVIRONMENT

H. NAKAHARA and H. KOZUKUE

Department of Hygiene
The Jikei University School of Medicine
Tokyo, Japan

In many bacteria, resistance to metallic ions is associated with plasmids (1-7). Penicillinase plasmids in *Staphylococcus aureus* can determine resistance to mercury, cadmium, arsenic, lead, antimony and zinc (1-4); and R plasmids mediating resistance to mercury, cobalt, nickel and arsenic have been observed in *Escherichia coli* (5-7). It is of interest that resistance to these metals is mediated by plasmids which also determine resistance to antibiotics. Most of these metals have recently been of concern as possible causes of environmental pollution. The role of R plasmids in drug resistance has been widely studied and extrachromosomal determinants are a main cause of the increase in number of drug-resistant bacteria. The factors selecting for metal resistant bacteria, however, have not yet been identified. We believe that metal-resistant microorganisms do not arise by chance, but that there must be selecting factors beyond mere drug resistance. One of the factors selecting for metal-resistant bacteria may be environmental pollution caused by these metals.

I. FREQUENCY OF METAL RESISTANCE IN BACTERIA FROM CLINICAL SOURCES

Previously, we tested the susceptibility to drugs and several metallic ions such as mercury, cadmium, arsenate and lead in clinical isolates of *E. coli, Klebsiella pneumoniae, Pseudomonas aeruginosa* and *S. aureus*, and observed that the frequencies of metal resistance were the same as, or higher than those of antibiotic resistances. We isolated 317 plasmids with mercury resistance from mercury resistant *E. coli* and *K. pneumoniae* (8).

We report here that the frequency of metal resistance in *E. coli* strains which were isolated from a river was higher than that of antibiotic resistance.

II. FREQUENCY OF METAL RESISTANCES IN *E.COLI* ISOLATED FROM AN AQUATIC ENVIRONMENT

We isolated 347 strains of *E. coli* from the Tama River in Tokyo. The Tama River is polluted by local factories which use metals. The distribution curves of susceptibility to mercury, cadmium and arsenate showed a clear-cut distinction between susceptible and resistant populations. The frequency of the metal-resistant strains was higher than the frequency of drug-resistant strains (Table 1). Concerning the frequency of metal resistance, a similar result was found in clinical isolates. But, in the frequency of drug resistance there was a great difference between aquatic strains and clinical strains. In addition, many strains were resistance to metals, but remained sensitive to the antibiotics (88% of *E. coli* isolates).

III. GENETIC PROPERTIES OF THE R PLASMIDS WITH Hg RESISTANCE

We checked 170 strains of *E. coli* resistant to mercury and examined the conjugal transferability of their resistance

Table 1. *Frequency of Metal Resistances and Drug Resistaces in Aquatic Isolates of E. coli*

		Clinical strains (1974-1977)	Aquatic strains (1979-1980)
Metal	Hg^{2+}	323 (57)[a]	170 (49)
	Cd^{2+}	524 (93)	258 (74)
	AsO_4^{3-}	342 (61)	292 (84)
Drug	Streptomycin	377 (67)	31 (8.9)
	Tetracycline	344 (61)	15 (4.3)
	Chloramphenicol	315 (56)	16 (4.6)
	Kanamycin	186 (33)	16 (4.6)
	Gentamicin	9 (1.8)	0 (0)
Total no. of strains		564	347

[a] *Figures in parentheses represent % of total isolates.*

Table 2. *Demonstration of R Plasmids Carrying Mercury Resistance in E. coli Isolates and Their Patterns of Resistance to Arsenate and Drugs*

Resistance pattern of R plasmids[a]	Number of E. coli strains with R plasmid
Hg^{2+}, AsO_4^{3-}	92
Hg^{2+}, SM	1
Hg^{2+}, AsO_4^{3-}, SM	6
Hg^{2+}, AsO_4^{3-}, KM	2
Hg^{2+}, AsO_4^{3-}, SM, TC	1
Hg^{2+}, AsO_4^{3-}, SM, KM	4
Hg^{2+}, AsO_4^{3-}, SM, CM	1
Hg^{2+}, SM, CM, KM	1
Hg^{2+}, AsO_4^{3-}, SM, TC, CM	2
Hg^{2+}, AsO_4^{3-}, SM, CM, KM	3
Hg^{2+}, AsO_4^{3-}, SM, TC, KM	1
Hg^{2+}, AsO_4^{3-}, SM, TC, CM, KM	3

[a] SM: Streptomycin, TC: tetracycline, CM: chloramphenicol KM: kanamycin, Hg^{2+}: $HgCl_2$, AsO_4^{3-}: Na_2HAsO_4

Table 3. *Mercury Volatilization Rates*

Strain	Resistance pattern of R plasmids	Rate of volatilization[a] Induced	Uninduced
J53	sensitive	0.01	0.01
J53(R222)	Hg^{2+}, SM, TC, CM	14.0	0.22
Clinical strains			
JE1111	Hg^{2+}, SM, TC, CM, KM	17.8	0.15
JE1217	Hg^{2+}, SM, TC, CM	14.8	0.15
Aquatic strains			
JET5	Hg^{2+}, SM, TC, CM, KM	9.9	0.15
JET18	Hg^{2+}	12.7	0.22

[a] nmole/min. per mg of cells with 5μM $^{203}Hg^{2+}$ and conditions as in ref. 6.

using two recipients. These recipients were as previously described (8). From the 170 isolates of mercury resistant *E. coli*, 117 mercury resistance plasmids were isolated, and of these 115 also conferred arsenate resistance. This is of interest as mercury and arsenic are environmental pollutants. Furthermore, R plasmids with resistance to mercury and arsenic but not to antibiotics were isolated most frequently (Table 2).

IV. MERCURY VOLATILIZATION

Starting with the discovery of Tonomura in 1969 that mercury resistance is due to enzymatic volatilization of Hg^0, many workers including Izaki and Silver have studied the reduction of Hg^{2+} to Hg^0 (3,6,9,10).

We checked the volatilization of mercury by strains containing these mercury resistance plasmids, by using radioactive ^{203}Hg. All of these aquatic isolates of *E. coli* have volatilization activity of Hg^{2+}. Also, all of these volatilization activity are inducible (Table 3).

ACKNOWLEDGMENTS

We are grateful to S. Silver and S. Mitsuhashi for helpful and constructive discussions. We thank M. Inoue for gifts of recipient strains.

REFERENCES

1. Richmond, M. H. & John, M. (1964) Nature *202*,1360-1361
2. Novick, R. P. & Roth, C. (1968) J. Bacteriol. *95*,1335-1342
3. Summers, A. O. & Silver, S. (1972) J. Bacteriol. *112*, 1228-1236
4. Kondo, I., Ishikawa, T. & Nakahara, H. (1974) J. Bacteriol. *117*, 1-7
5. Smith, D. H. (1967) Science *156*,1114-1116
6. Schottel, J. (1978) J. Biol. Chem. *253*,4341-4349
7. Summers, A. O. & Silver, S. (1978) Ann. Rev. Microbiol. *32*,637-672
8. Nakahara, H., Ishikawa, T., Sarai, Y., Kondo, I. & Mitsuhashi, S. (1977) Nature *266*,165-167
9. Furukawa, K., Suzuku, T. & Tonomura, K. (1969) Agric. Biol. Chem. *33*,128-130
10. Komura, I. & Izaki, K. (1971) J. Biochem. *70*,885-893

ROLE OF β-LACTAMASE INHIBITORS IN β-LACTAM-RESISTANT BACTERIA

T. Yokota and E. Azuma[1]

Department of Bacteriology
School of Medicine, Juntendo University
Tokyo, Japan

1. HISTORY OF β-LACTAMASE INHIBITORS

Since the main mechanism of β-lactam-resistance in bacteria was understood as the production of β-lactamase, attempts were made to find out effective β-lactamase inhibitors which can be used as a potentiator of β-lactam antibiotics against the resistant bacteria. Cloxacillin possessing a small *Ki* value to certain types of β-lactamse was combined once with ampicillin (ABPC) and expected their synergistic effect on β-lactamse-producing bacteria. The combined effect, however, was not pominent since the former drug could not penetrate well the outer membrane of gram-negative bacilli.

In 1977, Reading and Cole reported that clavulanic acid(CVA) discovered from a culture of *Streptomyces* exhibited a marked synergistic effect with ABPC or cephaloridine(CER) on β-lactamase-producing bacteria(1). Soon after that, sulbactam, penillanic acid sulfone, synthesized by English et al. was also confirmed to show the similar activity(2). Another β-lactamse inhibitor, izumenolide, was discovered in 1980. The chemical, however, was less practical because of the poor penetrability through the outer membrane(3).

Although CVA, sulbactam and izumenolide were rather weak for the own antimicrobial activities, carbapenem antibiotics, such as thienamycin(TM), N-formimidoyl TM(MK0787), PS-5, olivanic acids and carpetimycin, manifested both strong antibacterial and β-lactamase-inhibitory effects(4-7). In spite of unssolved problems, that is the susceptibility to humen dehydropeptidase-1 in the kidney, the latter group of antibiotics is interesting.

[1] *Present name and address: E. Maruyama, Asahikawa, Japan*

From the studies on β-lactamase inhibitors, some differences were recognized for their synergistic effect with β-lactams according to the kind of inhibitors and species of bacteria. To obtain some answers on that, the authors carried out experiments on interrelationships between tansient and permanent inhibitions of the enzyme and actual potentiation of β-lactams by various β-lactamse inhibitors.

2. *KI* VALUES OF β-LACTAMASE INHIBITORS DO NOT REPRESENT THE ACTIVITY AS β-LACTAM-POTENTIATOR

To elucidate the role of β-lactamase inhibitors, their transient inhibitory effects were compared on various types of β-lactamases as *Ki* values(Table 1). Halogenized isoxazolyl PCs, such as MCIPC and flucloxacillin(MFIPC), possessed small *Ki*s to cephalosporinase(CEPase)-type but the large values to penicillinase(PCase)-type β-lactamases. Contrary to the former, CVA and sulbactam exhibited small *Ki* values against PCase-type enzymes but not to CEPase type. On the other hand, carbapenem antibiotic, MK0787 manifested uniformly small *Ki* values to both PCase- and CEPase-type β-lactamase.

Since a CEPase-type β-lactamase produced chromosomally by *Proteus vulgaris* was transiently inhibited by various inhibitors with almost same *Ki* values, potentiation of CER by the inhibitors against the microbe was tested by the cross liquid medium dilution method as shown in Fig. 1.

Table 1. Ki Values of β-Lactamase Inhibitors to β-Lactamases

β-lactamase		*Ki in μM*				
Richmond	*Mitsuhashi*	*Sulbactam*	*CVA*	*MCIPC*	*MFIPC*	*MK0787*
*Ia**	*CEPase*	*78.3*	*496.*	*0.00089*	*0.0007*	*0.19*
Ic	*CXMase*	*1.52*	*0.53*	*1.21*	*0.70*	*0.16*
II	*PCase*	*0.41*	*0.089*	*45.8*	*43.5*	*8.47*
III	*PCase 1*	*0.69*	*0.30*	*20.8*	*12.7*	*2.30*
IV	*PCase*	*0.96*	*0.26*	*17.5*	*8.53*	*1.38*
V	*PCase 2*	*32.8*	*12.5*	*N.D.*	*N.D.*	*4.13*

**: Ia, Ic, II, III, IV, and V came from Ent. cloacae, P. vulgaris, P. mirabilis, E. coli(pRK1). Klebsiella and E. coli (pRE45), respectively.*

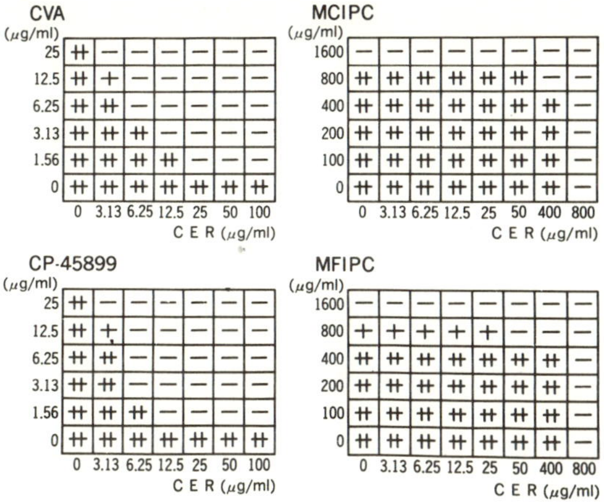

Fig. 1. Synergy of β-lactamase inhibitors with CER on P. vulgaris producing a CEPase-type β-lactamase

Contrary to CVA and sulbactam, both of which exhibited marked potentiation of the antibacterial activity of CER against the microbe, MCIPC and MFIPC showed only the additive effect on that even though *Ki* values of the formers were almost same as those of the latter drugs. The obtained results suggest that the transient inhibitory effects of β-lactamase inhibitors do not represent actual potentiating activity with β-lactams.

3. PERMANENT INACTIVATING ACTIVITIES OF β-LACTAMASE INHIBITORS

Permanent inactivation of the enzyme by β-latamase inhibitors was examined by two different methods.

A. Dilution Method

Crude β-lactamases were incubated at 30°C for 50 min without shaking with 5 to 50 μg/ml of the inhibitors. And the residual enzyme activies were measured by the macroiodometry after more than 200-fold dilution to avoid the influence of the inhibitor, using AMPC and CER as substrates for PCase and CEPase, respectively. The type Ia CEPase produced by *Enterobacter cloacae* Nek39 was permanently inactivated at 90 % by 5 μg/ml of sulbactam, 50 % by 10 ug/ml of MK0787 and only 10% by 10 μg/ml of CVA. It was confirmed that the *Ki* values are not related to the permanent inactivating activities of the inhibitors.

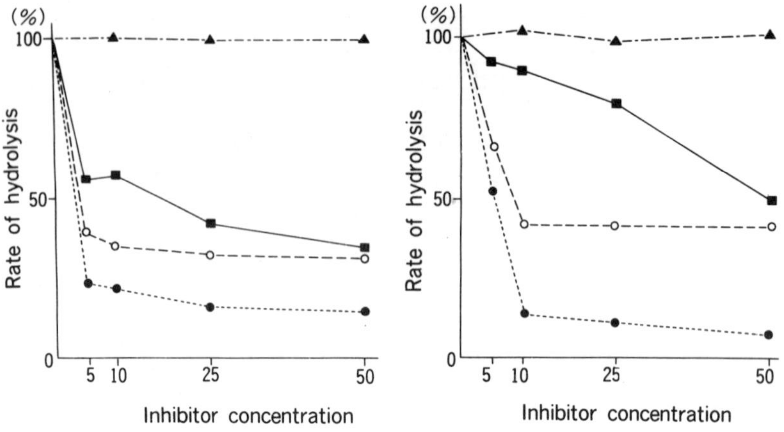

Fig. 2. Permanent inactivation of the type Ic(CXMase: left) and the type III(TEM: right) β-lactamases measured by the dilution method. Numbers of inhibitor concentration indicate μg/ml. Symbols: Triangle: MCIPC: square: sulbactam; open circle: MK0787; closed circle: CVA

As indicated in Fig. 2, the type Ic(CMXase) β-lactamase was strongly inactivated by CVA, MK0787 and sulbactam, although to the type III, inactivating activity of sulbactam was inferior than CVA and MK0787. MCIPC possessed no permanet inactivating effect on every types β-lactamases, i.e. Ia, Ic and III. Although the type II β-lactamase(PCase) was uniformly inactivated by the all inhibitors except MCIPC, the type IV(PCase) produced chromosomally by *Klebsiella* and the type V(OXA:PCase) controled by minor group of R plasmids were markedly inactivated by CVA and MK0787, respectively, and moderately by the others.

B. Isoelectric Focusing Method

Permanent inactivation of β-lactamases by the inhibitors was also investigated by the isoelectric focusing method followed by staining with a chromogenic β-lactam(Glaxo CEP 87/312), accoding to the report by Matthew et al.(9). The crude β-lactamases pretreated with the inhibitors at 30°C for 50 min were subjected to the electrophoresis and stained. By this method, β-lactamase and inhibitor were focused at different places, so that the residual enzyme activities could be demonstrated clearly without interference by the inhibitors.

As indicated in Fig. 3, CVA inactivated completly the main

fraction of 3 isozymes of the type Ia(CXMase) β-lactamase and moderately 2 other fractions accompanying with a change of the *pI*s. Whereas, sulbactam inactivated strongly the 2 subfractions and moderately the main fraction without *pI*-shift. The similar inactivation pattern of the type Ic enzyme was observed by the treatment with MK0787.

The permanent inactivation of type III(TEM) β-lactamase by CVA was characteristic. The all 3 isozymes were completely killed as indicated in Fig. 3. On the other hand, sulbactam inactivated strongly the main fraction and moderately the 2 subfractions. It may be of interest to note that MK0787 converted *pI*s of the isozymes successively, retaining the enzyme activity. Although 3 isozymes were recognized in the type Ia (CEPase) and type V(OXA:PCase) β-lactamases, only 2 isozymes were detected in the type II(PCase) and type IV(PCase). And inactivation of the each isozyme differed from one inhibitor to another. The most prominent inactivation of the type Ia, II, IV and V was caused by MK0787, CVA, CVA, and CVA, respectively.

4. PERMANENT INACTIVATION REFLECTS THE ACTIVITY OF β-LACTAMASE INHIBITORS AS A POTENTIATOR OF β-LACTAM ANTIBIOTICS

It was concluded that the synergistic effect of β-lactamase inhibitors on the resistant bacteria with β-lactam antibiotics was reflection of the penetrability and the permanent inactivating activity of an inhibitor. The details of mechanisms of the inactivation, however, differ from one inhibitor to another as described previously. Further studies may be necessary to elucidate molecular mechanisms of β-lactamase inhibitors, even though the reports by Fisher et al. and that by Brenner and Knowles were already appeared on CVA and sulbactam, respectively(10, 11).

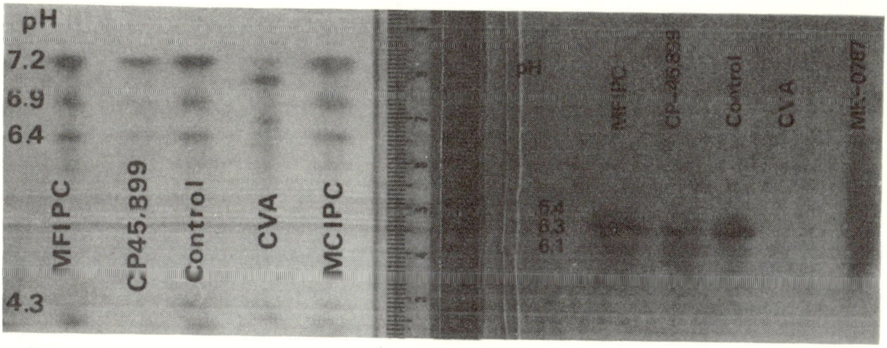

Fig. 3. Isoelectric focusing of the typeIc(left) and type III (right) β-lactamases treated with various inhibitors.

REFERENCES

1. Reading, C. & Cole, M.(1977) Antimicrob. Agents Chemother. *11*, 852-857
2. English, A.R., Retsema, J.A., Girard, A.E., Lynch, J.E. & Barth(1978) Antimicrob. Agents Chemother. *14*,414-419
3. Bush, K., Bonner, D.P., & Sykes, R.B.(1980) J. Antibiotics *33*,1262-1269
4. Kesado, T., Hashimizu, T., & Asahi, Y.(1980) Antimicrob. Agents Chemother. *17*,912-917
5. Okamura, K., Sakamoto, M., Fukagawa, Y.& Ishikura, T.(1979) J. Antibiotics *32*,280-286
6. Hood, J., Box,S. & Verrall, M(1979)J. Antibiotics *32*,295-293
7. Mori, T., Nakayama, M., Iwasaki, A., Kimura, S., Mizoguchi, T., Tanabe, S., Murakami, A., Watanabe, I., Okuchi, M., Ito, H., Saino, Y. & Kobayashi F.(1980) 20th Intersci.Conf. Antimicrob. Agents Chemother. New Orleans
8. Fisher, J., Charnas, R.L. & Knowles, J.R.(1978) Biochem. *17*,2180-2184
9. Matthew, M., Harria, A.M., Marshall, M.J., & Ross,G.W. (1975) J. Gen. Microbiol. *88*,169-178
10. Charnas, R.L., Fisher, J., & Knowless, J.R.(1978) Biochem. *17*,2185-2189
11. Brenner,D.G. & Knowless, J.R.(1981) Biochem. *20*, 3680-3687

MECHANISMS OF BACTERIAL RESISTANCES TO THE TOXIC HEAVY METALS ANTIMONY, ARSENIC, CADMIUM, MERCURY AND SILVER[1]

Simon Silver, Robert D. Perry[2]
Zofia Tynecka[3] and Thomas G. Kinscherf[4]

Biology Department
Washington University
St. Louis, Missouri, U.S.A.

I. INTRODUCTION

Bacteria of many species and from many sources have genes that confer resistances to specific toxic heavy metals. These resistances are often determined by genes on plasmids. The same resistance mechanisms occur in bacteria from soil, water, industrial and clinical sources. The mechanism of mercury and organomercurial resistance is the enzymatic detoxification of mercurials into volatile chemicals (methane, ethane, metallic mercury) which are rapidly lost from the growth medium. The genetic control of this resistance has been studied, and the enzymes responsible have been purified and characterized.

Cadmium and arsenate resistances are due to blocks in the net accumulation of these toxic materials. Efficient efflux systems cause the rapid excretion of Cd^{2+} and AsO_4^{3-}. The mechanisms of arsenite and of antimony resistance, which are usually found associated with arsenate resistance, are not known. Silver resistance is due to lowered affinity of the resistant cells for Ag^+; sensitivity is due to binding of Ag^+ more effectively to cells than to Cl^-.

[1] *Recent work from our laboratory was supported by grants from the U.S. National Science Foundation (PCM 79-03986) and the U.S. National Institutes of Health (AI15672).*

[2] *Present address: Department of Microbiology and Public Health, Michigan State University, East Lansing, Michigan, U.S.A.*

[3] *Present address: Department of Microbiology, Medical Academy, Lublin, Poland.*

[4] *Present address: Department of Biology, Stanford University, Stanford, California, U.S.A.*

Living cells relate to the chemical Periodic Table in three ways. Some elements are necessary for intracellular metabolism (1) while others, which abound in natural environments, are not generally used within the cell but can be used for extracellular structural or regulatory functions. Finally, some elements are toxic and have no useful function (2). Potassium and phosphorus are examples of the first class; calcium and chlorine are examples of the second class; and arsenic, mercury and cadmium are examples of toxic elements without biological function. This paper will deal with toxic elements and their compounds. The occurrence of toxic elements in natural environments can not be ignored since the levels of mercury, cadmium and arsenic are high enough to affect human health. The frequencies of bacterial resistances to heavy metal compounds are very high, especially among clinical isolates of bacteria resistant to various antibiotics because of the presence of resistance plasmids. It seems that bacteria have devised a highly specific resistance mechanism for every toxic heavy metal. This topic was reviewed four years ago (2). For this symposium, we shall provide a general picture and emphasize newer information.

II. MERCURY AND ORGANOMERCURIAL RESISTANCES

The mercury cycle in the environment is the best known case of microbial metabolism affecting the distribution and chemical form of a toxic heavy metal.

Microbial activity is associated with mercury transformations including methylation and demethylation of Hg^{2+} and oxidation and reduction of inorganic mercury. Since this discussion is limited to resistance mechanisms, we will deal only with the transformations from left to right (as drawn in Fig. 1). Metallic mercury is less toxic than ionic mercury in most biological systems and methylmercury is far more toxic than ionic Hg^{2+}. Thus, the transformations from right to left, although important to understanding microbial effects on toxic heavy metals, do not contribute to bacterial resistance to mercury or to organomercurials.

The earliest studies of plasmid-mediated resistance to Hg^{2+} were with a multiply drug-resistant *Escherichia coli* (3) and with a soil pseudomonad (4,5) for which a plasmid-determinant has still not been demonstrated. Plasmid-determined resistance to Hg^{2+} and to organomercurials occurs in both *Staphylococcus aureus* (6) and *E. coli* (7). The frequency of Hg^{2+} resistance determinants among clinical isolates can be over 50% (8,9) and among a collection of over 800 plasmids introduced into *E. coli* K-12 in London, about 25% conferred Hg^{2+} resistance (10).

$$CH_3Hg^+ \rightleftharpoons Hg^{2+} \rightleftharpoons Hg^{\circ}$$

Fig. 1. Environmental transformations of mercury.

More recently, mercuric and organomercurial-resistant strains with very similar properties have been found in a wide variety of bacterial species from soil, water and marine environments (11-15).

A small number of predictable resistance patterns for organomercurials was found among strains with plasmids (16): (a) In E. coli over 90% of the plasmids conferring mercury resistance also confer resistance to the organomercurials merbromin and fluorescein mercuric acetate (FMA) but to no other tested organomercurial (Fig. 2). These are "narrow spectrum" mercurial-resistance plasmids (16). The other 4% "broad spectrum" plasmids conferred additional resistances to phenylmercuric acetate (PMA) and thimerosal. (b) The plasmids in Pseudomonas aeruginosa also divided into "narrow" and "broad" spectrum with regard to resistance to organomercurials (17); however, about 50% fell into each class. Furthermore, the "narrow spectrum" Pseudomonas plasmids also confer resistance to p-hydroxymer-curibenzoate (pHMB) and the "broad spectrum" Pseudomonas plasmids show additional resistance to methylmercuric and ethylmercuric compounds

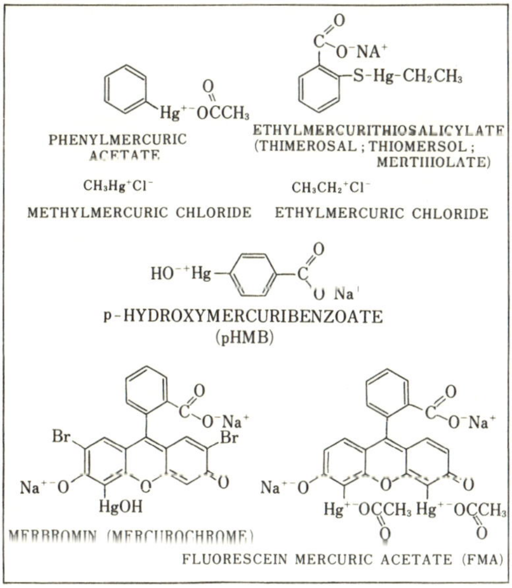

Fig. 2. Structures of the organmercurial compounds.

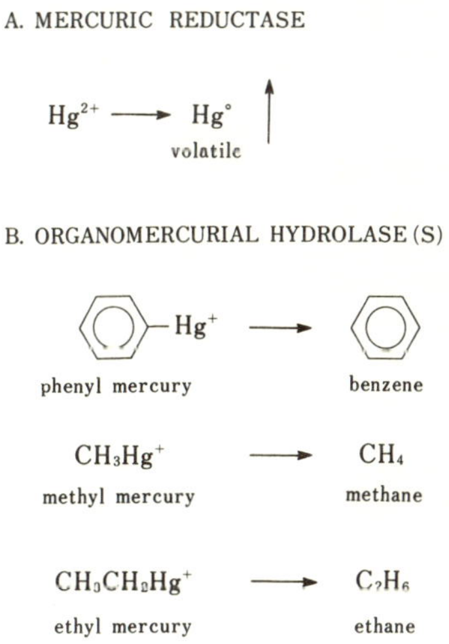

Fig. 3. Enzymatic detoxification of Hg(II) and organomercurials.

(16,17). (c) Only a single pattern has been reported with S. aureus plasmids (17,18). This pattern is different from those with the Gram-negative bacteria because all the S. aureus plasmids confer resistances to PMA, pHMB and FMA but not to thimerosal or to merbromin. In the last few months, the first thimerosal-resistant S. aureus was found (19).

The conclusion is not that there is a single pattern of resistance, but rather that there are a very limited number of patterns and that these have begun to be understood in terms of the biochemistry of the enzymes involved.

A. Enzymatic Mechanism of Mercury and Organomercurial Detoxification

Hg^{2+} resistance in bacteria results from enzymatic detoxification of mercury compounds leading to the volatilization of mercury from the growing culture. This was discovered independently in two laboratories in Japan (3,5,20,21) and in St. Louis (22). The volatile mercury was shown to be metallic Hg° in each case, and the enzyme responsible is called mercuric reductase.

Several organomercurials are also enzymatically detoxified to volatile compounds. These organomercurials include methylmercury, ethylmercury, PMA, pHMB and thimerosal (Figs. 2 and 3); benzene is produced from PMA, methane from methyl-mercury, and ethane from ethylmercury. The enzymes responsible for cleaving the Hg-C bond are organomercurial hydrolases. Tezuka and Tonomura (23,24) were able to separate two small soluble hydrolase enzymes. Both have molecular weights of about 19,000 and require thiol reagents such as thioglycolate. The two hydrolases were difficult to separate by chromatographic methods; but when this was accomplished (24), it was found that one enzyme cleaved PMA, pHMB and methylmercury, while the other enzyme cleaved only PMA and pHMB. With a plasmid-containing E. coli, there was no evidence for hydrolysis of pHMB (16), and Schottel (25) was unable to separate the two hydrolases. Nevertheless, kinetic analysis indicated that there were two enzymes active toward PMA and only one active toward methyl- and ethylmercury. The general properties of the enzymes from the soil pseudomonad and E. coli were similar except the E. coli organomercurial hydrolases had a somewhat greater molecular weight (25).

Mercuric reductase has been studied in greater detail both with plasmid-bearing E. coli (21, 25-27) and with the soil pseudomonad (20). The enzyme is strictly NADPH-dependent, and one NADPH is oxidized per Hg(II) reduced (25,27). The molecular weight and subunit structure of the mercuric reductase enzyme from various sources is currently under reevaluation. Furukawa and Tonomura (28) reported that the enzyme from the soil pseudomonad had a molecular weight of about 65,000 (from movement through a gel filtration column). Schottel (25), however, reported that the enzyme from a particular plasmid that originated in *Serratia* was a trimer of 170,000 daltons containing identical monomer subunits, each approximately the size of the *Pseudomonas* enzyme. Each subunit contained a single bound FAD, for a total of 3 FAD's per 170,000 molecular weight. Recently, the situation has become more complex, as Fox and Walsh (27) reported that the molecular weight of still another mercuric reductase (this time from transposon Tn*501*, which originated in a clinical *P. aeruginosa* isolate) was about 125,000, and the enzyme appeared to be dimeric. We have confirmed the dimeric structure of the mercuric reductase of Tn*501* using both gel filtration and migration through polyacrylamide gels of varying porosity as measures of molecule size(26).

Antibodies have been prepared against purified mercuric reductases coded by two plasmids in *E. coli* (26). All reductases obtained from different Gram-negative sources reacted with these antibodies, as shown by inhibition of enzyme activity and by formation of precipitin bands on double-diffusion gels. The enzymes fell into two major subclasses, based on only partial immunological identity.

The prototype of the first enzyme class is coded by transposon Tn*501*, the first well-studied mercuric resistance transposon (29). It is the Tn*501* enzyme that appears to be dimeric (26,27). This enzyme class also includes mercuric reductases governed by a variety of plasmids found in clinical isolates of enteric bacteria and *P. aeruginosa*, in marine pseudomonads, and in *Pseudomonas putida* (the MER plasmid). The MER plasmid harbors a transposon, Tn*1861*, which appears indistinguishable (11) from Tn501 (29). One strong conclusion from studies of plasmid-determined mercuric resistance is that the same system appears widely in clinical isolates and in bacteria from other environments. However, newer Hg^{2+}-(and in one case phenylmercuric-) resistance transposons from soil microbes show different patterns of digestion by restriction endonuclease enzymes (15), indicating that although the systems are related immunologically, they do not have identical DNA sequences.

The second immunological subgroup of the Gram negative mercuric reductases has as its prototype the enzyme coded by plasmid R100, one of the earliest and most thoroughly studied of the antibiotic resistance plasmids. It is with plasmid R100 that the genetic structure of the mercuric resistance operon was studied in detail (30,31). The enzyme from R100 appears to be trimeric (26) as was the enzyme from plasmid R831 studied by Schottel (25). This subgroup includes enzymes from plasmids of a wide variety of incompatability groups and also the enzyme determined by a second *Pseudomonas* mercury transposon, Tn*502* (26; V. Stanisich, personal communication). Although all of the mercuric reductases from Gram-negative bacteria were immunologically related, the antibodies prepared against the two classes of Gram-negative enzymes did not cross react with mercuric reductases from S. aureus strains and marine and soil bacilli. These enzymes from Gram-positive sources showed similar functional requirements to those from the Gram-negative bacteria (18,26), yet they are immunologically distinct.

To summarize briefly the current understanding of plasmid-determined mercuric and organomercurial resistances: (a) They occur widely in both Gram positive and Gram negative species and are the best understood of all plasmid-coded heavy metal resistances. (b) Resistance is due to enzymatic detoxification of the mercurials to volatile compounds of lesser toxicity. (c) The enzymes responsible (mercuric reductases and organomercurial hydrolases) have been purified and studied in vitro.

III. ARSENIC AND ANTIMONY RESISTANCES

Arsenic and antimony resistances are governed by plasmids that also code for antibiotic and other heavy metal resistances (6,32,33). Arsenate, arsenite and antimony(III) resistances are coded for by an inducible operon-like system in both S. aureus and E. coli (34). Each of the three ions induces all three resistances. In E. coli, Bi(III) is a gratuitons inducer of arsenate resistance, since it causes reduced $^{74}AsO_4^{3-}$ accumulation even though the plasmid system does not confer Bi(III) resistance. S. aureus has a genetically separate plasmid-mediated Bi(III) resistance determinant (6,35) of unknown mechanism. Not all oxyanions induce the arsenate system; in E. coli harboring plasmid R773, highly toxic VO_4^{3-} does not induce resistance to itself or to AsO_4^{3-} (data not shown).

The mechanism of arsenate resistance is a reduced accumulation of arsenate by induced resistant cells (Fig. 4). Arsenate is normally accumulated via the cellular phosphate transport systems, of which many bacteria appear to have two (1). Phosphate protects cells from arsenate toxicity, just as high Mn(II) protects sensitive S. aureus from Cd(II) toxicity (Perry, unpublished; see below). The distinction between arsenate and arsenite resistances was shown by finding that phosphate did not protect against arsenite inhibition of growth (34). Genetic studies with S. aureus plasmids have also demonstrated that the arsenate resistance gene is different from but closely linked to the arsenite resistance gene; additionally, arsenite and antimony resistances may be determined by separate genes (35). The presence of the resistance plasmid does not alter the kinetic parameters of the cellular phosphate transport systems, even the K_i for arsenate as a competitive inhibitor of phosphate transport is unchanged. This finding, coupled with direct evidence for plasmid-governed energy-dependent efflux of arsenate (Fig. 5) indicated that the block on uptake of

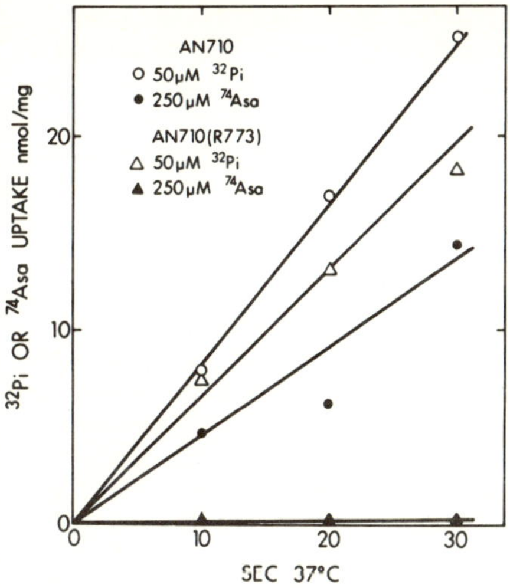

Fig. 4. Arsenate and phosphate uptake by sensitive (AN710) and resistant (AN710(R773)) E. coli (34).

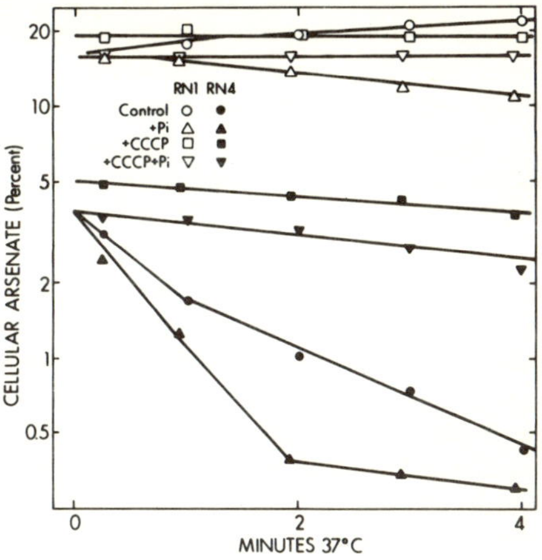

Fig. 5. Accelerated energy-dependent efflux of arsenate by resistant S. aureus cells (RN4) but not by the sensitive cells (RN1). Cultures were grown, induced, "loaded" with $^{74}AsO_4^{3-}$ and efflux was initiated by dilution as described in ref. 36. 40 μM CCCP and 5 mM PO_4^{3-} (Pi) were added as indicated (from ref. 36).

arsenate in Figure 4 may result from rapid efflux. The energy dependence of the efflux process was shown by its sensitivity to "uncouplers" such as CCCP and tetrachlorosalicylanilide and ionophore antibiotics such as nigericin and monensin (36). Hg^{2+} and pHMB also inhibit the efflux system, and the Hg^{2+} inhibition is readily reversed by mercaptoethanol (36). The tentative conclusion is that AsO_4^{3-} efflux is directly coupled to a pH-sensitive ATPase-like transport system, similar to the Na^+ efflux system found in Streptococcus faecalis (37). The inhibition by uncouplers (at low pH only) and the inhibition by nigericin but not by valinomycin are consistent with this hypothesis (37).

A current question about the arsenate efflux system concerns its specificity. Arsenate generally functions as a phosphate analogue and is accumulated by bacteria via phosphate transport systems (1,34). The arsenate-resistance efflux system should not excrete both phosphate and arsenate, since the cells would then become phosphate starved, a situation no more advantageous than being arsenate inhibited! A basic conclusion from our work on arsenate (and also on Cd^{2+} resistance; see below) is that toxic heavy metals often get into cells by means of transport systems for normally required nutrients (1,2). Energy-dependent efflux systems functioning as resistance mechanisms should be highly specific for the toxic anion or cation to prevent loss of the required nutrient.

The mechanism(s) of arsenite and of antimony resistances are not known. Arsenicals and antimonials are toxic by virtue of inhibiting thiol-containing enzymes (38). Some dithiol reagents such as BAL (British anti-Lewisite) protect against arsenical and antimonial toxicity. Growing resistant cells did not excrete soluble thiol compounds into the medium to bind arsenite and antimony, since pre-growth of resistant cells in medium containing these toxic ions does not allow subsequent growth of sensitive or of uninduced resistant cells (34). Arsenite is not oxidized to the less toxic arsenate by plasmid-bearing E. coli or S. aureus (34). The absence of "detoxification" measured by experiments inocculating sensitive cells into medium "preconditioned" by growth of resistant cells eliminates all other possible mechanisms involving changes in extracellular chemical states. Although plasmid-mediated resistance to As(III) and Sb(III) does not involve chemical transformations, bacteria and fungi are known to oxidize, reduce, methylate, and demethylate arsenicals and antimonials (most recently summarized in ref. 39). For plasmid-mediated resistance, only the untested hypotheses of an alteration in uptake or a change in a key intracellular target are left.

IV. CADMIUM AND ZINC RESISTANCE

Plasmid-determined cadmium resistance has been found only in S. aureus (6). In some clinical collections, Cd(II) resistance is the most common of the S. aureus plasmid resistances, exceeding in frequency both mercury and penicillin resistances (8). Gram-negative cells without plasmids are just as resistant to Cd(II) as are staphylococci with plasmids (9), probably because of relatively reduced Cd(II) uptake (Silver, unpublished). However, there are occasional Gram-negative bacteria that are sensitive or even "hypersensitive" to Cd(II) (12; T. Barkay, unpublished data). The basis of Cd(II) sensitivity and resistance in other bacterial species is a subject in need of more effort.

In staphylococci, Cd(II) is accumulated by a membrane transport system utilizing the cross-membrane electrical potential (40,41). This uptake system is highly specific for Cd(II) and Mn(II) (40,42) with respective K_m's of 10 μM and 16 μM in whole cells (41) and 0.2 μM and 0.95 μM in membrane vesicles (40).

Two separate plasmid genes are responsible for the Cd(II) resistance of S. aureus strains (6). The cadA and cadB genes

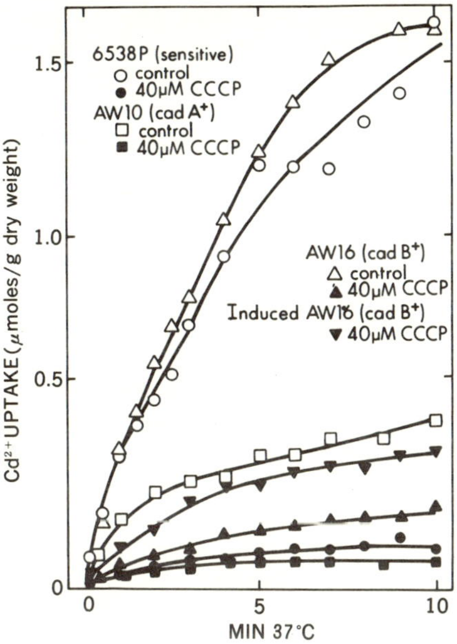

Fig. 6. Cadmium uptake and binding by S. aureus strains 6538P (sensitive), AW10 (cadA$^+$ resistant) and AW16 (cadB$^+$ resistant). Cells were grown in brain heart infusion broth, harvested, washed and resuspended. 40 μM CCCP was added 5 min prior to the addition of 1 μM ^{109}CdCl$_2$ at 37°C. AW16 was induced by exposure to 1 μM non-radioactive CdCl$_2$ during logarithmic growth.

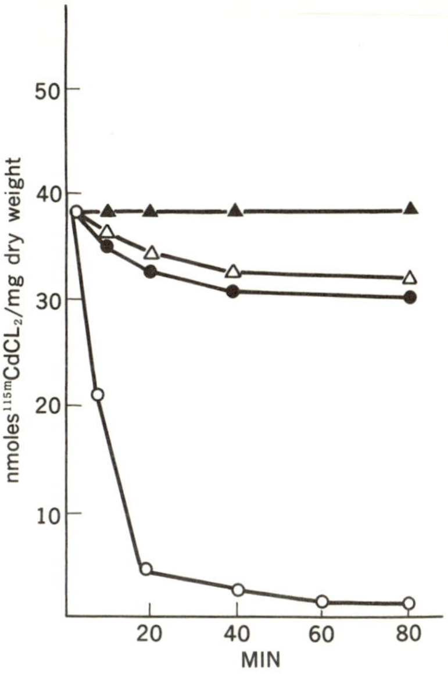

Fig. 7. Retention of ^{115m}Cd^{2+} by cadmium-resistant S. aureus. Cell suspensions were preincubated with 1 mM ^{115m}Cd^{2+} for 10 min at 37°C. Cells were washed free of Cd^{2+} by filtration with 4°C broth, and then ^{115}Cd^{2+} efflux was assayed at 37°C. Symbols: ○, control cells at 37°C; ▲, cells at 4°C; △, cells with 10 mM dinitrophenol; and ●, cells with 100 μM DCCD (ref. 46).

confer, respectively, 100-fold and 10-fold increases in Cd(II) resistance (6,42,43). When both genes are present, the effect of cadA masks the cadB gene effect; cadA$^+$ cadB$^+$ strains are no more resistant than are cadA$^+$ cadB$^-$ strains (43). Both genes confer increased Zn(II) resistance similar to that for Cd(II) (40). This and the apparent genetic linkage of Cd(II) and Zn(II) resistance (6) indicate that the cadA and cadB genes are also responsible for Zn(II) resistance.

Cd(II) resistance is due to a constitutive reduction in Cd(II) accumulation by resistant cells (41,42,44,45). The cadA gene product causes this reduced Cd(II) accumulation (40,42, Fig. 6). Recently, it has been shown that reduced Cd(II) accumulation is due to a plasmid-encoded efflux system which rapidly excretes

Cd(II) (46, Fig. 7). This efflux system is coded for by the cadA gene. $cadA^+$ resistant strains (but neither sensitive nor $cadB^+$ resistant strains) possess this efflux system (40). Although not directly demonstrated, it seems plausible that the cadA gene product might also cause Zn(II) efflux. However, the presence of the $cadA^+$ gene does not reduce Mn(II) uptake nor cause rapid efflux of accumulated Mn(II) (40).

The cadA-encoded efflux system is energy dependent. Cd(II) efflux was abolished by dinitrophenol and at low temperature (4°C) (46, Fig. 7). Although efflux directly energized by ATP has not been definitely eliminated, inhibitor studies indicate that the efflux system is a $Cd^{2+}/2H^+$ antiport. The current model for Cd(II) uptake and efflux is presented in Fig. 8.

A $cadB^+$ resistant strain failed to show significant differences in energy-dependent uptake by whole cells (Fig. 6) and membrane vesicles (40). Additionally, the kinetic parameters of Cd(II) and Mn(II) uptake by sensitive and $cadB^+$ resistant membrane vesicles were nearly identical (40). The only significant difference between sensitive and $cadB^+$ resistant cells was the extent of energy independent Cd(II) binding (40, Fig. 6). When sensitive and $cadB^+$ resistant cells were exposed to 1 μM Cd(II) during logarithmic growth, the energy-independent Cd(II) binding by the induced and noninduced $cadB^+$ strain was increased 3.2- and 1.9-fold respectively over that observed in the sensitive strain (40). Energy-independent Mn(II) binding was unaffected. Since L-forms of the $cadB^+$ resistant strain still express Cd(II) and Zn(II) resistance, it is unlikely that the gene product is located in the cell wall (40). Our working hypothesis that the cadB gene product is an inducible (or amplifiable) Cd(II)- and Zn(II)-binding component is tentative and must await confirmation by subsequent work.

V. SILVER RESISTANCE

Microbial silver toxicity is found in situations of industrial pollution and is especially associated with the use of photographic film. In hospitals, silver salts are the preferred antimicrobial agents for burns covering large areas (47). It is thus not surprising that silver-resistant bacteria have been found in urban and industrial polluted sites (48) or that silver resistant bacteria (49-51) and silver resistance plasmids have recently been described (52). R.W. Hedges (personal communication) produced a recombinant between plasmid R1 and a Ag^+ resistance plasmid from *Citrobacter* (51) and introduced it into an *E. coli* K-12 strain. Silver resistance is constitutive in *E. coli*. The plasmid-determined resistance is very great, and the ratio of minimum-inhibitory concentrations (resistant: sensitive) can be greater than 100:1 (53). The level of resistance is strongly dependent upon available halide ions; without Cl^-, there

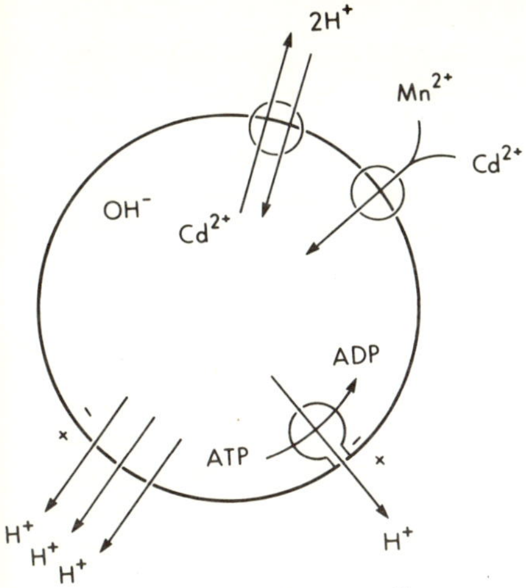

Fig. 8. Model for Cd^{2+} uptake and efflux systems (46).

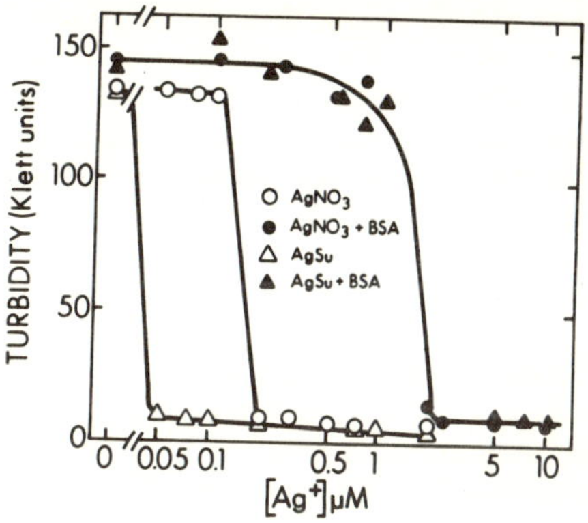

Fig. 9. Effect of bovine serum albumin (BSA; 0.5 mg/ml) on sensitivity of E. coli K-12 to $AgNO_3$ and silver sulfadiazine (AgSu). Silver et al., manuscript in preparation.

was relatively little difference between cells with or without the resistance plasmid (53). At concentrations far below those used for Cl^-, Br^- and I^- conferred resistances on both sensitive and resistant cells. Not only halides that precipitate Ag^+ affect Ag^+ toxicity, but proteins such as bovine serum albumin which can bind Ag^+ salts also increase the resistance of both sensitive cells (Fig. 9) and plasmid-containing resistant cells (data not shown). In the presence of serum albumin, the resistant cells had a 20-fold higher minimum inhibitory concentration for $AgNO_3$ than did the sensitive cells. However, the resistant cells (data not shown) were about equally resistant to silver sulfadiazine with or without serum albumin -- but more resistant than sensitive cells. The complex interplay between the bacterial cells and the potential Ag^+-binding materials (Cl^- and proteins) on burn surfaces must be responsible for the success or failure of Ag^+ treatment of burns, and in these complex interactions there must be an answer to the apparently greater utility of Ag^+ sulfadiazine than $AgNO_3$ in such treatments (47). The function of sulfadiazine in topical preparations is not to inhibit bacterial growth directly; the concentrations released are too low (47), but rather to bind silver in a form subject to slow release. Both sensitive and resistant cells bind Ag^+ tightly and are killed by effects on cell respiration and other cell surface functions. Once bound extracellularly, Ag^+ enters the cell and is found in high speed centrifugal supernatant fluids (unpublished data). Our current hypothesis is that the sensitive cells bind Ag^+ so tightly that they extract it from AgCl and proteins, whereas cells with the resistance plasmid do not compete successfully with Ag^+-halide precipitates for Ag^+. Although some Ag^+-resistant isolates have determinants of sulfadiazine resistance as well, these determinants can be on separate plasmids (Hedges, personal communication).

VI. OTHER HEAVY METAL RESISTANCES

Even though there are many other plasmid heavy metal resistances (2, 48), nothing today is known about the mechanisms of resistance to bismuth, boron, cobalt, nickel or tellurium ions. Chromate resistance in a pseudomonad isolated from river sediment seems to be due to reduction of toxic Cr(VI) to less toxic Cr(III) (Bopp and Ehrlich, Abstract Q111, 1980 American Society for Microbiology Meetings), and this resistance appears to be plasmid determined (A.M. Chakrabarty, personal communication). Caution on this point is needed. Bacteria capable of oxidizing toxic As(III) to less toxic As(V) are also known; however, this has not turned out to be the mechanism of plasmid-governed resistance.

Over the last ten years, our laboratory has been studying the mechanisms and genetics of plasmid-mediated heavy metal resistances. In the absence of any obvious source of direct selection, one may ask why these resistance (on plasmids) occur in clinical strains with such high frequencies. Selective agents in hospital and "normal human" environments are only beginning to be examined. W. Witte and R. Lacey (personal communications) have suggested that heavy metal resistances may have been selected on plasmids in earlier times, and they are merely carried along today "for a free ride" with selection for antibiotic resistances. However, we doubt that there is such thing as "a free ride" as far as these determinants are concerned. Where they exist, there is likely to be selection by either non-human or human sources of heavy metals. Radford et al. (15) found Hg^{2+}-resistant soil microbes in agricultural soil, with no known human mercurial input. In such settings, the prevalence of resistant microbes may be very low, but may come into much greater quantitative prominance after industrial or agricultural pollution (4,5,12,14,15). This situation may be closely analogous to that with antibiotic-resistance plasmids, which are found in low frequencies in antibiotic-virgin populations (54,55), but which become dominant with extensive human use of antibiotics

REFERENCES

1. Silver, S. (1978) in Bacterial Transport (Rosen, B.P., ed.) pp. 221-324, Marcel Dekker Inc., New York

2. Summers, A.O. & Silver S. (1978) Annu. Rev. Microbiol. 32, 637-672
3. Komura, I. & Izaki, K. (1971) J. Biochem. 70, 885-893
4. Tonomura, K., Nakagami, T., Futai, F. & Maeda, K. (1968) J. Ferment. Technol. 46, 506-512
5. Furukawa, K, Suzuki, T. & Tonomura, K. (1969) Agric. Biol. Chem. 33, 128-130
6. Novick, R.P. & Roth, C. (1968) J. Bacteriol. 95, 1335-1342
7. Smith, D.H. (1967) Science 156, 1114-1116
8. Nakahara, H., Ishikawa, T., Sarai, Y. & Kondo, I. (1977) Zentralbl. Bakteriol. Parasitenkd. Infektionskr. Hyg. 1 Abt. Orig. A.237, 470-476
9. Nakahara, H., Ishikawa, T., Sarai, Y., Kondo, I., Kozukue, H. & Silver, S. (1977) Appl. Environ. Microbiol. 33, 975-976
10. Schottel, J., Mandal, A., Clark, D., Silver, S. & Hedges, R.W. (1974) Nature 251: 335-337
11. Friello, D.A. & Chakrabarty, A.M. (1980) in Plasmids and Transposons: Environmental Effects and Maintenance Mechanisms (Suttard, C. & Rozee, K.R., eds.) pp. 249-260, Academic Press, New York
12. Olson, B.H., Barkay, T. & Colwell, R.R. (1979) Appl. Environ. Microbiol. 38, 478-485
13. Olson, G.J., Iverson, W.P. & Brinckman, F.E. (1981) Current Microbiol. 5, 115-118
14. Timoney, J.F., Port, J., Giles, J., & Spanier, J. (1978) Appl. Environ. Microbiol. 36, 465-472
15. Radford, A.J., Oliver, J., Kelly, W.J. & Reanney, D.C. (1981) J. Bacteriol. 148(2), November issue
16. Weiss, A.A., Schottel, J.L., Clark, D.L., Beller, R.G. & Silver, S. (1978) in Microbiology -1978 (Schlessinger, D., ed.) pp. 121-124, American Society for Microbiology, Washinton, D.C.
17. Clark, D.L., Weiss, A.A. & Silver, S. (1977) J. Bacteriol. 132, 186-196
18. Weiss, A.A., Murphy, S.D. & Silver, S. (1977) J. Bacteriol. 132, 197-208
19. Porter, F.D., Silver, S. & Nakahara, H. (1982) Antimicrob. Agents Chemother., in preparation
20. Furukawa, K. & Tonomura, K. (1972) Agric. Biol. Chem. 36, 217-226
21. Izaki, K., Tashiro, Y. & Funaba, T. (1974) J. Biochem. 75, 591-599
22. Summers, A.O. & Silver, S. (1972) J. Bacteriol 112, 1228-1236
23. Tezuka, T. & Tonomura, K. (1976) J. Biochem. 80, 79-87
24. Tezuka, T. & Tonomura, K. (1978) J. Bacteriol. 135, 138-143
25. Schottel, J.L. (1978) J. Biol. Chem. 253, 4341-4349
26. Kinscherf, T.G. & Silver, S. (1982) Proc. Natl. Acad. Sci. U.S.A., submitted

27. Fox, B. & Walsh, C. (1982) J. Biol. Chem., submitted
28. Furukawa, K. & Tonomura, K. (1971) Agric. Biol. Chem. 35, 604-610
29. Bennett, P.M., Grinsted, J., Choi, C.L. & Richmond, M.H. (1978) Mol. Gen. Genet. 159, 101-106
30. Nakahara, H., Silver, S., Miki, T. & Rownd, R.H. (1979) J. Bacteriol. 140, 161-166
31. Foster, T.J., Nakahara, H., Weiss, A.A. & Silver, S. (1979) J. Bacteriol. 140, 167-181
32. Hedges, R.W. & Baumberg, S. (1973) J. Bacteriol. 115, 459-460
33. Smith, H.W. (1978) J. Gen. Microbiol. 109, 49-56
34. Silver, S., Budd, K., Leahy, K.M., Shaw, W.V., Hammond, D., Novick, R.P., Willsky, G.R., Malamy, M.H. & Rosenberg, H. (1981) J. Bacteriol. 146, 983-996
35. Novick, R.P., Murphy, E., Gryczan, T.J., Baron, E. & Edelman, I. (1979) Plasmid 2, 109-129
36. Silver, S. & Keach, D. (1982) Proc. Natl. Acad. Sci. U.S.A. submitted
37. Heefner, D.L. & Harold, F.M. (1980) J. Biol. Chem. 255, 11396-11402
38. Albert, A. (1973) in Selective Toxicity, Fifth Edition, pp. 392-397 Chapman and Hall, London
39. Pickett, A.W., McBride, B.C., Cullen, W.R. & Manji, H. (1981) Canad. J. Microbiol., in press
40. Perry, R.D. & Silver, S. (1982) J. Bacteriol. 149, in press
41. Tynecka, Z., Gos, Z. & Zajac, J. (1981) J. Bacteriol. 147, 305-312.
42. Weiss, A.A., Silver, S. & Kinscherf, T.G. (1978) Antimicrob. Ag. Chemother. 14, 856-865.
43. Smith, K.D. & Novick, R.P. (1972) J. Bacteriol. 112, 761-772
44. Tynecka, Z., Zajac, J. & Gos, Z. (1975) Acta Microbiol. Pol. 7, 11-20
45. Chopra, I. (1975) Antimicrob. Agents Chemother. 7, 8-14
46. Tynecka, Z., Gos, Z. & Zajac, J. (1981) J. Bacteriol. 147, 313-319
47. Fox, C.L., Jr. (1965) Arch. Surg. 96, 184-188
48. Summers, A.O., Jacoby, G.A., Swartz, M.N., McHugh, G. & Sutton, L. (1978) in Microbiology (D. Schlessinger, ed.) American Society for Microbiology, Washington, D.C. pp. 128-131
49. Annear, D.I., Mee, B.J. & Bailey, M. (1976) J. Clin. Path. 29, 441-443
50. Bridges, K., Kidson, A., Lowbury, E.J.L. & Wilkins, M.D. (1979) Brit. Med. J. 1, 446-449
51. Hendry, A.T. & Stewart, I.O. (1979) Canad. J. Microbiol. 25, 915-921

52. McHugh, G.L., Moellering, R.C., Hopkins, C.C. & Swartz, M.N. (1975) Lancet 1, 235-240
53. Silver, S. (1981) in Molecular Biology, Pathogenicity and Ecology of Bacterial Plasmids (Levy, S.B., Clowes, R.C. & Koenig, E.L., eds.) pp. 179-189. Plenum Press, N.Y.
54. Maré, I.J. (1968) Nature 220, 1046-1047
55. Gardner, P., Smith, D.H. Beer, H. & Moellering, R.C., Jr. (1969) Lancet 2, 774-776

EPIDEMIOLOGY

EVOLUTION OF SUPPORT FOR PLASMID RESEARCH BY THE NATIONAL INSTITUTE OF ALLERGY AND INFECTIOUS DISEASES

Irving P. Delappe

National Institute of Allergy and Infectious Diseases
National Institutes of Health
Bethesda, Maryland, U.S.A.

It is now almost 30 years ago since Joshua Lederberg brilliantly deduced the presence of extrachromosomal genetic elements that exercise a profound influence on reproduction in certain bacteria and allow them to mate as an alternative to simple binary fission (1). He created the term "plasmid" for these elements.

Transmissibility of drug resistance among different species of Enterobacteriaceae by cell-to-cell contact (conjugation) was discovered in Japan based on outbreaks of bacillary dysentery in the mid-1950's (2, 3, 4). Until his untimely death from cancer the National Institute of Allergy and Infectious Diseases supported the work of another pioneer in this field, Dr. Tsutomu Watanabe (5, 6).

The program on Mechanisms of Resistance to Antimicrobial Agents in the National Institute of Allergy and Infectious Diseases had its inception in 1965 as the consequence of a report prepared for the Institute on drug resistance. From its origin, this program has consisted of an amalgam of unsolicited grants for the most part involving basic research in this area. The program includes research of a high degree of relevance to other programs of interest to the Institute, such as hospital-associated infections of staphylococcal and streptococcal origin and sexually transmitted diseases of bacterial origin. Substantial clinical and epidemiological evidence indicates that resistance to antimicrobial agents is to a great extent plasmid-mediated.

An International Committee on Plasmid Nomenclature was formed in November 1972 as a result of a National Science Foundation-sponsored United States-Japan Cooperative Science Program Conference on Bacterial Plasmids. A final report of this committee established a uniform nomenclature involving definitions and terminology in plasmid research (7). In the course of the committee's deliberations, and as a result of its discussions with other plasmid workers during the preparation of the above report, the need for a plasmid reference center became increasingly apparent.

Dr. Esther Lederberg's research grant application for such a center was submitted with the full support and encouragement of the Bacterial Plasmid Nomenclature Committee, all but one of whom were recipients of research grants from the National Institute of Allergy and Infectious Diseases. In April 1975, the NIH Microbial Chemistry Study Section reviewed this unsolicited application submitted by Dr. Lederberg of Stanford Medical School for the establishment of a Plasmid Reference Center (PRC). The study section recommended approval for one year instead of the three-year period requested by the applicant. Its reason for doing so was that the applicant would be performing a service function which ultimately should be supported by a funding mechanism other than a research grant. The NIAID Council concurred with the study section's reasoning and approved it for one year in the amount of $75,181. The grant was initiated on June 1, 1976, and terminated on June 1, 1978 (subsequent to May 31, 1977, attributable to Dr. Lederberg's request for two extensions of time without additional funds). Later she competed successfully for a solicited contract on this occasion, and was awarded $337,903 for continuation of the PRC. This support commenced on September 29, 1977, was adjusted for overlap with the research grant, and did not terminate until February 2, 1981. The contract was not renewed because funds for contracts in general were sharply reduced at that time and have not been increased since then. Since the mid-1950's and prior to the advent of collections surpassing the Stanford PRC in size, two examples of which are the collections of Drs. Mitsuhashi (in excess of 50,000 plasmids) and Datta (approximately 8,000 plasmids), plasmid researchers routinely have been exchanging plasmids. This was an additional reason for nonrenewal of the contract.

Prior to, and during the period of our support of the Stanford PRC, it performed a valuable service to plasmid research. Subsequently, it was successful in obtaining support from an alternate source (the National Science Foundation) and continues to carry on its highly important function in the rapidly expanding field of plasmid research.

An announcement in the first issue of Plasmid indicated the establishment and proposed development of The (Stanford) Plasmid Reference Center (8). Its aims included the collection, verification, nomenclature and registration, cataloging, storage, and distribution of plasmid strains. During its period of support by an NIAID contract the PRC achieved many of these objectives (9).

The collection consists of over 800 accessions, a number of which were acquired during the term of the NIH contract. These comprise plasmids carried by enteric bacteria, preponderantly *E. coli* and their plasmid-free host strains together with a set of *Staphylococcus aureus* prototype plasmids. Cloning vectors contributed by Drs. Stanley N. Cohen and Herbert Boyer are included also. Quality control (verification) is based on routine examinations for plasmids and chromosomal attributes in order to confirm known markers and add new ones if they are present. Symbols for newly discovered plasmid attributes may be registered with the Center for inclusion in future compilations. To avoid duplication of plasmid names, a registry of plasmid prefix designations is maintained. Over 250 permutations of the proposed code (pXY prefix) are still available. A registry of Tn allocations is also maintained (10).

The Cold Spring Harbor 1977 tabulation is the principal catalog for the collection because it contains all available data and published references, although it has been modified by the PRC (11). It has been used for acquisitions and has been available since the second year of the contract. Also, each new PRC acquisition is assigned a serial number unique to the PRC. In addition a Plasmid Data Sheet is written for each plasmid giving its history and the results of verification of markers, etc. by the PRC. One of these was filed with the Research Resources Branch of NIAID for each plasmid acquired under the contract.

The primary repository for long-term storage of the plasmid collection, including those members obtained under the NIAID contract, is at Stanford University. Four 1/2 ml. lyophilized samples of each plasmid obtained under our contract are stored in the Washington, D.C. area. Duplicate cultures of these are located at the University of Iowa. The *Staphylococcus aureus* plasmids received from Dr. Richard Novick can not be lyophilized and have been preserved in glycerol-lactate broth in a -70° freezer at Stanford and are duplicated in stab agar cultures at the University of Nebraska. Certain other enteric plasmids including *E. coli* have been stored in glycerol at -70° and lyophils of some of these sent to the University of Calgary in Canada.

In addition to answering requests for individual plasmids, the Stanford PRC has also assembled and distributed a number of helpful kits:

(1) *Escherichia coli* incompatibility (Inc) tester.
(2) Molecular weight determination.
(3) Transposable elements.
(4) B-lactamase coding plasmids and others coding for tetracycline resistance.
(5) Representative colicinogenic strains.

Certain of these kits have been subject to periodic modifications based on new information received by the PRC.

Overall during the period of contract support, the Stanford PRC answered requests from 794 individuals for 2,829 cultures (12).

ACKNOWLEDGMENT

The National Institute of Allergy and Infectious Diseases acknowledges its debt to Dr. Esther M. Lederberg from whom it received a great deal of the information on which this presentation is based.

REFERENCES

1. Lederberg, J.(1952) Cell genetics and hereditary symbiosis. *Physiol. Rev. 32*,403-430.
2. Ochiai, I., Yamanaka, T., Kimura, K. & Sawada, O.(1959) Studies on the inheritance of drug resistance between *Shigella* strains and *Escherichia coli* strains. Nippon Iji Shimpo *1861*,34-36.
3. Akiba, T.(1959) Mechanism of development of resistance in *Shigella* in Proc. of the 15th Gen. Mtg. of the Japan. Med. Assoc. *5*,299-305.
4. Akiba, T., Koyama, K., Ishiki, Y., Kimura, S. & Fukushima, T.(1960) On the Mechanism of the development of multiple drug-resistant clones of *Shigella. Japan J. Microbiol. 4*, 219-227.
5. Watanabe, T. & Fukasawa, T.(1961) Episome-mediated transfer of drug resistance in Enterobacteriaceae. I. Transfer of resistance factors by conjugation. *J. Bacteriol. 81*,669-678.
6. Watanabe, T. & Fukasawa, T.(1961) Episome-mediated transfer of drug resistance in Enterobacteriaceae. II. Elimination of resistance factors with acridine dyes. *J. Bacteriol. 81*, 679-683.
7. Novick, R.P., Clowes, R.C., Cohen, S.N., Curtiss, R., III, Datta, N., & Falkow, S.(1976) Uniform nomenclature for bacterial plasmids: A proposal. *Bacteriol. Rev. 40*,168-189.
8. Lederberg, E.M.(1977) Announcement, The Plasmid Reference Center. *Plasmid 1*,123.
9. Lederberg, E.M., Personal communication.
10. Campbell, A., Starlinger, P., Berg, D.E., Botstein, D., Lederberg, E.M., Novick, R.P. & Szybalski, W.(1979) Nomenclature of transposable elements in prokaryotes. *Plasmid 2*, 466-473.
11. Jacob, A.E., Shapiro, J.A., Yamamoto, L., Smith, D.I., Cohen, S.N. & Berg, D.(1977) Plasmids studied in *Escherichia coli* and other enteric bacteria. in DNA Insertion Elements, Plasmids, and Episomes, 607-656 (A.I. Bukhari, J.A. Shapiro & S.L. Adhya, eds.) Cold Spring Harbor Laboratory.
12. Lederberg, E.M., Personal communication.

RESISTANCE TO AMINOGLYCOSIDE ANTIBIOTICS AND CONJUGATIVE R PLASMIDS IN *SERRATIA MARCESCENS*

R. Katoh, T. Ikeda, M. Kimura, K. Nakata
K. Kawahara and S. Kimura

Department of Bacteriology
Teikyo University School of Medicine
Tokyo, Japan

In recent years, *S.marcescens* has been recognized as a causative microorganism of nosocomial infections(1,2,3,4). It has been reported that these clinical strains have shown a steady increase in their resistance to aminoglycosides and that conjugative R plasmids account for the majority of the resistant strains(5,6,7,8). This paper presents the results of a 1980 survey on aminoglycoside-resistance and conjugative R plasmids in the clinical strains of *S.marcescens*. Aminoglycoside modifying enzymes mediated by these R plasmids are also discussed.

AMINOGLYCOSIDE-RESISTANCE IN *SERRATIA MARCESCENS*

A total of 239 strains of *S.marcescens* isolated from 5 hospitals were investigated in 1980. Of these, 147 were isolated from urine, 35 from sputum, 27 from pus and 30 from other specimens. The susceptibility of *S.marcescens* to antibiotics was determined on Mueller Hinton medium(Difco) by an agar dilution method and was expressed as the maximum allowable concentration of growth(MAC). Aminoglycosides used were kanamycin A (KM), ribostamycin(RSM), dibekacin(DKB), amikacin(AMK), tobramycin(TOB), gentamicin(GM), paromomycin(PRM), neomycin B(NM), butirosin A(BUT), lividomycin A(LVDM) and sisomicin(SISO). The criterion of resistance of *S.marcescens* was as follows: $\geqq$6.3μg/ml for GM and $\geqq$12.5μg/ml for the other aminoglycosides.

Of the 239 strains, 85.4% were resistant to one or more aminoglycosides and 60.3% were resistant to more than 5 aminoglycosides. There was little differece in the percentages of those strains resistant to one or more aminoglycosides by source of specimens, but remarkable differences were found in the resistant strains to more than 5 aminoglycosides: urine 71.4%, sputum 34.3%, pus 44.4% and others 40.8%.

As shown in Table 1, the strains resistant to RSM were those most frequently isolated. About 16% of the strains were single-resistant to RSM and this single-resistance was unique with the antibiotic among the other aminoglycosides. The strains resistant to DKB,KM,TOB or BUT were found at an intermediate rate. GM and LVDM were the most effective antibiotics and AMK followed these antibiotics.

CONJUGATIVE R PLASMIDS MEDIATING AMINOGLYCOSIDE-RESISTANCE

Mating experiments were carried out with the resistant strains as donors to examine the existence of conjugative R plasmids. Details of these experiments were described by Eda et al(5). The susceptibility of transconjugants was determined in the same way as that of *S.marcescens*. The criterion of resistance of transconjugants was as follows: ≧12.5μg/ml for KM,RSM,NM,PRM,BUT and LVDM, ≧6.3μg/ml for TOB,GM and DKB, ≧3.1μg/ml for AMK and SISO.

The conjugative R plasmids carrying aminoglycoside-resistance were detected in 40.2% of the strains resistant to one or more aminoglycosides and 50.0% of the strains resistant to more than 5 aminoglycosides. R plasmids were most frequently detected in the strains isolated from urine.

As shown in Table 1, 80% of the strains resistant to GM carried conjugative R plasmids mediating GM-resistance. The R plasmids mediating DKB,TOB,KM,PRM,LVDM or NM-resistance were detected at an intermediate rate. The detection-rate of R plasmids mediating the resistance to AMK or BUT was low.

Table 1. Resistant Strains and Conjugative R plasmids

Antibiotics	*Resistant strains*		*Strains Carrying R Plasmid*	
	Number	*Percentage*	*Number*	*Percentage**
RSM	*198*	*82.8*	*69*	*34.8*
BUT	*111*	*46.4*	*22*	*19.8*
KM	*148*	*61.9*	*82*	*55.4*
DKB	*150*	*62.7*	*69*	*46.0*
TOB	*124*	*51.8*	*64*	*51.6*
AMK	*79*	*33.0*	*14*	*17.7*
GM	*65*	*27.1*	*52*	*80.0*
SISO	*86*	*35.9*	*58*	*67.4*
PRM	*86*	*35.9*	*50*	*58.1*
LVDM	*65*	*27.1*	*38*	*58.4*
NM	*85*	*35.5*	*48*	*56.4*

**Percentage of conjugative R plasmids in resistant strains*

RESISTANT PATTERN AND MODIFYING ENZYMES OF TRANSCONJUGANTS

The percentage of the resistant strains of *S.marcescens* to each aminoglycoside may relate to the resistant patterns of the R plasmids and the phenotypes of modifying enzymes mediated by those plasmids. The modifying enzymes in transconjugants were presumed from both resistant patterns and the mode of modification determined by radioisotopic assay(9) or microbiological assay(10).

As shown in Table 2, APH(3')I and APH(3')II were prevalent enzymes, however these enzymes were responsible only for the resistance to classical aminoglycosides and BUT. The aminoglycosides often used in the treatment of urinary tract and septic infections(AMK,GM,TOB and DKB) were refractory to these enxymes. AAD(2") mediating the resistance to KM,GM,TOB, DKB and SISO was found in 18.2% of the total R plasmids in 2 hospitals. The R plasmids mediating the same phenotype as

Table 2. Resistant Patterns and Possible Modifying Enzymes of Transconjugants

Hospitals	Resistant Pattern*	Enzyme	Number of Transconjugants
Y.	K,N,R,B,P	APH(3')II	10
	K,N,R, P,L	APH(3')I	3
	K, G,T,D,S	N.D.**	4
	K, R, T,D	N.D.	5
To.	K,N,R,B,P	APH(3')II	1
	K,N,R, P,L	APH(3')I	3
	K, G,T,D,S	N.D.	1
Te.	K,N,R, P,L	APH(3')I	3
	K, G,T,D,S	N.D.	12
	K,N,R, P,L,G,T,D,S	APH(3')I + N.D.	25
	K, R,B, G,T,D,S,A	AAC(6')IV + N.D.	2
S.	K, R,B, T,D,S°,A	AAC(6')IV	7
	K, G,T,D,S	AAD(2")	3
	K,N,R, P,L,G,T,D,S	APH(3')I + AAD(2")	5
	K,N,R,B,P,L,G,T,D,S,A	(3')I + (2") + (6")IV	1
D.	K,N,R,B,P	APH(3')II	1
	K, G,T,D,S	AAD(2")	8
	K, R,B° T,D,S,A	AAC(6')IV	3
	K,N,R,B,P, T,D,S,A	APH(3')II + AAC(6')IV	1
	K,N,R, P,L,G,T,D,S	APH(3')I + AAD(2")	1

*K:KM,R:RSM,B:BUT,P:PRM,L:LVDM,G:GM,T:TOB,S:SISO,A:AMK,N:NM,
S°, B°:some transconjugants may show susceptiblity to it .
**N.D..No detectable enzymatic activity.

AAD(2") were frequently found in 3 other hospitals although no enzymatic activity in these transconjugants could be detected by microbilogical assay. AAC(6')IV responsible for the resistance to AMK,KM,RSM,BUT,TOB,DKB and SISO was found in 14% of the total R plasmids in 3 hospitals. It is noteworthy that both AAD(2") and AAC(6')IV are responsible for the resistance to TOB,DKB,SISO and KM. Although the percentage of the resistant strains to GM or AMK was relatively low, it was suggested that the effectiveness of GM or AMK to *S.marcescens* might be quickly abolished by the selection of R plasmids mediating AAD(2") or AAC(6')IV through the use of not only GM or AMK but also TOB,DKB,SISO and KM. It was also suggested that the strains resistant to both GM and AMK might be selected by the use of TOB,DKB,SISO and KM.

As the detection-rate of R plasmids mediating AMK-resistance was low, we examined the existence of AMK-modifying enzymes in those representative strains of *S.marcescens* highly resistant to AMK(≧50μg/ml). Most of the cell free extracts of the strains could inactivate AMK by acetylation. This result suggests that the AMK-acetylating enzymes, which do not associate with conjugative R plamids, are attributable to one of the mechanisms of AMK-resistance in *S.marcescens*.

ACKNOWLEDGMENTS

The authors are grateful to those who kindly supplied the strains of *S.marcescens* from Tokyo University Hospital, Tokyo Medical and Dental University Hospital, Yokohama City University Hospital, Saiseikai Utsunomiya Hospital and Teikyo University hospital.

REFERENCES

1. Davis et al (1970) *JAMA 214,* 2190-2192
2. Shimizu (1975) *J.Jap.Associ.Infec.Dis. 49,* 77-79
3. Cooksey et al (1975) *Antimicob.Agents.Chemother.* 7,396-399
4. Schaberg et at (1976) *J.Infec.Dis. 134,* 181-188
5. Eda et al (1980) in Antibiotic Resistance; Transposition & Other Mechanisms 285-291, Czechoslovak Medical Press Prague
6. Ueda et at (1980) Chemotherapy *28,* 1-8
7. Verbist et al (1978) *J.Antimicrob.Chemother. 4,*47-55
8. Price et al (1981) *J.Antimicrob.Chemother. 8 sup A,* 89-105
9. Haas et al (1975) in Methods in Enzymology, *43,* 611-628 Academic Press Inc. New York
10. Kawabe et al (1975) *Antimicrob.Agents.Chemother. 7,* 494-499

TRANSFER RESISTANCE OF CLINDAMYCIN AND TETRACYCLINE BETWEEN BACTEROIDES

A.Umemura
K.Watanabe
K.Ueno

Institute of Anaerobic Bacteriology
Gifu Univercity School of Medicine
Tsukasa-Machi 40,Gifu,Japan

INTRODUCTION

Recently clindamycin-resistant strains have been increasing among the Bacteroides fragilis group in Japan and tetracycline resistant strains of B.fragilis have also appeared in increasing frequency.Privitera et al,(1979) and Tally et al,(1979)have already reported that resistance of clindamycin was determined by a transferrable extrachromosomal DNA,there are also several reports on a transferable tetracycline resistance of C.perfringens and others.Using the membrane mating method suggested by Tally et al,we succeeded in a transfer experiment of clindamycin and tetracycline resistance among the B.fragilis group.

MATERIALS AND METHODS

A.Bacterial Strains

Seven strains of B.fragilis(GAI-0605,GAI-0626,GAI-0624, GAI-0753,GAI-1213,GAI-0764,GAI-1214)resistant to 15ug/ml clindamycin were employed as donor strains of clindamycin resistance.B.fragilis GAI-1213 which was resistant to 6.5 ug/ml tetracycline was selected as a donor strain of resistance of tetracycline.Recipient strains used were two strains of B.thetaiotaomicron,GAI-0449RFP and GAI-0914RFP, which were derivatives from the parent strains and susceptible to both 15ug/ml clindamycin and 6.5ug/ml tetracycline.These strains were isolated from the clinical specimens and identified according to the V.P.I.manual, except B.fragilis GAI-0449 which was kindly distributed by Werner.

B. Transfer Experiment

The membrane mating method,described by Tally et al, was employed.The broth mating method was also attempted.As selective markers,arabinose and rifampicin were used.

C. Trial of Elimination of Plasmid

Acridine orange,proflavine and ethidium bromide were used for this purpose.

D. Susceptibility Test

Susceptibility of B.fragilis to antimicrobial was determined by the agar dilution method using GAM agar.

E. Demonstration of Extrachromosomal DNA

Demonstration of extrachromosomal DNA was performed according to the method reported by Meyers et al,(1976).

RESULTS

A. Transfer Experiment of Clindamycin Resistance

In eight of fourteen mating pairs,clindamycin resistance was transferred to B.thetaiotaomicron from B.fragilis by the membrane mating method(Table 1).In the case of B.fragilis GAI-0605 x B.thetaiotaomicron GAI-0449RFP mating,the transfer frequency was 10^{-4} per donor by the membrane method. Transcipient strains obtained had the same biochemical characteristics as the parent strain,GAI-0449RFP,were resistant to lincomycin,josamycin and erythromycin at same time.These transfer experiments never succeeded by the broth mating method.

B. Transfer Experiment of Tetracycline Resistance

Tetracycline resistance of B.fragilis GAI-1213 was also

Table 1. Screening of transferrable clindamycin resistance among strains of B.fragilis and B.thetaiotaomicron

Donor		Recipient		Transfer
B.fragilis	GAI 0605	B.thetaiotaomicron	GAI 0449RFP	+
		"	GAI 0914RFP	-
"	GAI 0626	"	GAI 0449RFP	-
		"	GAI 0914RFP	+
"	GAI 0624	"	GAI 0449RFP	-
		"	GAI 0914RFP	-
"	GAI 0753	"	GAI 0449RFP	+
		"	GAI 0914RFP	-
"	GAI 1213	"	GAI 0449RFP	+
		"	GAI 0914RFP	+
"	GAI 0764	"	GAI 0449RFP	+
		"	GAI 0914RFP	-
"	GAI 1214	"	GAI 0449RFP	+
		"	GAI 0914RFP	+

transferred to B.thetaiotaomicron GAI-0449RFP by the membrane mating method. Its frequency was 5×10^{-6} per donor without induction and 2.5×10^{-5} per donor with induction. Six transcipients obtained showed the same biochemical characteristics as the parent strain. Three of the six transcipients were resistant to macrolides such as clindamycin josamycin and erythromycin simultaneously.

C. Elimination Experiment of Plasmid

A subinhibitory concentration of acridin orange, proflavin and ethidium bromide did not eliminate the plasmid from the clindamycin resistant strain used as the donor strain in our repeated experiments.

D. Demonstration of Extrachromosomal DNA

By agarose gel electrophoresis, two extrachromosomal bands were demonstrated in the donor strain GAI-0605 and all of the transcipient strains tested.

DISCUSSION AND CONCLUSION

The transferable clindamycin- and tetracycline-resistance in B.fragilis was demonstrated by both the membrane mating method and agarose gel electrophoresis as it was by Privitera et al(1979) and Tally et al(1979).For success mating,the cell to cell contact and selection of the recipient strain seemed to be very important.We found that the clindamycin- and tetracycline-resistance were transferred at same time in some cases but the data of agarose gel electrophoresis seemed to indicate these resistances were transferred separately.Induction may be useful for increasing the transfer frequency of tetracycline resistance.

REFERENCES

1.Privitera, G. Dublanchet, A. & Sebald, M.(1979) J.Infectious Disease Vol.139, pp.97-101
2.Tally, F.P. Snydman, D.R. Gorbach, S.L. & Malamy, M.H. (1979) J.Infectious Disease Vol.139, PP.83-88
3.Meyers, J.A. Sanchez, D. Elwell, L.P. & Falkow, S. (1976) J.Bacteriology Vol.127, pp.1529-1537

R PLASMIDS DETECTED IN FISH-PATHOGENIC BACTERIA, *PASTEURELLA PISCICIDA*

T. Aoki, T. Kitao and Y. Mitoma

Department of Fisheries, Faculty of Agriculture, Miyazaki University, Miyazaki, Japan

The multiple drug resistant strains of fish-pathogenic bacteria have increased and transferable R plasmids have been detected with high frequency in various fish-pathogenic bacteria in fish farms(1). Recently, drug resistant strains of *Pasteurella piscicida*, which is well known as the causative agent of Pseudotuberculosis in cultured marine fish, yellow-tail(*Seriola quinqueradiata*) in Japan(2,3) and Pasteurellosis in white perch(*Morone americanus*) in the U. S. A.(4), have emerged in marine fish farms(5). We have already reported the detection of transferable R plasmids from naturally occurring strains(5). The detected R plasmids had the common drug resistance markers of chloramphenicol(Cm), tetracycline(Tc), kanamycin(Km) and sulfonamide(Su). The growth of phage Tl was not inhibited by these R plasmids.

The large number of R plasmids detected in fish-pathogenic bacteria, *Edwardsiella tarda* and *Aeromonas hydrophila*, were classified into incompatibility(Inc) group A-C(1). The other R plasmids from marine *Vibrio* sp. and *V. anguillarum* were classified into Inc group E(1). We previously described a new Inc group R plasmid detected in *A. salmonicida*(6). The genus of *Pasteurella* is quite different from the above mentioned fish-pathogenic bacteria. It is quite interesting to classify the R plasmids from *P. piscicida* into an Inc group from the epidemiological as well as evolutional point of view.

pJA8001 and pJA8005 in the detected R plasmids were used for the determination of the Inc group. The standard R plasmids used were selected for their possible combinations of drug resistant markers on the R plasmids from *P. piscicida* strains(Table 1). Entry exclusion was observed in the combinations of pJA8001 and pJA8005 with standard R plasmids of Inc group A-C, although entry exclusion was not observed in any other combinations. However both R plasmids in any of the transconjugants were stable after successive culturing in a drug-free medium. This fact indicated that both were compatible with all the tested standard R plasmids: FI, FII, A-C, F, H_1, Iα, I_1, I_2, K, L, M, N, O, B-O, P, W, X, and the

Table 1. *Incompatibility grouping of R plasmids pJA8001 and pJA8005*

Standard R plasmid (Incompatibility group)	Resistance marker[a]	Transductants with both markers/tested transconjugants	
		pJA8001	pJA8005
pIP162/2(FI)	Tc Ap	10/10	10/10
222R3(FII)	Cm Sm Su	20/20	20/20
W8015(A-C)	Cm Sm Ap Su	10/10	10/10
R40a(A-C)	Km Ap Su	10/10	10/10
pJA4318(E)	Cm Sm Su	10/10	10/10
pIP166/1(H_1)	Cm Sm Su	10/10	10/10
R64-11(Iα)	Tc Sm	10/10	10/10
pIP186(I_1)	Su Tp	10/10	10/10
pIP175(I_2)	Ap	10/10	10/10
R387(K)	Cm Sm	10/10	10/10
R471a(L)	Ap	10/10	10/10
R446b(M)	Tc Sm	20/20	20/20
N-3(N)	Tc Sm Su	10/10	10/10
R16(O)	Tc Sm Ap Su	10/10	10/10
pIP185(B-O)	Km Tp	10/10	10/10
RP4(P)	Tc Km Ap	10/10	10/10
R7K(W)	Sm Ap	10/10	10/10
R6K(X)	Ap	10/10	10/10
pAr32(New group)	Cm Sm Su	10/10	10/10

[a] *Abbreviations: Cm, chloramphenicol; Tc, tetracycline; Sm, streptomycin; Km, kanamycin; Ap, aminobenzyl penicillin; Su, sulfonamide; Tp, trimethoprim.*

proposed new Inc group(6). These R plasmids detected from *P. piscicida* were untypable and could possibly belong to a new Inc group.

Escherichia coli AB1157 *recA* strains carrying pJA8001 and pJA8005 which had been transferred from original strains of *P. piscicida* were used for isolation of their plasmids DNAs. The plasmids used had different sizes from the RP4, R40a, R1, and ColE1Tn5 used as references. The R plasmid DNA was isolated by the Miki method(7). Agarose gel electrophoresis of R plasmid DNAs of pJA8001 and pJA8005 are shown in Fig. 1.

Plasmid DNAs of pJA8001 and pJA8005 were digested by the endonucleases: BamHI, EcoRI, HindIII, PstI and SalI. EcoRI digestion of pJA8005 produced 3 fragments. The molecular weights of these fragments were 26.7, 9.7, and 5.7 x 10^6 daltons(Fig. 2). The sum of these fragments was 42.1 x 10^6 daltons, although the molecular size of the R plasmid DNA calculated was about 80 x 10^6 daltons as shown in the

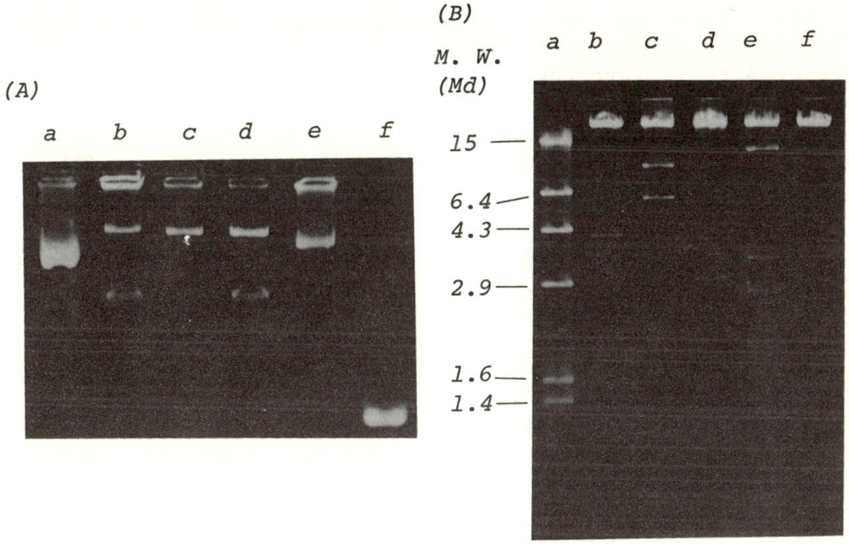

Fig. 1. *Agarose gel electrophoresis of plasmid DNA pJA8001 and pJA8005*

The purified DNA and the DNA fragments were electrophoresed on 0.8% agarose gel and visualized after ethidium bromide staining. (A) pJA8001(c), pJA8005(d), molecular weight standards of RP4(35 M dalton)(a), R40a(96 M dalton)(b), R1(62 M dalton)(e), and ColE1Tn5(7.8 M dalton)(f).
(B) Digestion patterns of pJA8005 with endonucleases in gel electrophoresis. (a) Reference phage DNA λcI85757(received from Dr. Sakaki) digested by HindIII. The size of λ fragments are shown on left of the photograph. (b) digested by BamHI, (c) digested by EcoRI, (d) digested by HindIII, (e) digested by PstI and (e) digested by SalI.

photograph in Fig. 1. This value was twice as much as the sum of the fragments and the R plasmid would have a dimeric form in the *E. coli* cells. The R plasmid DNA was cleaved by BamHI to yield 2 fragments, by HindIII to yield 2 fragments, by PstI to yield 8 fragments, and by SalI to yield 5 fragments (Fig. 1). The estimated molecular weight of the R plasmid equalled the total of the fragments digested BamHI, HindIII, PstI, and SalI. Digestion of pJA8001 with the same endonucleases gave fragments identical to those of pJA8005. This fact suggested that both R plasmids could have same origin, although the strain habouring these plasmids were isolated in different areas. The restriction pattern of R plasmid will be a useful tool for identification of R plasmid.

There have been many reports of the detection of various transposable elements of drug resistant determination(8). We attempted to isolate the transposon of the drug resistant determinant from the R plasmids pJA8001 and pJA8005 using cosmid(ColE1-*cos*λ-*guaA*) for acceptor of the tranposon, according to the convenient methods developed by Maeda *et al.*(9) and Shimada *et al.*(10). Both pJA8001 and pJA8005 were conjugally transferred separately to the *E. coli* KS2127 *recA* mutant isolated from HfrH, which had deletions of a *guaA-guaB* region and carried cosmid, and lysogenized with λBAM. The heat-induced lysates infected the KS1616 of *E. coli* K12 with the deletion of *gal-att*λ*-bio* and *guaA-guaB*. No drug resistant clones of the transductant were found. However transposons are thought to play a large part in such rapid evolution. The origin of the drug resistant gene and the physical propeties of the R plasmid from *P. piscicida* are now under investigation in our laboratory.

REFERENCES

1. Aoki, T., Kitao, T. & Arai, T.(1977) in Plasmids-Medical and Theoretical Aspects(Mitsuhashi, S. *et al.*, ed.) pp.39 -45, Spring-Verlag, Berlin
2. Kimura, M. & Kitao, T.(1971) Fish Pathol. *6*,8-14(in Japanese.)
3. Kusuda, R., Kawai, K. & Matsui, T.(1978) Fish Pathol. *13*, 79-83(in Japanese.)
4. Janssen, W. A. & Surgalla, M. J.(1968) J. Bacteriol. *96*, 1906-1610
5. Aoki, T., Kitao, T., Itabashi, T. & Mitoma, Y.(1981) J. Fish Diseas.(in press)
6. Aoki, T., Kitao, T. & Arai, T.(1979) in Microbial Drug Resistance(Mitsuhashi, S., ed.) Vol. *2*,219-222, Japan Scientific Societies Press, Tokyo
7. Miki, T., Easton, A. M. & Rownd, R. H.(1980) J. Bacteriol. *141*,87-99
8. Calos, M. P. & Miller J. H.(1980) Cell *20*,579-595
9. Maeda, S., Shimada, K. & Takagi Y.(1977) in DNA insertion Elements, Plasmids, and Episomes(Bukhari, A. I. *et al.*, ed.) pp.543-548, Cold Spring Habor Laboratory
10. Shimada, K., Umene, K., Nakamura, T. & Takagi, Y.(1979) in Cold Spring Habor Symposium on Quantitative Biology, Vol. *XLIII*, pp.991-997, Cold Spring Habor Laboratory

STABILITY OF *VIBRIO ANGUILLARUM* R PLASMIDS IN *VIBRIO PARAHAEMOLYTICUS* AND *VIBRIO CHOLERAE*

F. Hayashi

Gunma Prefectural
Fisheries Experimental Station
Maebashi, Gunma, Japan.

T. Nakajima, M. Suzuki, K. Harada, M. Inoue* and S. Mitsuhashi

Department of Microbiology and
*Laboratory of Drug Resistance in Bacteria**
School of Medicine, Gunma University
Maebashi, Gunma, Japan

Enterobacteriaceae R plasmids have a wide host range and are transferred to many gram-negative bacteria, but not to *Pseudomonas aeruginosa*. They are quite unstable in *Vibrio cholerae*. Most of the R plasmids isolated from *P. aeruginosa* cannot be transferred to *Escherichia coli* (1).

Vibrio parahaemolyticus is known as a human pathogen and causes food poisoning and the strains of this species are sensitive to many antibiotics. On the other hand, ayu *(Plecoglosses altivelis)* usually live in clean rivers and no drug-resistant strains of fish pathogenic *Vibrio anguillarum* have been isolated from freshwater fish in Japan for many years.

However, most of these strains have recently become resistant to many antibiotics. During a survey taken between 1973 and 1978 of 50 strains of diseased ayu in geographically dispersed fish-breeding pond in Japan, 48 antibiotic- resistant strains of *V. anguillarum* were found. Among these, 34 strains (71%) were found to have transferable plasmids encoding resistance to tetracycline(Tc), chloramphenicol(Cm) and sulfanilamide (Sa). It should be noted the all plasmids from *V. anguillarum* strains isolated in various ponds had the same resistance patterns to Tc, Cm and Sa. The isolation frequency of these R plasmids was the same as reported by Aoki et al (2). This incidence of R plasmids with the same resistance pattern is assumed to be the result of selective pressure exerted by the same chemotherapeutic agents, since the breeders continued to treat the / ponds during this survey using Tc, Cm, nalidixic acid(Na) and Sa. Fishes at 6 of the ponds were obtained from the same breeding- pond when they were fry. Accordingly, we assumed that the 33 drug-resistant plasmids isolated from the 6 ponds were of the same origin, and that one plasmid was of a different origin.

Two plasmids, Rms418 from the 6 ponds and Rms419 from the other, were randomly selected from 34 drug-resistant plasmids of *V. anguillarum* strains, and the ability of the plasmids to transfer to *V. parahaemolyticus*, *V. cholerae* and *E. coli* strains was examined.

All of the resistance markers on the plasmids were transferred to the four recipient strains at a frequency ranging from about 2.7×10^{-4} to 3.0×10^{-7} per donor strain. The two drug-resistant plasmids in *V. parahaemolyticus* were also reciprocally transferred to the three recipient strains at

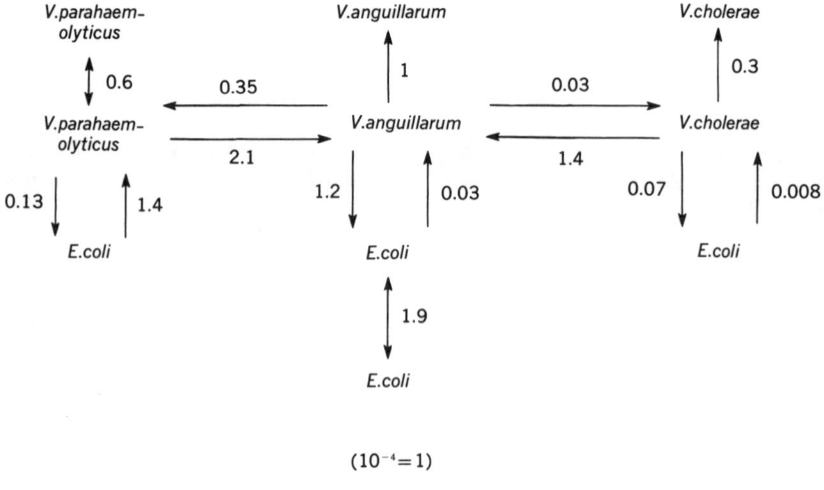

Fig. 1. Transfer frequency of R plasmids[a]

[a] *Transfer frequency was expressed by the number of transconjugants per donor cell. Equal parts from the donor and recipient strains at the middle exponential phase were mixed for mating and incubated at 30 or 37 C. After 2 hr of incubation, the appropriately diluted mixed culture was spread on HI agar plates containing Na(50μg/ml) or Rf(100μg/ml)and one of the selective drugs, i. e., Tc(12.5μg/ml) or Cm(12.5μg/ml). The liquid culture for V. anguillarum and R plasmid from V. anguillarum were incubated at 30°C.*

almost the same frequencies as those of *V. anguillarum*. In contrast, the plasmid transferability from *V. anguillarum* to *V. cholerae* was about 10 times lower than that from *V. anguillarum* to *V. parahaemolyticus* (Fig. 1).

The other 32 plasmids were very similar in transferability. When both *V. parahaemolyticus* and *V. cholerae* had received *V. anguillarum* R plasmids, transfer frequency of these R plasmids increased to *V. anguillarum* strains but not to *E. coli*

Incompatibility testing of R plasmids isolated from *V. anguillarum* strains was performed with representatives of the known plasmid incompatibility groups(3-4). In reciprocal transfer experiments, the transfer of Rms418 or Rms 419 was reduced about 100 times by the presence of R40a in the recipient strain, and almost all of the incoming plasmid was eliminated from transconjugants(Table 1). From these results, we conclude that Rms418 and Rms419 plasmids belong to incompatibility group C.

Our results confirmed the belief that the R plasmid set of *V. anguillarum* is dominated by plasmids of incompatibility

Table 1. Incompatibility testing of R plasmids[a]

Donor carrying	Recipient carrying	Selection	Frequency	No. of cells carrying: Donor plasmid	No. of cells carrying: Recipient plasmid
R40a	Rms418	Km/Rf-r[b]	1.8×10^{-6}	56/100	44/100
"	Rms419	"	6.7×10^{-6}	40/100	60/100
"	-	"	9.5×10^{-8}	100/100	
Rms418	-	Tc/Rf-r[c]	1.4×10^{-5}	100/100	
"	R40a	"	3.5×10^{-7}	4/100	96/100
Rms419	-	"	3.0×10^{-5}	100/100	
"	R40a	"	1.0×10^{-7}		100/100

[a] *Incompatibility experiments were performed by introducing one plasmid into a strain carrying another and selecting for the donor plasmid. Transconjugants were purified on a nonselective medium and tested by replica plating for the presence of bothe the incoming and resident plasmids. E. coli K12 J53-1(nal) and E. coli C(rif) were used as donor and recipient, respectively.*

[b] *25μg of Km per ml was used for the selection of R40a plasmid.*

[c] *12.5μg of Tc per ml was used for the selection of Rms418 and Rms419 plasmids.*

group C. Group C plasmids have been found in a wide range of genera including *Pseudomonas*, *E. coli*, *Proteus*, *Providencia*, *S. typhimurium*, *K. pneumoniae* and *S. marcescens*(5-6).

The average molecular weight of plasmids calculated from their DNA length was from 103 x 10^6 to 124 x 10^6 daltons.

Yokota et al (7) introduced fi^+ and fi^- plasmids into a strain of *V. cholerae* and found that most were very unstably inherited. Rms418, Rms419 and R40a plasmids belonging to group C were stably inherited in *V. parahaemolyticus* and overnight cultures grown in drug-free medium regularly carried more than 99.9 % R^+ cells. When the cells were stocked in cooked meat medium, however, these plasmids were unstable only in *E. coli* (Table 2).

Seafood poisoning from *V. parahaemolyticus* is common in Japan because of the custom of eating raw seafood. *Salmonella*, *V. parahaemolyticus* and *E. coli* strains are frequently isolated from patients showing symptoms of bacillary dysentery. In spite of the marked development of isolation techniques for plasmid study in gram-negative rods, drug-resistant *V. parahaemolyticus* strains have not been isolated. However, it should be noted that the R plasmids from *V. anguillarum* strains are stably maintained in *V. parahaemolyticus* and *V. cholerae*. This suggests that drug-resistant *V. parahaemolyticus* and *V.*

Table 2. Spontaneous loss of R plasmids either in BHI broth or cooked meat medium

Strains	*Segregation in*	
	BHI[a]	*Cooked meat*[b]
V. anguillarum		
HFV33 Rms418	0/100	0/100
HFV34 Rms419	0/100	0/100
V. cholerae		
NIH41 Rms418	0/100	0/100
Rms419	0/100	0/100
V. parahaemolyticus		
GN11382 Rms418	0/100	0/100
Rms419	0/100	0/100
E. coli ML1410		
Rms418	0/100	103/105
Rms419	2/100	66/66

[a] *Overnight culture.*

[b] *Stocked for three months*

cholerae may increase in number if this type of R plasmid spreads naturally from fish-breeding ponds to sea water as a result of water pollution.

This work was supported in part by grants 311201(S.M) and 977401(M.I) from the Ministry of Science Education of Japan.

REFERENCES

1. Mitsuhashi, S. (1977) Epidemiology of R factors. pp25-48 University of Tokyo Press, Tokyo.
2. Aoki, T., S. Egusa, and T. Arai. (1974) Antimicrob. Ag. Chemoth. *6*,534-538
3. Coetzee, J.N., N.Datta and R.W.Hedges. (1972) J.Gen.Microbiol. *72*,543-552
4. Dennison, S. (1972) J.Bacteriol. *109*,416-422
5. Witchity,J.L., and G.R.Gerbaud. (1972) Ann. Inot. Pasteur Paris *123*,333-339
6. Shapiro,J.A. (1977) In DNA insertion elements, plasmids, and episomes. pp601-704. Cold Soring Harbor Laboratory.
7. Yokota, T., T. Kasuga, M. Kaneko, and S. Kuwahara. (1972) J. Bacteriol. *109*,440-442

R PLASMID WITH CARBADOX RESISTANCE FROM *ESCHERICHIA COLI* OF PORCINE ORIGIN

K.Ohmae and S.Yonezawa

Antibiotic division, National Veterinary Assay Laboratory Kokubunji, Tokyo, Japan

N.Terakado

Biological Products Division, National Institute of Animal Health Tsukuba, Ibaraki, Japan

Carbadox is a synthetic antibacterial agent, chemically classified as a quinoxaline-di-N-oxide, and quindoxin, olaquindox were known as the same agents (8) . Although the action mechanism is not completely clarified, it is known that the antibacterial activity is remarkably enhanced under anaerobic conditions (5). In Japan, Cdx was approved in 1969 for animal use in preventing or treating porcine dysentery(2,10,11). Subsequently,the frequency of its use has been steadily increasing . However, it was unknown whether bacteria have acquired resistance to Cdx or not.

The present investigation is concerned with the presence of Cdx resistant-strains of *Escherichia coli* of porcine origin and the demonstration that this resistance is associated with R plasmids from such strains.

SUSCEPTIBILITY TO CARBADOX

Table 1 shows the antibacterial activities of carbadox(Cdx) against the 72 *E.coli* isolates(3). Strains were isolated from fecal samples in 1980 on hog farm. Cdx had been used on this farm during the periods of fattening. The MIC under aerobic conditions ranged from 6.25 to above 200 µg/ml,presenting a diphasic distribution pattern, with12.5 µg/ml as one peak and above 200 µg/ml as the other. In contrasts, under anaerobic conditions, the MIC exhibited a triphasic distribution pattern, with peaks at 0.78 to 1.56, 6.25 and 50 µg/ml, respectively. Strains resistance to Cdx had MICs of more than 50 µg/ml under aerobic conditions or more than 12.5 µg/ml under anaerobically. The resistance patterns of Cdx resistance strains to seven drugs, tetracycline (Tc), chloramphenicol (Cm), streptomycin (Sm), spectinomycin (Spc), sulfadimethoxin (Su), kanamycin (Km) and ampicillin (Apc),were examined. Resistance to combination of Tc,Sm,Spc,Su,Km and Apc was found in 18 of the 24 Cdx resistant strains (75%). Then, five Cdx resistance strains were chosen, and examined whether Cdx resistance is associated with R plasmid or not.

CONJUGAL TRANSMISSION OF CARBADOX

Transmissibility of drug resistance was examined by mixed cultivation to *E.coli* K-12 ML1410 (9). The MIC of Cdx for *E.coli* ML1410,which was used as the recipient,was 6.25 µg/ml under aerobic condition and 0.2 µg/ml under anaerobic condition. Transconjugants which possessed two or three different resistance patterns were obtained with each donor (Table 2). The resulting transconjugants, though Cdx resistance, had lower Cdx MICs either under aerobic or anaerobic conditions, than those of the donor strains. Cdx resistance was almost always transmitted jointly with Sm Spc Apc resistance, but only Cdx resistance strains were not obtained.

DETECTION OF PLASMID DNA

The transconjugant obtained from strain NV 13 was used to detect plasmid DNA (13). Agarose gel electrophoresis (4) revealed a single band with molecular weight of about 28 x 10^6 (Fig.1). A transformation experiment was also carried out with this DNA as donor(1). A total of 25 transformants were obtained with different initial selections. The MIC of Cdx was 6.25 ug/ml under anaerobic conditions and all transformants possessed Sm Spc Apc resistance. The MIC of Cdx was 0.2µg/ml under anaerobic conditions for the *E.coli* C used as the recipient. Then,when the Cdx-resistant transformants were used as doner, the character of Cdx-resistance was found to be further transmissible by conjugation. All the transconjugants obtained were found to possess Cdx Sm Spc Apc resistance, and MIC of Cdx was 6.25 µg/ml under anaerobically. This result demonstrated that a transformant obtained with plasmid DNA as donor material could subsequently transmit the identical resistance pattern, Cdx Sm Spc Apc, as one unit by conjugation to another strain.

GENETIC PROPERTIES OF THE R PLASMID CARRING CARBADOX RESISTANCE

The fi character (12) of the R plasmid was examined by com-

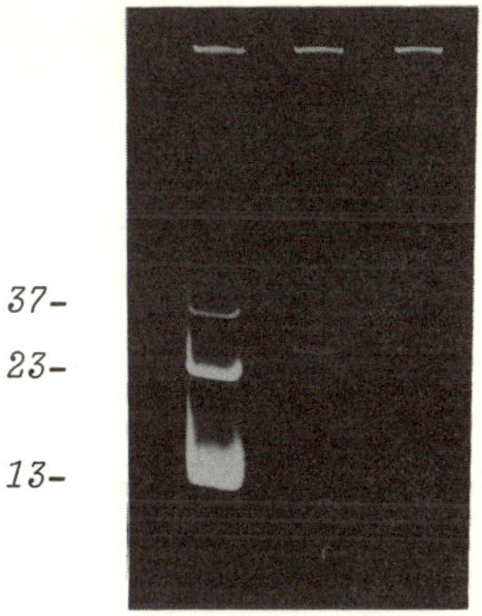

Fig. 1 . Agarose gel electrophoresis of DNAs from transconjugant E.coli ML 1410 and transformant E.coli C. From left to right: marker plasmid DNA , E.coli ML 1410(pNV 13), E.coli C(pNV13). Numbers indicate sizes (megadalton) of bands.

paring the plating efficiency of the male-specific f2 phage, using *E.coli* W 1895, an Hfr derivative of K-12 as indicator strain. All of 6 R plasmids derived from *E.coli* reduced the plating efficiency of the f2 phage, indicating that these R plasmid are fi negative type. The phage restriction (12) by the R plasmids was examined by the double layer method, using *E.coli* W 1895, as an indicater strain. In this experiment, phage λ was employed. None of 6 R plasmids tested reduced the plating efficiency of the phages employed,indicating that these R plasmids did not confer the character of phage restriction on their host bacteria. On the other hand,all R plasmids examined,were not calssified into any incompatibility group.

HOST RANGE OF pNV 13

The host range of pNV 13, a representative Cdx-resistant R plasmid, was transmissible to *K.pneumoniae, C.freundii, Sal. typhi, Sal. typhimurium* and *Sh.flexneri* at 24 hours conjugation. And the levels of drug resistance to Cdx in enterobacteria carrying pNV 13 was 6.25 to 12.5 μg/ml under anaerobic conditions.

Table 1. *Antibacterial Activities of Carbadox against the 72 E.coli Isolated from Pigs*

Culture condition	*No. of isolates with MIC (μg/ml) of:*										
	0.39	*0.78*	*1.56*	*3.13*	*6.25*	*12.5*	*25*	*50*	*100*	*200*	*200*
Aerobic	*0*	*0*	*0*	*0*	*12*	*18*	*6*	*3*	*7*	*11*	*15*
Anaerobic	*6*	*12*	*12*	*6*	*12*	*1*	*7*	*15*	*1*	*0*	*0*

Table 2. *Conjugal Transfer of* Cdx^r *to E.coli K-12 ML 1410*

Donor strain tested			*Character of transconjugant strain*		
Designation	*Cdx MIC(μg/ml)*		*Resistance pattern*	*Cdx MIC(μg/ml)*	
	Aerobic	*Anaerobic*		*Aerobic*	*Anaerobic*
NV 1	*200*	*100*	*Cdx Sm Spc Su Apc*	*12.5*	*6.25*
			Cdx Sm Spc Apc	*12.5*	*6.25*
			Sm Spc Su	*6.25*	*0.2*
NV 7	*100*	*50*	*Cdx Sm Spc Apc*	*12.5*	*6.25*
			Sm Spc Su	*6.25*	*0.2*
NV 13	*200*	*50*	*Cdx Sm Spc Apc*	*12.5*	*6.25*
			Sm Spc Su	*6.25*	*0.2*
NV 14	*50*	*50*	*Cdx Sm Spc Apc*	*12.5*	*6.25*
			Sm Spc Su	*6.25*	*0.2*
NV 16	*100*	*50*	*Cdx Sm Spc Apc*	*12.5*	*6.25*
			Sm Spc Su	*6.25*	*0.2*

DISCUSSION

There are many kind of drug resistance controlled by R plasmid. The present study adds resistance to additional antibacterial agent to the list and the first report of such resistance and its plasmid nature. What is clear from the present investigation is that Cdx resistance in *E.coli* was always associated with resistance to other drugs (Sm Spc Apc) and was always present on a conjugative R plasmid. No R plasmid was detected which carried Cdx resistance only. Since Cdx is routinely used on the farm from which the Cdx resistance strains were isolated, it is not difficult to see how this may have provided the high selective pressure for the high frequency of appearance of Cdx resistance organisms. As is clear from the work of Rubens et al. (6,7), it is reasonable to consider that a transposable gene,or transposon, participates predominantly in the addition of a gene(s) for a new resistance mechanism that has not been encountered previously. Therefore, it is highly possible that a gene(s) for Cdx resistance may have been transposed to an existing R plasmid. Studies are now under way to verify this possibility. And the biochemical mechanism of the resistance to Cdx cotrolled by an R plasmid is unknown, but it will be of interest to know if it is related to the above.

REFERENCES

1. Cohen,S.N.,Chang,A.C.Y. & Hsu,L. (]972) Proc.Natl.Acad. Sci.U.S.A.,69,2110-2114
2. Davis,J.W.,Libke,K.G. & Kornegay,E.T. (1968) J.Am.Vet. Med.Assoc.,153,1181-1184
3. Ishiyama,S.,Ueda,Y.,Kuwabara,S.,Kosakai,N.,Koya,G.,Konno, M. & Fujii,R. (1968) chemotherapy(Tokyo)16,98-99
4. Meyers,J.A.,Sanchez,D.,Elwell,P. & Falkow,S. (1976) J. Bacteriol.127,1529-1537
5. Ohno,Y.,Saito,K.,Kashiwazaki,M. & Namioka.S. (1974) J.Jpn. Vet.Med.Assoc. 27, 451-453
6. Rubens,C.E.,McNeill,W.F. & Farra,W.E.Jr. (1979) J.Bacteriol.139,877-882
7. Rubens,C.E.,McNeill,W.F. & Farra,W.E.Jr. (1979) J.Bacteriol.140,713-719
8. Suter,W.,Rosselet,A. & Knusel,F. (1978) Antimicrob.Agents chemother.13,770-783
9. Terakado,N. & Mitsuhashi,S. (1974) Antimicrob.Agents chemother.6,836-840
10. Thrasher,G.W.,Shively,J.E.,Askelson,C.E.,Babcock,W.E., & Chalquest.R.R. (1969) J.Anim.Sci.28,208-215
11. Thrasher.G.W.,Shively,J.E.,Askelson,C.E.,Babcok,W.E. & Chalquest,R.R. (1970) J.Anim.Sci.31,333-338
12. Watamabe,T.,Nishida,H.,Ogata,C.,Arai,T. & Sata,S. (1964) J.Bacteriol.,88,716-726

ANTIBIOTIC RESISTANCE AND R PLASMIDS AMONG CLINICAL ISOLATES OF *SALMONELLA* IN JAPAN, 1966-1979

R. Nakaya, S. Horiuchi, N. Goto, N. Okamura, T. Chida, H. Shibaoka, A. Shoji, K. Hasegawa, T. Nagai[1]

Department of Microbiology
Tokyo Medical and Dental University School of Medicine
Tokyo, Japan

S. Sakai, T. Ito, K. Saito, and M. Ohashi[2]

Tokyo Metropolitan Research Laboratory of Public Health
Tokyo, Japan

An increase or a decrease in the incidence of antibiotic resistant *Salmonella* isolates has been reported from various countries (1-10). To study temporal changes in the antibiotic resistance and R plasmids of *Salmonella* in Japan, a total of 3,179 strains of human non-typhoid *Salmonella* isolated mainly in Tokyo from 1966 to 1979 were tested for drug resistance and distribution of R plasmids, that extends our previous studies (1-3). Comparisons were made by dividing the period of study into the early (1966-1973) and the late (1974-1979). Susceptibilities to ampicillin(Ap), chloramphenicol(Cm), kanamycin(Km), streptomycin(Sm), and tetracycline(Tc) were tested on the strains in the early period, while in the late period cephaloridine(Cer) and gentamicin were incorporated in the study. Sulfisoxazole, nalidixic acid, and rifampicin were also used to determine sensitivity.

I. CHANGES IN INCIDENCES AND PATTERNS OF ANTIBIOTIC RESISTANCE AND CONJUGATIVE R PLASMIDS AMONG THE ISOLATES OF *SALMONELLA*

The incidence of resistance to one or more of the antibiotics used decreased significantly from 87% in the early period of study to 53% in the late period, resulting from the drastic decrease in the incidence of singly Sm resistant strains since 1974 (see Fig. 1). In contrast, the incidence of Ap, Cm, and Km resistance was increased in the late period (see Fig. 1). The incidence of resistant isolates bearing conjugative R plasmids was not changed in both periods of study (26 and 24%), although more versatile resistance patterns were observed in the late period.

Single Sm resistance was predominant among the isolates until 1973 and a pattern Sm Tc followed in the next. In the late pariod, the most predominant pattern was replaced by Tc and the second place was taken over by Km Sm Tc. Tc resistance was ranked at the first place among the resistance patterns of R plasmids in the early period, whereas in the late period a trend of divergence became remarkable due to the association of Km, Ap, and Cm resistance markers (Table 1).

II. CHANGES IN INCIDENCE OF SEROTYPES OF *SALMONELLA* WITH ANTIBIOTIC RESISTANCE AND CONJUGATIVE R PLASMIDS

Salmonella typhumurium was the most predominant serotypes among the resistant isolates (59%) throughout the study period. The percentage, however, decreased from 77% in the early period to 51% in the late period. The percentage of R^+ isolates among the resistant *S. typhumurium* decreased from 84% to 68%.

Table 1. *Predominant Patterns of Antibiotic Resistance of R Plasmids Carried by Salmonella Isolated in 1966-1973 and in 1974-1979*

Most common resistance pattern	*Number of R^+ isolates*		
	1966-1973 (n = 320)[a]	*1974-1979 (n = 470)*	*1966-1979 (n = 790)*
Tc	*173*	*89*	*262*
Sm Tc	*44*	*60*	*104*
Km Sm Tc	*16*	*56*	*72*
Sm Su Tc	*27*	*14*	*41*
Km Tc	*2*	*39*	*41*
Cm Sm Su Tc	*14*	*24*	*38*
Sm	*12*	*26*	*38*
Ap (Cer) Km Sm Tc	*7*	*26*	*33*
Ap (Cer) Cm Sm Tc	*0*	*20*	*20*
Km Sm Su Tc	*3*	*7*	*10*

[a]*Total number of isolates carrying conjugative R plasmids.*

III. CHANGES IN FI(F) TYPE OF R PLASMIDS

Fi^+(F) type of R plasmids was increased from 4.9% in the early period to 22.5% in the late period.

IV. PRESENCE OF CRYPTIC PLASMIDS IN THE ISOLATES OF *SALMONELLA* WITH NONCONJUGATIVE DRUG RESISTANCE

A total of 109 isolates that were resistant but nonconjugative were arbitrarily selected from the isolates in 1978 and were subjected to analysis for the presence of plasmid DNA by agarose gel electrophoresis (11). It was revealed that 55 strains harbored plasmid DNA (Table 2). Among these, 19 out of 25 strains resistant to three to seven drugs (76%) possessed plasmid DNA whose molecular weights distributed in two discrete groups. On the other hand, 36 out of 84 strains resistant one to two drugs (43%) contained plasmid DNA which covered a wide range of molecular weights. It is not known whether these cryptic plasmids mediate resistance, but at least some may be nonconjugative R plasmids.

V. RELATIONSHIP BETWEEN ANNUAL PRODUCE OF ANTIBIOTICS AND RESISTANCE OF *SALMONELLA*

It can be implicated that the prevalence of resistance and R plasmids in *Salmonella* isolates is due to the usage of a vast amount of chemotherapeutic agents in human patients, domestic animals, and animal feed additives. We constructed the charts that illustrate the relationship between the annual produce of individual antibiotic in Japan and the incidences of resistance and R plasmids among the isolates (Fig. 1). It seems that the usage of antibiotics above certain critical amount over several years would promote the emergence of resistance and R plasmids in the clinical isolates of *Salmonella*. On the other hand, with respect to Sm, Km, and Tc, the critical reduction of the annual produce for several years coincided with the gradual decrease in the prevalence of the resistant and R^+ isolates. *Salmonella* organisms are not exceptional bacteria that change their attitude toward antimicrobials in reponse to the selective pressure against them.

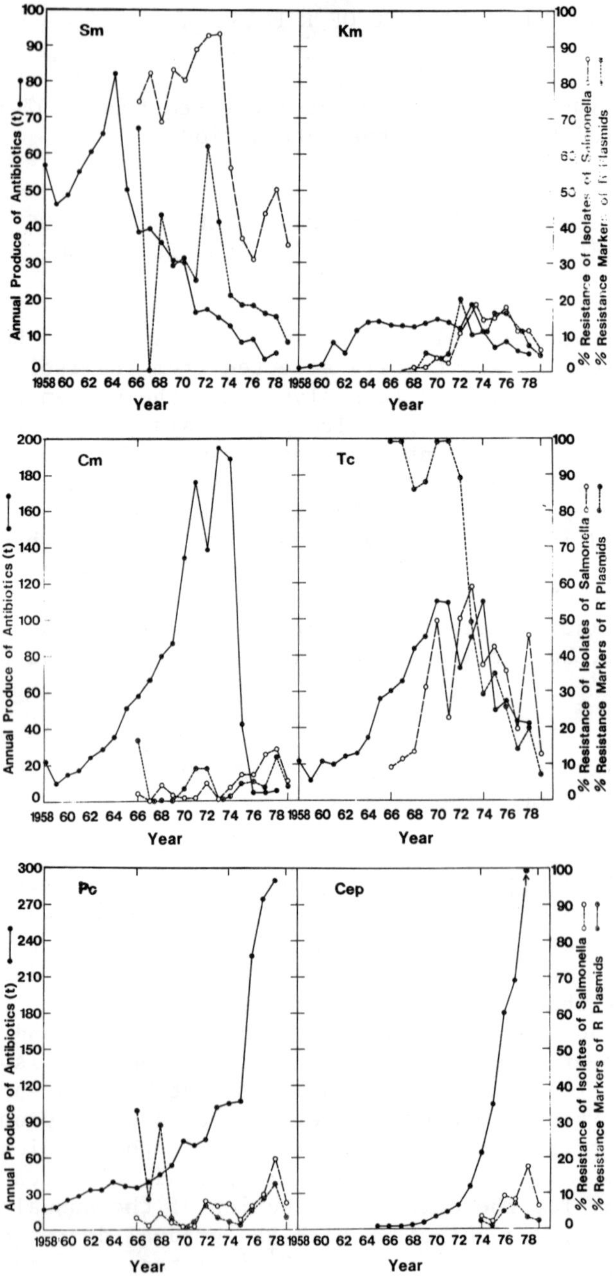

Fig. 1. Relationship between the annual produce of antibiotics in Japan and the incidences of resistance and R plasmids among the isolates of Salmonella.

Table 2. *Incidence of Plasmid DNA and Number of Drug Resistance Markers of the Non-Conjugative, Resistant Strains of Salmonella Isolated in 1978*

Resistant to	*No. of strains tested*	*No. of plasmid-bearing strains (%)*
1 drugs	*53*	*23 (43)*
2 drugs	*31*	*13 (42)*
3 drugs	*16*	*11 (69)*
4 drugs	*2*	*1 (50)*
5 drugs	*2*	*2 (100)*
6 drugs	*4*	*4 (100)*
7 drugs	*1*	*1 (100)*
Total	*109*	*55 (50)*

REFERENCES

1. Nakaya, R., Yoshida, Y. & Terawaki, Y.(1975) in Microbial Drug Resistance (Mitsuhashi, S. & Hashimoto, H., ed.) pp.237-252, University of Tokyo Press, Tokyo
2. Nakaya, R., Horiuchi, S., Yoshida, Y., Sakai, S., Ito, T., Saito, K., Terayama, T., Zen-Yoji, H. & Onogawa, T.(1979) in Microbial Drug Resistance (Mitsuhashi, S., ed.) Vol.2, pp.207-209, Japan Scientific Societies Press, Tokyo
3. Nakaya, R.(1979) Igaku no Ayumi *111*,905-910
4. Marsik, F. J., Parsi, J. T. & Blenden, D. C.(1975) J. Infect. Dis.*132*,296-303
5. Tanaka, T., Ikemura, K., Tsunoda, M., Sasagawa, I. & Mitsuhashi, S.(1976) Antimicrob. Agents Chemother. *9*,61-64
6. Chun, D., Seol, S. Y., Cho, D. T. & Tak, R.(1977) Antimicrob. Agents Chemother. *11*,209-213
7. Barros, F., Korzeniowski, O. M., Sande, M. A., Martins, K., Santos, L. C. & Rocha, H.(1979) Antimicrob. Agents Chemother. *11*,1071-1073
8. Duck, P. D., Dillon, J. R., Lior, H. & Eidus, L.(1978) Can. J. Microbiol. *24*,1358-1365
9. van Leeuwen, W. J., van Embden, J., Guinée, P., Kampelmacher, E. H., Manten, A., van Schothorst, M., & Voogd, C. E.(1979) Antimicrob. Agents Chemother. *16*,237-239
10. Ryder, R. W., Blake, P. A., Murlin, A. C., Carter, G. P., Pollard, R. A., Merson, M. H., Allen, S. D., & Brenner, D. J.(1980) J. Infect. Dis. *142*,485-491
11. Eckhardt, T.(1978) Plasmid *1*,584-588

BIOCHEMISTRY : NEW DRUG

DL-8280, A NEW SYNTHETIC ANTIMICROBIAL AGENT: *IN VITRO* AND *IN VIVO* ANTIMICROBIAL POTENCY AGAINST CLINICAL ISOLATES RESISTANT TO NALIDIXIC ACID, PIPEMIDIC ACID AND GENTAMICIN

Yoshiharu Matsuura
Kenichi Sato
Matsuhisa Inoue
Susumu Mitsuhashi

Department of Microbiology
Laboratory of Drug Resistance in Bacteria
School of Medicine
Gunma University
Maebashi, Japan

More than 10 years ago, nalidixic acid (NA) was developed as an orally effective antimicrobial agent (1). NA was active mainly against gram-negative enteric bacteria and was used for the treatment of bacterial infection, especially urinary tract infections. Recently, many NA analogs with more potent activity and broader spectra have been developed. Among these analogs, pipemidic acid (PPA) was the first drug that possessed antipseudomonal activity (2), and was followed by AM-715 (AM), which is more active to a surprising extent against either gram-negative bacteria including *Ps. aeruginosa* or against gram-positive bacteria than PPA (3). DL-8280 (DL), 9-fluoro-3-methyl-10-(4-methyl-1-piperazinyl)-7-oxo-2,3-dihydro-7H-pyrido 1,2,3-de 1,4-benzoxazine-6-carboxylic acid, a new NA analog. DL had an activity almost equal to that of AM against gram-negative bacteria and 2 to 8 times the activity of AM against gram-positive bacteria. Furthermore, this compound was also highly active against obligatory anaerobes.

This paper deals with the *in vitro* and *in vivo* antimicrobial activity of DL, particularly against clinical isolates resistant to NA, PPA, gentamicin (GM) and clindamycin (CLDM), in comparison with that of NA, PPA, AM and GM. The chemical structure of DL is shown in Fig. 1.

I. ANTIMICROBIAL ACTIVITIES AGAINST NA-, PPA-, GM- AND CLDM-RESISTANT BACTERIA

The antimicrobial activities of DL against NA, PPA, GM and

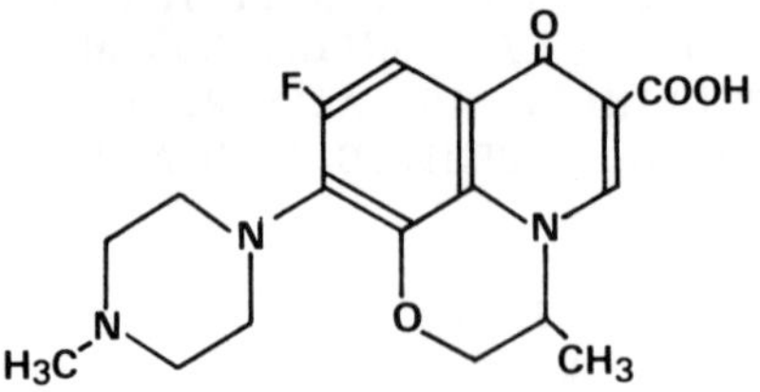

Fig. 1. Chemical structure of DL-8280.

CLDM-resistant bacteria are shown in Fig. 2. Against NA, PPA and GM-resistant bacteria, DL was almost as active as AM. However, some strains were highly resistant to AM but not to DL. Against CLDM-resistant anaerobes including *B. fragilis*, *C. difficile* and *C. perfringens*, DL showed an excellent activity far exceeding that of AM and PPA. DL destroyed them completely at a concentration of 12.5 μg/ml or less. Among the GM-resistant strains, *Ps. cepacia*, *Ps maltophilia* and staphylococci showed a high susceptibility to DL, as shown in Fig. 3. DL was 4 to 8 times more active than AM.

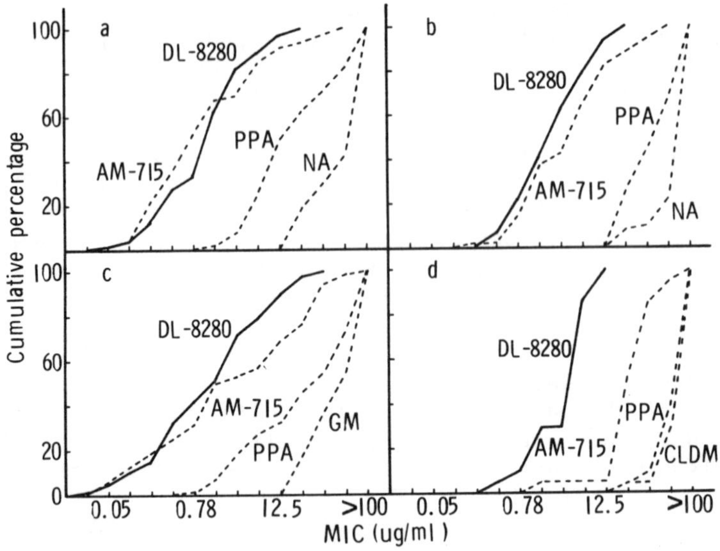

Fig. 2. Antimicrobial activity of DL, AM, PPA and NA. (a) NA-resistant bacteria (141 strains), (b) PPA-resistant bacteria (73 strains), (C) GM-resistant bacteria (256 strains), (d) CLDM-resistant bacteria (21 strains).

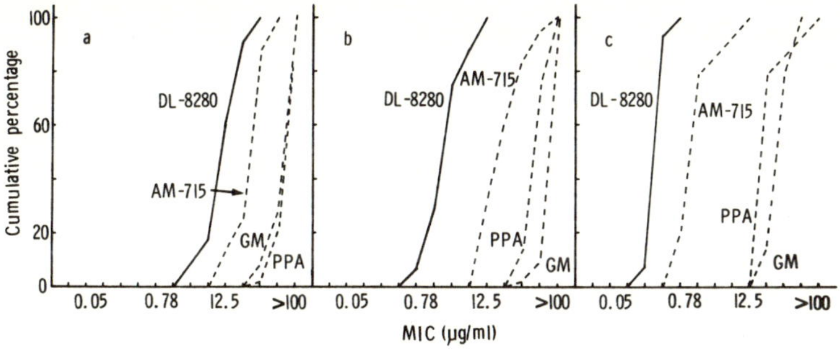

Fig. 3. Antimicrobial activity against GM-resistant bacteria: (a) Ps. cepacia (52 strains), (b) Ps. maltophilia (44 strains), (c) Staphylococcus spp. (14 strains).

II. BACTERICIDAL ACTIVITY

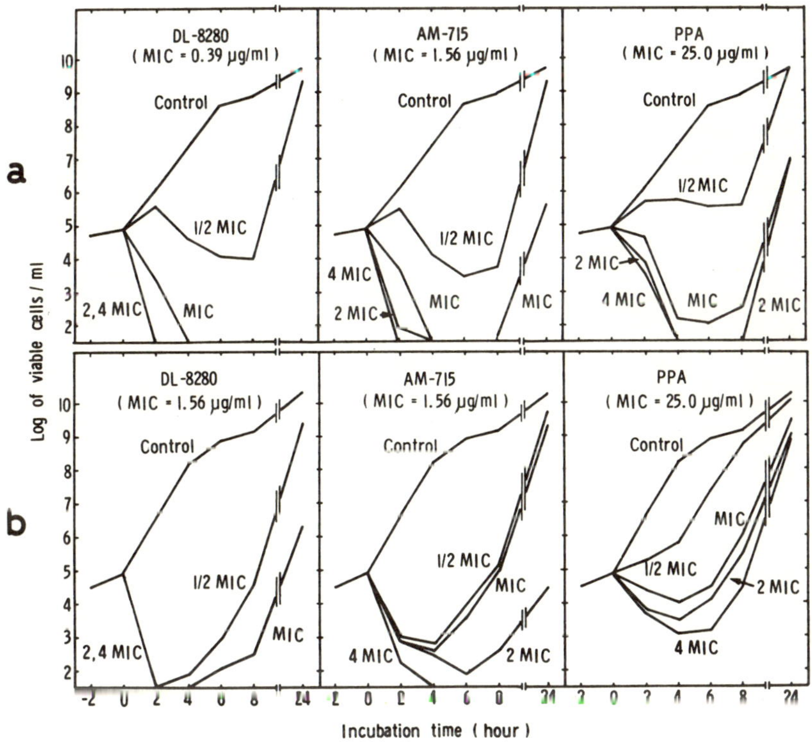

Fig. 4 Bactericidal activity of DL, AM and PPA (a) S. aureus Smith, (b) S. marcescens GN7577.

The bactericidal activity against *S. aureus* Smith and *S. marcescens* GN7577 is shown in Fig. 4. An overnight culture was diluted to 10^4 cells/ml and incubated with shaking at 37 °C. When the bacterial cell count reached about 10^5 cells/ml, the drugs were added for examination of their bactericidal activities. Against *S. aureus*, DL showed excellent bactericidal activity. MIC of DL killed them all within 4 h of incubation, while at the MIC of AM, regrowth was observed after 8 h of incubation. PPA could not kill them even at twice the MICs. Against *S. marcescens*, DL showed potent bactericidal activity far exceeding that of AM and PPA. At twice the MICs, DL destroyed them all within 2 h of incubation, while twice the MICs of AM could not. PPA could not destroy them even at four times the MICs.

III. *IN VIVO* ANTIMICROBIAL ACTIVITY

These *in vitro* antimicrobial activities of DL were reflected to its *in vivo* effectiveness as shown in Table 1. Against *S. marcescens* infections, oral treatment with DL was more effective than that with AM and PPA, but less so than subcutaneous treatment with GM. Against *Ps. aeruginosa* infection, DL was most active, even though this strain was less susceptible to DL than to AM.

These *in vivo* effectiveness of DL, as compared with AM, is in part attributable to its excellent bactericidal potency as well as to its high absorbability (data to be published). Recent studies of the oral absorbability of DL in animals

Table 1. *In Vivo Antimicrobial Activity of DL, AM, PPA and GM*[a]

Test organism	*Challenge dose (log/mouse)*	*Drug*	*MIC(μg/ml)*	*ED$_{50}$(mg/Kg)*
S. marcescens GN7577	*5.0*	*DL-8280*	*0.78*	*5.25*
		AM-715	*0.78*	*19.05*
		PPA	*25.0*	*>200*
		GM	*0.39*	*1.25*
Ps. aeruginosa GN11189	*6.8*	*DL-8280*	*1.56*	*27.54*
		AM-715	*0.78*	*63.09*
		PPA	*12.5*	*>200*
		GM	*1.56*	*32.36*

[a] *A group of 10 male mice were treated orally with DL, AM and PPA or subcutaneously with GM immediately after intraperitoneal infection.*

revealed that it was readily absorbed and well distributed to renal, hepatic and pulmonary tissues. In human studies, a level of DL equal to that in the serum was detected in the saliva, and more than 80 % of the dose had been excreted in the urine in an unchanged form by 8 h after dosing. By 48 h, 95 to 100 % of the dose had passed into the urine. Thus, the high urinary level and excellent bactericidal activity of DL are sure to contribute to its effectiveness against urinary tract infections. Furthermore, no significant side-effect was observed at any time during the study.

REFERENCES

1. Lesher, G.Y., Froelich, E.H., Gruett, M.D., Bailey, J.H., and Brundage, R.P. (1962) J. Med. Pharm. Chem. *5*, 1063-1065
2. Shimizu, M., Takae, S., Nakamura, S., Takae, H., Minani, A., Nakata, K., Inoue, S., Ishiyama, M., and Kubo, Y. (1975) Antimicrob. Agents Chemother. *8*, 132-138
3. Ito, A., Hirai, K., Inoue, M., Koga, H., Suzue, S., Irikura, T., and Mitsuhashi, S. (1980) Antimicrob. agents Chemother. *17*, 103-108

IN VITRO ANTIBACTERIAL ACTIVITY OF E-0702, A NEW SEMISYNTHETIC CEPHALOSPORIN

Kanemasa Katsu, Matsuhisa Inoue and Susumu Mitsuhashi

Laboratory of Drug Resistance in Bacteria
School of Medicine, Gunma University
Maebashi, Japan

Many broad spectrum cephalosporins have been developed to date. E-0702, (6R,7R)-3-[(1-carboxymethyl-1H-tetrazol-5-yl) thiomethyl]-7-[(2R)-2-(6,7-dihydroxy-4-oxo-4H-1-benzopyran-3-carboxamido)-2-(4-hydroxyphenyl)acetamido]-8-oxo-5-thia-1-azabicyclo[4.2.0]oct-2-ene-2-carboxylic acid disodium salt is a new parenteral semisynthetic cephalosporin derivative. This paper deals with the *in vitro* antibacterial activity of E-0702, its stability to β-lactamases, and its affinity for penicillin-binding proteins.

ANTIBACTERIAL ACTIVITY OF E-0702

The susceptibility distribution of E-0702 was examined against about 1400 clinical isolates of gram-positive and gram-negative bacteria according to the standard method of the Japan Society of Chemotherapy at an inoculum size of 10^6 colony-forming units (CFU) per ml (1). As shown in Table 1, *in vitro* activity of E-0702 against gram-negative bacteria except *Neisseria gonorrhoeae*, was more potent than Cefoperazone (CFP). The concentrations of E-0702 required to inhibit the growth of 50% of the total number of tested strains (MIC_{50}) were 0.09 µg/ml against *Pseudomonas aeruginosa*, 0.39 against *Ps. maltophilia*, 1.26 against *Ps. cepacia*, 0.29 against *Acinetobacter calcoaceticus*, 0.006 against *Citrobacter freundii* and 0.001 against *Salmonella* species. E-0702 was 9 to 400-fold more potent than CFP and Cefotaxime (CTX) against those strains. Furthermore, it was also highly active against *Escherichia coli*, *Klebsiella pneumoniae*, *Serratia marcescens*, *Enterobacter cloacae*, *Proteus mirabilis* and indole positive *Proteus* species. The MIC_{50} values of E-0702 against those isolates were 0.01, 0.02, 0.06, 0.27, 0.05 and 0.09 µg/ml, respectively, indicating that the compound was more potent than CFP and comparable to CTX in its efficacy. On the other

Table 1. *Antibacterial activity of E-0702*

Species	No. of strains	Drug concentration (μg/ml)[a]					
		MIC_{50}			MIC_{90}		
		E-0702	CFP	CTX	E-0702	CFP	CTX
S. aureus	100	4.15	0.97	0.61	5.80	1.50	1.25
S. epidermidis	50	7.00	1.47	1.38	28.0	9.50	12.5
N. gonorrhoeae	53	0.09	0.02	0.006	0.50	0.07	0.04
E. coli	200	0.01	0.16	0.03	0.08	2.10	0.04
E. cloacae	100	0.27	0.37	0.11	2.20	17.5	8.70
K. pneumoniae	100	0.02	0.34	0.03	0.19	2.70	0.07
C. freundii	50	0.006	0.10	0.18	0.40	17.0	25.0
Salmonella sp.	50	0.001	0.43	0.06	0.005	1.00	0.16
P. mirabilis	50	0.05	0.82	0.02	0.57	1.56	0.04
P. morganii	50	0.14	0.92	0.02	1.18	5.00	0.20
P. vulgaris	50	0.08	1.00	0.02	0.80	2.35	0.04
P. rettgeri	25	0.06	0.64	0.02	0.24	7.40	0.09
P. inconstans	21	0.02	1.38	0.05	0.39	6.25	0.20
S. marcescens	100	0.06	0.88	0.22	56.0	38.2	2.30
A. calcoaceticus	100	0.29	26.7	7.45	0.98	75.0	20.0
Ps. aeruginosa	200	0.09	3.30	10.0	0.45	8.80	24.5
Ps. cepacia	50	1.26	21.0	10.8	37.5	94.0	42.5
Ps. maltophilia	50	0.39	8.50	31.5	19.8	44.5	100

[a] *MIC_{50} and MIC_{90} represent the concentrations required to inhibit the growth of 50 and 90%, respectively, of the total number of strains used.*
CFP, cefoperazone; CTX, cefotaxime.

hand, E-0702 was less active than CFP and CTX against gram-positive bacteria such as *Staphylococcus aureus* and *S. epidermidis*.

The bactericidal activities of E-0702 against *Ps. aeruginosa* PAO 1 and *S. marcescens* IAM1184 were examined in Sensitivity test broth containing serial two-fold dilutions of the antibiotic. A clear decrease in viable bacteria was observed at 0.1 µg/ml against *Ps. aeruginosa* PAO 1 and *S. marcescens* IAM 1184, which was lower concentration than those of CFP. At these concentrations, however, regrowth of both strains was observed after twenty-four hours of incubation.

The minimal bactericidal concentrations of E-0702 were identical to or two times higher than the MICs against *E. coli* and *K. pneumoniae*, and 4 times higher against *S. marcescens* and *Ps. aeruginosa*.

The effect of inoculum size on the MICs of E-0702 against *E. coli*, *K. pneumoniae*, *S. marcescens* and *Ps. aeruginosa* was studied with the inocula varying between 10^3 and 10^8 CFU/ml. With inocula between 10^5 and 10^8 CFU/ml a great difference was observed with all species.

STABILITY TO BETA LACTAMASE

Stability of E-0702 to hydrolysis by various types of β-lactamase was determined using spectrophotometric assay (2). The maximum rate of hydrolysis (Vmax) is shown in Table 2. E-0702 was stable to various types of cephalosporinase (CSase) produced by *E. coli*, *E. cloacae*, *C. freundii*, *Ps. aeruginosa* and *P. rettgeri*. The compound also showed a high stability to penicillinase types II and III (oxacillin-hydrolyzing PCase), and type IV (carbenicillin-hydrolyzing PCase). But E-0702 was partially hydrolyzed by PCase type I (TEM type PCase), *P. morganii* CSase and the cefuroximases (3) produced by *P. vulgaris*, *Ps. cepacia* and *Bacteroides fragilis*.

AFFINITY FOR PENICILLIN-BINDING PROTEINS

According to B.G. Spratt (4), inner membrane protein fractions were prepared from *E. coli* JE1011 and *Ps. aeruginosa* NCTC10490. Affinity of E-0702 for penicillin-binding proteins (PBPs) was detected by measuring competition for binding of ^{14}C-penicillin G to the PBPs. E-0702 showed extremely high affinities for PBP-1A, -2, -3 and -1Bs of *E. coli*, and PBP-1A, -3 and -2 of *Ps. aeruginosa*. These proteins have been demonstrated to be essential and to have discrete function in cell

Table 2. *Substrate profile of cephalosporinase*

Organism	*Relative rate of hydrolysis*[a]					
	E-0702	*CER*	*CFP*	*CTX*	*CEZ*	*PCG*
E. coli GN5482	1	100	4	1	311	63
E. cloacae GN7471	1	100	1	1	100	12
C. freundii GN346	1	100	2	1	116	3
Ps. aeruginosa GN10362	1	100	5	1	222	29
P. rettgeri GN4430	1	100	1	1	99	8
P. morganii GN5407	20	100	1	1	20	16
S. marcescens GN10857	11	100	3	1	198	3
Ps. cepacia GN11164	153	100	10	174	156	161
P. vulgaris GN7919	50	100	15	29	387	20
B. fragilis GN11478	20	100	8	7	60	50

[a]*Hydrolysis of each substrate by CSase and CXase is expressed as a relative rate of hydrolysis taking the absolute rate of CER hydrolysis as 100.*
CER, cephaloridine; CFP, cefoperazone; CTX, cefotaxime; CEZ, cefazolin; PCG, penicillin G.

elongation (5,6), shape (6,7) and division (8). E-0702 also showed high affinity for PBP-1B of *Ps. aeruginosa* and low affinities for PBP-4, -5 and -6 of *E. coli* and PBP-5 of *Ps. aeruginosa*. The affinity of E-0702 for the essential PBPs is accounted for by its high antibacterial activity.

REFERENCE

1. Japan Society of Chemotherapy (1981) Chemotherapy *29*,76-79
2. Samuni, A.(1957) Anal. Biochem. *63*,17-26
3. Mitsuhashi, S. & Inoue, M.(1981) in Beta-Lactam Antibiotics (Mitsuhashi,S.,ed.) pp.41-56, Japan Sci. Soc. Press, Tokyo/ Springer-Verlag, Berlin, Heiderberg, New York
4. Spratt, B.G.(1977) Eur. J. Biochem. *72*,341-352
5. Tamaki, S., Nakajima, S. & Matsuhashi, M.(1977) Proc. Natl. Acad. Sci.(USA) *74*,5472-5476
6. Suzuki, H., Nishimura, Y. & Hirota, Y.(1978) Proc. Natl. Acad. Sci.(USA) *75*,664-668
7. Spratt, B.G.(1975) Proc. Natl. Acad. Sci.(USA) *72*,2999-3003
8. Spratt, B.G.(1977) J. Bacteriol. *131*,293-305

NOVEL NALIDIXIC ACID-RESISTANCE MUTATIONS RELATING TO DNA GYRASE ACTIVITY

S. INOUE, J. YAMAGISHI, S. NAKAMURA, Y. FURUTANI and M. SHIMIZU

Research Laboratories, Dainippon Pharmaceutical Co., Ltd., Suita, Osaka, Japan

Nalidixic acid and related compounds are a group of synthetic antibacterial agents having a pyridonecarboxylic acid moiety as a common chemical structure, which are mainly active against gram-negative bacteria (1-5). Cross-resistance among the agents is so incomplete that bacteria highly resistant to nalidixic acid are inhibited by some others at relatively low concentrations (4). During the study on the mode of the incomplete cross-resistance, we found two new types of nalidixic acid-resistance mutations designated as *nal-21* or *nal-31* (formerly *nalC*) and *nal-24* (formerly *nalD*) (6), both of which were located at about 82 min on the *E. coli* K-12 chromosome and affected sensitivity of the bacterial DNA synthesizing system to the agents.

In this paper, we will discribe the results showing that the *nal* mutations are related to DNA gyrase.

Nalidixic acid (NA) (7), oxolinic acid (OA) (8), piromidic acid (PA) (9), pipemidic acid (PPA) (10) and AT-2266 (11) were prepared in our laboratory and novobiocin (NB) was purchased from Sigma Chemical Co..

Drug susceptibility of the *E. coli* strains used as determined by colony formation inhibition assay is shown in Table 1.

Compared to the wild-type strain, KL-16, the *gyrA-51* mutant, N-51, was resistant to all of the pyridonecarboxylic acids. The *nal-31* mutant, N-31, caused resistance to the acid-type compounds such as NA, OA and PA but at the same time hypersusceptibility to the amphoterite-type compounds such as PPA and AT-2266. The *nal-24* mutant, N-24, conferred resistance to all of the pyridonecarboxylic acids though the levels of resistance were lower than those in N-51. N-31 and N-24 but not N-51 were twice more resistant to NB. Similar resistance patterns were obtained in bacterial DNA synthesis (Table 1). So, the *nal-31* and *nal-24* mutations seemed to be related to bacterial DNA synthesis.

Table 1. Effect of pyridonecarboxylic acids and novobiocin on cell growth, DNA synthesis and DNA gyrase activity in vivo of E. coli KL-16 and its nalidixic acid-resistant mutants

Drug	Minimum drug concn. inhibiting colony formation*				50% inhibitory dose (μg/ml) DNA synthesis**				50% inhibitory dose (μg/ml) DNA gyrase activity***			
	KL-16	N-51	N-31	N-24	KL-16	N-51	N-31	N-24	KL-16	N-51	N-31	N-24
PPA	1.56	12.5	0.39	6.25	3.0	45	0.5	8.4	54	1350	15	1000
PA	12.5	>400	100	100	7.0	800	34	76	420	>800	>800	>800
NA	3.13	400	100	50	2.5	400	61	52	78	>3200	2600	1200
OA	0.2	6.25	3.13	0.78	0.40	23	1.9	1.6	21	>400	76	56
AT-2266	0.1	1.56	0.025	0.2	0.30	3.5	0.18	1.0	7.8	84	0.86	7.2
NB	50	50	100	100	-	-	-	-	170	150	300	280

Abbreviations : PPA ; pipemidic acid, PA ; piromidic acid, NA ; nalidixic acid, OA ; oxolinic acid, NB ; novobiocin.

* *Minimum drug concentration (μg/ml) reducing colonies formed by more than 99%.*

** *^{3}H-thymidine incorporation in intact cells.*

*** *Supercoiling activity of λ DNA inside cells.*

KL-16 ; E. coli KL-16 (parent strain), N-51 ; a gyrA-51 mutant of E. coli KL-16, N-31 ; a nal-31 mutant of E. coli KL-16, N-24 ; a nal-24 mutant of E. coli KL-16.

DNA gyrase activity *in vivo* was examined by the supercoiling reaction of [^{14}C]-labeled $\lambda cI_{857}S_7$ DNA inside the λ lysogenic *E. coli* strains in the presence of drugs as shown in Table 1. In KL-16, 50% gyrase inhibitory dose at which supercoiled λ DNA was reduced by 50% of control was low in the order of AT-2266, OA, PPA, NA, NB and PA. This order is similar to that found in colony formation inhibition assay and in DNA synthesis inhibition with respect to pyridonecarboxylic acids, though the drug concentrations required for gyrase inhibition were one or two orders of magnitude higher except for NB. In N-51, gyrase activity was highly resistant to all of the pyridonecarboxylic acids but not to NB when compared with KL-16. This result is consistent with the previous reports (12) saying that *gyrA* mutations change gyrase resistant to NA or OA but not to NB. In N-31, gyrase activity was resistant to the acid-type compounds, OA, NA and PA but on the contrary hypersusceptible to the amphoterite-type ones, AT-2266 and PPA. Thus, the unique resistance pattern of the *nal-31* mutation was observed in gyrase inhibition too. N-31 was about twice as resistant to NB as KL-16. In N-24, gyrase activity was resistant to OA, PPA, NA and PA but not to AT-2266. This strain was slightly resistant to NB as N-31 was.

These results indicate that *nal-31* and *nal-24* are mutations on a gene or genes relating to gyrase activity *in vivo*.

P1 transduction using N-31 and N-24 as donors and a temperature-sensitive *dnaA* mutant, LC257, as a recipient revealed that the *nal-31* and *nal-24* mutations are very close to the *dnaA* gene (data not shown).

Specialized transduction analysis was carried out using transducing λ phages carrying chromosomal fragments around the *dnaA* gene. These phages (13) were kindly supplied by Dr. H. Yamagishi. As shown in Table 2, λ*tna* carrying a

Table 2. Specialized transduction of nal and tnaA genes by transducing λ phages

Phage strain	Chromosomal DNA carried by phage	Transducing activity		
	*tnaA rimA dnaA recF gyrB**	*nal-31*$^+$	*nal-24*$^+$	*tnaA*$^+$
λ*tna*	6.2 kb (*tnaA*–*rimA*)	-	-	+
λ*dnaA* #4-Ap52	4.3 kb (*tnaA*); 2.6 kb (*gyrB**)	+	-	+
λ*dnaA* #4	13.8 kb (*tnaA*–*gyrB**)	+	-	+

* partially deleted.
# 153-4 *tnaA nal-31* or # 153-1 *traA nal-24* was infected with one of the phages and selected for TnaA$^+$ or disappearance of hypersusceptibility to PPA and checked for unselected markers.

chromosomal fragment of 6.2 kb long corresponding to the *tnaA rimA* region showed transducing activity of *tnaA* but not of *nal-31* and *nal-24*. λ*dnaA*#4 carrying a wild-type chromosomal fragment of 13.8 kb long corresponding to the region from *tnaA* to *gyrB* showed transducing activity of *nal-31* and *tnaA* but not of *nal-24*. λ*dnaA*#4-Ap52 is a deletion mutant of λ*dnaA*#4 bearing 4.3 and 2.6 kb chromosomal fragments corresponding to the *tnaA* and *gyrB* regions respectively and showed transducing activity of *nal-31* and *tnaA*.

These results suggest that *nal-31* but not *nal-24* is a mutation on the chromosomal fragment of 2.6 kb pairs long corresponding to the *gyrB* region commonly carried by λ*dnaA*#4 and λ*dnaA*#4-Ap52. As the molecular size of this chromosomal fragment is too small for that expected from the molecular weight of gyrase subunit B protomer (14), the *gyrB* gene included in this fragment seems to be partially deleted. It is possible that *nal-24* is a mutation on the deleted part of the *gyrB* gene.

These biochemical and genetic results indicate that *nal-31* and *nal-24* are mutations on the gene relating to DNA gyrase activity *in vivo*, and one of them, *nal-31* is a mutation probably on the *gyrB* gene. Pyridonecarboxylic acids may act on gyrase subunit B as well as subunit A.

REFERENCES

1. Deitz, W. H., J. H. Bailey and E. J. Froelich (1964) *Antimicrob. Agents Chemother. 1963*, 583-587
2. Sammes, P. G. (1980) Topics in Antibiotic Chemistry Vol.3, pp.13-38, John Wiley and Sons
3. Shimizu, M., S. Nakamura and Y. Takase (1971) *Antimicrob. Agents Chemother. 1970*, 117-122
4. Shimizu, M., Y. Takase, S. Nakamura, H. Katae, A. Minami,

K. Nakata, S. Inoue, M. Ishiyama and Y. Kubo (1975) *Antimicrob. Agents Chemother. 8,* 132-138

5. Shimizu, M., Y. Takase, S. Nakamura, H. Katae, S. Inoue, A. Minami, K. Nakata and Y. Sakaguchi : Current Chemotherapy and Infectious Disease, Proceeding of the 11th ICC and 19th ICAAC, ed. J. D. Nelson and C. Grassi, Am. Soc. Microbiol., Washington, D.C. Vol.1, pp.458-460 (1980)
6. Inoue, S., T. Ohue, J. Yamagishi, S. Nakamura and M. Shimizu (1978) *Antimicrob. Agents Chemother. 14,* 240-245
7. Lesher, G. Y., E. J. Froelich, M. D. Gruett, J. H. Bailey and R. P. Brundage (1962) *J. Med. Pharm. Chem. 5,* 1063-1065
8. Kaminsky, D. and R. I. Meltzer (1968) *J. Med. Chem. 11,* 160-163
9. Minami, S., T. Shono and J. Matsumoto (1971) *Chem. Pharm. Bull. 19,* 1426-1432
10. Matsumoto, J. and S. Minami (1975) *J. Med. Chem. 18,* 74-79
11. Matsumoto, J., T. Miyamoto, A. Minamida, Y. Nishimura, H. Egawa and H. Nishimura : Current Chemotherapy and Infectious Disease, Proceeding of the 11th ICC and 19th ICAAC, ed. J. D. Nelson and C. Grassi, Am. Soc. Microbiol. , Washington, D.C. Vol.1, pp.454-456
12. Gellert, M., K. Mizuuchi, M. H. O'Dea, T. Itoh and J. Tomizawa (1977) *Proc. Natl. Acad. Sci. U.S.A. 74,* 4772-4776
13. Murakami, A., H. Inokuchi, Y. Hirota, H. Ozeki and H. Yamagishi (1980) *Molec. Gen. Genet. 180,* 235-247
14. Higgins, N. P., C. L. Peebles, A. Sugino and N. R. Cossarelli (1978) *Proc. Natl. Acad. Sci. U.S.A. 75,* 1773-1777

MODE OF ACTION OF VIOMYCIN

T. Yamada

Research Institute for Microbial Diseases
Osaka University, Suita 565, Japan

K. H. Nierhaus

Max-Planck Institut für Molekulare Genetik
Berlin, Federal Republic of Germany

T. Teshima and T. Shiba

Department of Science, Faculty of Science,
Osaka University, Toyonaka 560, Japan

Y. Mizuguchi

Department of Microbiology, School of Medicine,
University of Occupational and Environmental Health,
Kitakyushu 807, Japan

T. Yamanouchi

Research Institute for Microbial Diseases
Osaka University, Suita 565, Japan

We describe herein (A) progress in the study of viomycin (Vim) action, (B) structure and activity relationship of Vim, (C) binding of the drug to ribosomal RNA, and (D) inhibition by Vim of *in vitro* assembly of ribosomal subunits.

A. PROGRESS IN THE STUDY OF VIOMYCIN ACTION

Vim is a peptide antibiotic which differs from most antibiotics in that it is more active against mycobacteria than other bacterial species in the culture. The fact that Vim inhibits protein synthesis was first shown in intact

Mycobacterium avium (1). Later, its inhibition of poly(U) directed polyphenylalanine synthesis was shown in an *Escherichia coli* cell-free system (2). However, *E. coli* is rather resistant to Vim in culture and therefore no successful attempts to isolate mutants with altered ribosomes from *E. coli* have yet been reported. Under these circumstances, it was reasonable to expect that ribosomal resistant mutants from mycobacteria might be obtained. This was found to be the case. There are two types of mutants, some with altered 50S subunits and others with altered 30S subunits (3). Genetic study showed that there are two loci for Vim resistance on host chromosomes. One affects a component of 50S subunits and other that of the 30S subunits (4, 5). Interestingly, ribosomes with high level Vim resistance are always associated with streptomycin (SM) or kanamycin resistance (6). Binding experiments have suggested that there is interaction between the binding sites of Vim and SM on ribosomes (7). The inhibition of the translocation step of protein synthesis by Vim has been reported by other groups (8, 9). We found that Vim stabilizes 70S couples and promotes association of ribosomal subunits, Vim-resistant strains were not affected by Vim with respect to dissociation and 70S couple formation (10), and it has later reported that Vim resistance is conferred by RNA of either of the ribosomal subunits (11). More recently, we found that Vim binds to ribosomal RNA and blocks the *in vitro* assembly of 50S subunits (12).

B. THE STRUCTURE AND ACTIVITY RELATIONSHIP OF VIOMYCIN

Vim is a kind of tuberactinomycin (Tum) which consists of 6 amino acids and is named Tum B (Fig. 1). Many analogs of Tum were prepared either by total synthesis, semi-synthesis, or chemical modification by Shiba's group. From the results of minimal inhibitory concentration of these analogs, a correlation between the structure and activity of the drug was discussed previously (13). The activity inhibiting polypeptide synthesis on ribosomes was tested in a cell-free system (14), and essentially the same conclusion was obtained with a few exceptions as described in the following. An amino acid in a branch part is not essential, but modifies the activity arising from the cyclic peptide part, tuberactinamine N. Elimination of hydroxy groups of ser residues does not lead to a significant loss of activity. Guanidino amino acid seems to be the most important moiety. The presence of the ureido group or C=C double bond in the 3-ureidodehydroalanine residue is not necessary for the activity,

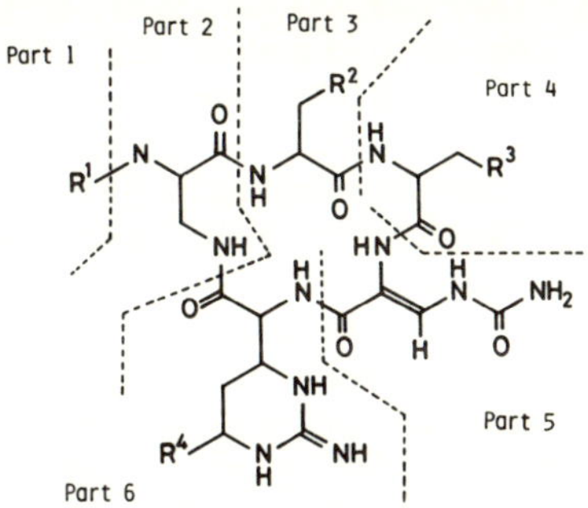

	R1	R2	R3	R4
Tum A	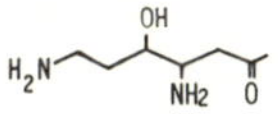	OH	OH	OH
Tum B (Vim)	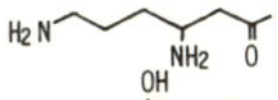	OH	OH	OH
Tum N	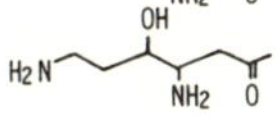	OH	OH	H
Tum O	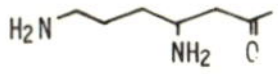	OH	OH	H
Tua N	H	OH	OH	H
Cpm IA	H	OH	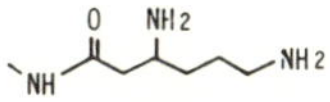	H
Cpm IB	H	H	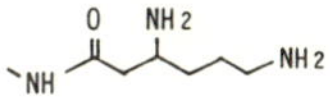	H
Cpm IIA	H	OH	NH_2	H
Cpm IIB	H	H	NH_2	H
Pseudo Cpm IB	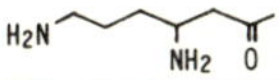	H	NH_2	H

Fig. 1. Structure of viomycin

but this residue is necessary for labelling the antibiotic. We were able to label Tum O by incubating it with ^{14}C-urea, and using the ^{14}C-Tum O thus obtained, the binding experiments were performed.

C. BINDING OF THE DRUG TO RIBOSOMES

The binding of ^{14}C-Tum O to ribosomes, ribosomal RNA, and ribosomal proteins was measured by means of equilibrium dialysis. One compartment was filled with 100 μl of each sample derived from *E. coli* and other with ^{14}C-Tum O.

Dialysis reached equilibrium after 19 hr at 4 C; then aliquots were withdrawn from each chamber and counted in a scintillation counter. From the difference between the two chembers, the number of moles of antibiotic bound to each component was calculated. The following results were obtained. The binding increases with decreasing concentration of Mg^{++} and NH_4^+, suggesting multiple binding sites on ribosomes. We asked what components of ribosomes were responsible for the binding. To find this, we used 10 mM Tris-HCl(pH 7.8) buffer containing 8 mM Mg^{++} and 100 mM NH_4^+. One A_{260} unit of 50S subunits bound 54 p moles of the drug, whereas the total proteins from 50S particles bound only 4 p moles. RNA from 50S bound 74 p moles; one A_{260} unit of 30S bound 80 p moles of the drug; total proteins bound nothing. In contrast, 16S RNA bound 102 p moles of the antibiotic. Core particles with decreasing protein content were produced from both subunits by incubation with an increasing concentration of LiCl. The 50S derived cores showed a slight increase of binding after washing off the first proteins, whereas 30S derived cores showed about the same binding capacity. Clearly, the influence of proteins on the conformation of RNA does not contribute to RNA-dependent binding of the drug (12).

D. INHIBITION BY VIOMYCIN OF *IN VITRO* ASSEMBLY OF RIBOSOMAL SUBUNITS.

Since the drug binds to RNA, it is conceivable that it would also influence assembly of the ribosomal subunits. To test this possibility, the following experiments were performed. The 50S subunits from *E. coli* were reconstituted by incubating RNA and total proteins in the presence of various concentrations of Vim first at 44 C for 30 min. The concentration of Mg^{++} was 4 mM. Following this, the concentration of Mg^{++} was adjusted to 20 mM, and incubation was continued at 50 C for another 90 min. As a control, native 50S subunits were subjected to the same procedure in the presence of various concentrations of the drug. The samples were dialyzed against 10 mM Tris-HCl buffer(pH 7.8) containing high salts in order to remove excess amounts of the drug. To the 50S particles thus obtained, native 30S subunits were added and the poly(U)-directed polyphenylalanine synthesis was measured. The results may be summarized as follows. The reconstitution of ribosomes was severely affected, even at a low concentration of Vim. The activity of native 50S subunits in the control was hardly affected, indicating that the dramatic reduction of activity seen with reconstituted particles may not be due to a transfer of the drug from the incubation system to poly(U) system (12).

Vim-resistant mutants of *Mycobacterium tuberculosis* were isolated both from patients and by test tube selection. The biochemical analysis of these is in progress.

REFERENCES

1. Tsukamura, M. (1960) J. Biochem. (Tokyo) *47*, 685-688
2. Davies, J. E., Gorini, L. and Davis, B. D. (1965) Mol. Pharmacol. *1*, 93-106
3. Yamada, T., Masuda, K., Shoji, K. and Hori M. (1972) J. Bacteriol. *112*, 1-6
4. Mizuguchi, Y., Suga, K., Masuda, K. and Yamada, T. (1974) Japan J. Microbiol. *18*, 457-462
5. Yamada, T., Masuda, K., Mizuguchi, Y. and Suga, K. (1976) Antimicrob. Agents. Chemother. *9*, 817-823
6. Yamada, T., Masuda, K., Shoji, K. and Hori, M. (1974) Antimicrob. Ag. Chemother. *6*, 46-53
7. Masuda, K. and Yamada, T. (1976) Biochim. Biophys. Acta *438*, 333-339
8. Liou, Y. F. and Tanaka, N. (1976) Biochem. Biophys. Res. Commun. *71*, 477-483
9. Modolell, J. and Vazquez, D. (1977) Europ. J. Biochem. *81*,491-497
10. Yamada, T. and Nierhaus, K. H. (1978) Molec. Gen. Genet. *161*, 261-265
11. Yamada, T., Mizuguchi, Y., Nierhaus, K. H. and Wittmann, H. G. (1978) Nature *275*, 460-461
12. Yamada, T., Teshima, T., Shiba, T. and Nierhaus, K. H. (1980) Nucl. Acids Res. *8*, 5767-5777
13. Shiba, T. (1978) Bioactive Peptides Produced by Microorganisms, Edited by Umezawa, H., Takita, T. and Shiba, T. Kodansha Ltd,Tokyo;Halsted Press,New York p.71
14. Yamada, T., Teshima, T. and Shiba, T. (1981) Antimicrob. Agents. Chemother. (in press)

Subject Index

Author Index